Contents

Electronic Circuit Guidebook
Volume 4:
Electro-Optics

Written By
Joseph J. Carr

A Division of Howard W. Sams & Company
A Bell Atlantic Company
Indianapolis, IN

PROMPT© Publications is an imprint of Howard W. Sams & Company, A Bell Atlantic Company, 2647 Waterfront Parkway, E. Dr., Indianapolis, IN 46214-2041.

International Standard Book Number: 0-7906-1132-5
Library of Congress Card Catalog Number: 97-65787

Acquisitions Editor: Candace M. Hall
Editor: Loretta L. Leisure
Assistant Editors: Pat Brady, Natalie F. Harris
Layout Design: Loretta L. Leisure
Typesetting: Loretta L. Leisure
Cover Design: Phil Velikan
Graphics Conversion: Terry Varvel
Illustrations and Other Materials: Courtesy of the Author

Trademark Acknowledgments:
All product illustrations, product names and logos are trademarks of their respective manufacturers. All terms in this book that are known or suspected to be trademarks or services have been appropriately capitalized. PROMPT© Publications, Howard W. Sams & Company, and Bell Atlantic cannot attest to the accuracy of this information. Use of an illustration, term or logo in this book should not be regarded as affecting the validity of any trademark or service mark.

PRINTED IN THE UNITED STATES OF AMERICA

9 8 7 6 5 4 3 2 1

Preface

Electro-optics is the marriage of electronics and optics to form a wide variety of instruments, circuits, and devices with a large number of uses. For example, the "electric eye" used by the shopkeeper to alert her when customers arrive is an electro-optical device; the burglar alarm that alerts her when other-than-customers arrive after hours is also an electro-optical (E-O) device.

This book is about E-O *sensors*, i.e., those electronic transducers that convert light waves into a proportional voltage, current, or resistance. The coverage of the sensors will be wide enough to allow you to understand the physics behind the theory of operation of the device, and also the circuits used to make these sensors into useful devices. We will take a look at the photoelectric effect, photoconductivity, photovoltaics and PN junction photodiodes and phototransistors.

Also examined is the operation of lenses, mirrors, prisms and other optical elements. These discussions are keyed to a couple of chapters on light physics, which you will need to understand before studying E-O instruments.

Other devices including optoisolators, optical fibers, X-ray sensors, TV sensors and some additional optical instruments such as telescopes and scientific instruments are also discussed in detail. The circuits for using sensors in E-O instruments, and for interfacing them with computers, are also discussed in sufficient depth to allow you to apply them to practical situations.

The author wishes to acknowledge the assistance of editor Jim Rounds of SAMS, Steve Bigelow, and Bob Clark—who provided immense wisdom and advice.

Joseph J. Carr
P.O. Box 1099
Falls Church, VA 22041
E-mail: carrjj@aol.com

Chapter 1

Introduction to Light and Electro-Optics

Light is a natural phenomenon that is both mysterious and perplexing to those who do not understand it, yet light is also one of mankind's most common experiences. Most humans, indeed nearly all animals, are at least familiar with light--even most plants cannot do without it. We intuitively know that light is "something" because its absence is noted: it's dark at night and in closed rooms. Most of the light on Earth comes from the Sun. Every year some 745 quadrillion kilowatt-hours of light energy fall on the Earth's surface, yet that immense amount of light energy represents less than one billionth of the total energy emitted by the Sun over the same period.

Light is a form of energy; energy is the capacity to do work, and is commonly measured in any of several units: gram-calories, watt-seconds, Joules, and kilowatt-hours. *Power*, the rate at which energy is used, is another concept used in light physics. The *power density* of light is the amount of light that falls on a unit area. This power per unit area is measured in milliwatts per square centimeter (mW/cm^2) or kilowatts per square meter (kW/m^2). The amount of sunlight reaching the Earth is approximately 136 mW/cm^2, of which 90 mW/cm^2 reaches the surface at noontime on a clear day (at moderate latitudes); atmospheric absorption and scattering account for the loss. Some solar power and weather researchers use a unit of measure for the sunlight reaching the surface called the *Langley* (1 Langley = 11.62 watt-hours per square meter).

A Brief History of Light

Light has interested people from ancient times. In the third verse of the first book of the Bible (Gen. 1:3) the text says: "And God said: 'Let there be light,' and there was light." The ancient Greeks, not to be outdone by the Hebrews, had their own ideas. Two schools of thought concerning light arose in ancient Greece. The *emission* school thought that light was some "thing" or substance emitted by all physical objects, emission that could stimulate the eye. The *tactile* school believed that the eye did the emitting; that is, the eye sent out invisible tentacles that touched and registered on all physical objects. Other ancient cultures, and some primitive cultures today, have what we would consider odd ideas concerning light. Modern ideas about light would be a real shock to the "ancients" of even the nineteenth century! Indeed, they shocked most scientists early in the twentieth century, and often disquiet students today.

As science progressed, other ideas arose concerning the nature of light. These theories became increasingly more sophisticated, and closer to our modern understanding, as scientific knowledge increased over the centuries.

The Early Scientists

A seventeenth century British scientist named Robert Hooke came close to understanding something of the nature of light in his book Micrographia, published in 1665. Hooke decided that light was a series of small, wave-like vibrations (**Figure 1-1a**). These vibrations, according to Hooke's idea, varied in frequency for the different colors. Another model (**Figure 1-1b**) considered light as "corpuscles" or "particles."

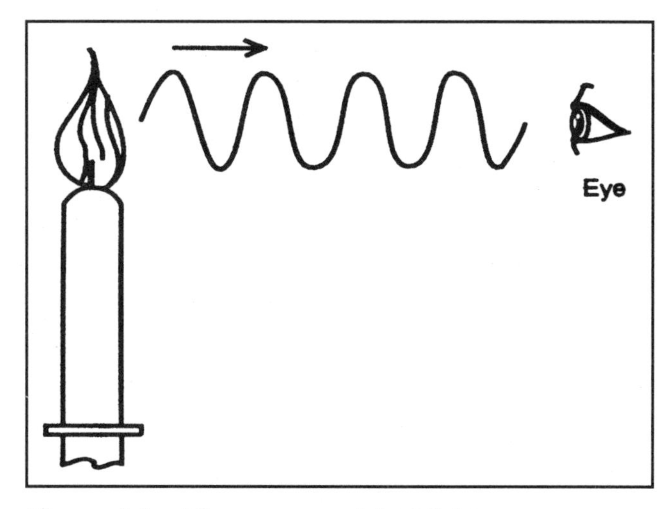

Figure 1-1a. The wave model of light.

Figure 1-1b. The particle model of light.

Sir Isaac Newton, a contemporary of Hooke, is famous for his work on gravity and for inventing one form of the calculus. But he also made substantial contributions to our understanding of the nature of light. Newton's famous experiments involving color separation in a prism (**Figure 1-2**) demonstrated that white light was composed of light of different colors. We know now that the prism bends or refracts light rays by different amounts according to the wavelength, implying the velocity through a refractive material is different for different wavelengths. Newton accepted the experiment as proof of a corpuscular or "particle" nature of light. The results of this experiment were forwarded to the Royal Society in London in 1672.

In 1690, Dutch scientist Christian Huygens proved that light has the property of polarity, a wave-like behavior. When joined with Hooke's theory of waves, the wave theory of light was born, which served to fuel the "particle/wave" debate for centuries to come. Huygens' model is what we now call the transverse theory of light. *Transverse waves* are those in which the vibrations (oscillations) are perpendicular to the line of travel. The *electric* and *magnetic* wave components of light seem to vibrate sideways. For more than three centuries the debate raged between the disciples of Newton and those of Hooke over the nature of light: particles or waves? In the twentieth century, however, a few surprising findings rendered them both right.

The theories of light, particle and wave, continued to compete with each other throughout the seventeenth and eighteenth centuries. While the nature of light was in controversy, there was one fact that was not: light traveled at a very fast velocity — faster than anything else in nature. Today, we recognize

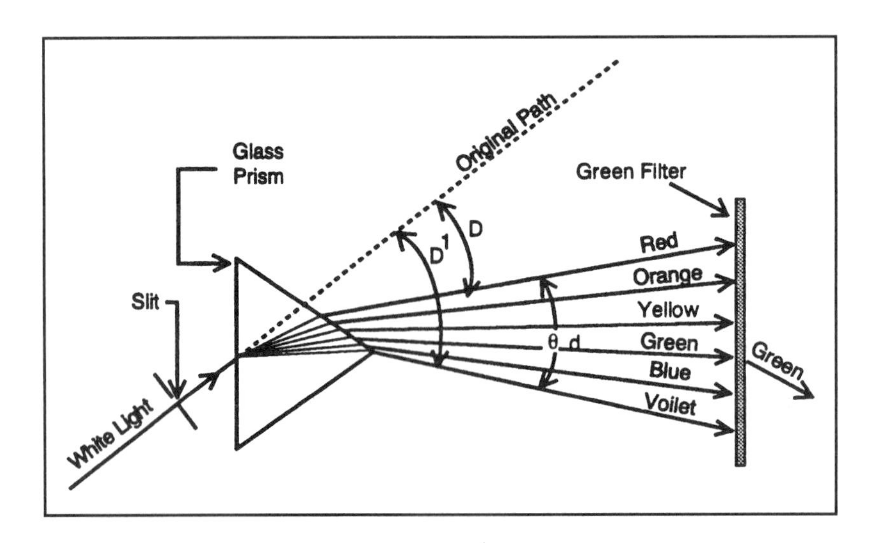

Figure 1-2. The spectrum of different light wavelengths shown dispersed through a prism.

that the speed of light is a universal "speed limit." Much of the scientific work on light concerned just how fast it traveled. At first, it was believed that the speed of light was infinite, but that proved to be not the case.

In 1675, Danish astronomer Olaus Römer proved that the speed of light was not infinite, but rather was merely great enough to seem so. Römer's observations of the orbits of the moons of Jupiter suggested that the speed of light was finite. He calculated that light would take 22 minutes to travel a distance equal to the diameter of the Earth's orbit.

About the same time that Römer was observing the moons of Jupiter and making his calculations, two French astronomers, Richer and Cassini, measured the diameter of the Earth's orbit. In 1678, Huygens combined the Richer-Cassini value for the diameter of the Earth's orbit with the transit time of Römer to come up with the first rough calculation of the speed of light as 140,000 miles per second.

Light actually takes closer to 16.7 minutes to traverse the Earth's orbit, not 22 minutes as measured by Römer, which accounts for the error (140,000 mi/s versus 186,000 mi/s). Nonetheless, given that the calculations were made on the basis of the crude astronomical and time measurements available to him from the work of Römer, Richer, and Cassini, Huygens was remarkably close to correct.

Science made tremendous advances in light studies during the nineteenth century. In 1860, Scottish scientist James Clerk Maxwell derived a series of equations that described the propagation of light waves (now called *electromagnetic waves*). From Maxwell came a spectrum table of wavelengths and frequencies for visible, infrared, and ultraviolet light. Heinrich Hertz, one of the fathers of modern radio, generated waves of a much lower frequency during his famous 1887 radio experiments. Studies very quickly revealed that these "Hertzian waves" were the equivalent of Maxwell's electromagnetic waves at extremely long wavelengths. While the wavelengths of light are measured in nanometers, the wavelengths of Hertzian waves are measured in centimeters, meters, or even kilometers.

Despite the advances in other areas of light theory, at the end of the nineteenth century scientists still did not have a precise grasp of the speed of light. It was not until 1926 that American physicist Albert Michelson measured the speed of light at 299,796,000 ± 4,000 meters per second, or about 186,290 miles per second, which is very near the currently accepted value. Michelson's experiment involved using a precisely surveyed light path between his laboratory at the Mount Wilson Observatory in California, and Mount San Antonio 35.4 kilometers away. He placed a stationary mirror on the distant mountain, and used a rotating octagonal mirror arrangement to measure the transit time to and from it.

As the turn of the twentieth century neared, scientists began to grapple with some perplexing issues that showed up in the experiments of nineteenth century scientists. As far back as 1839, Alexandre Becqueral discovered that a chemical battery worked more effectively when it was exposed to a strong light. Radio pioneer Hertz discovered another form of this *photoelectric effect* when he noticed electrical changes in his radio apparatus when a metal plate used in the generation of the spark that created the signal was exposed to ultraviolet light. Shortly thereafter, Willoughby Smith discovered that selenium changed its dc (direct current) resistance when light was shining on the metal. All of these discoveries set the stage for a monumental breakthrough that came at the dawn of the twentieth century.

On December 14, 1900, a thin, not-yet-graying physicist named Max Planck walked onto the stage of a Berlin auditorium to address fellow scientists of the German Physical Society. The paper he presented that night had an immediate and long-lasting impact on science; indeed, on the history of this century. It marked the birth of the theory of quantum mechanics. The New Physics that grew out of Planck's discovery supplanted the Old Physics of Isaac Newton, and revolutionized the way we look at the world. This period is now called the Second Scientific Revolution.

Old Physics had been immensely successful. It grew more and more competent from the sixteenth century onwards in a seemingly never ending cascade of important new discoveries. By the late nineteenth century, scientists were supremely confident in their ability to ultimately explain reality, predict the behavior of physical phenomena and control nature. Their world was simple, mechanical, and approachable through an elegant mathematics.

In the last decades of the nineteenth century, scientific experimenters stumbled on many phenomena that simply could not be explained in the familiar terms of Newtonian science. The new theory of energy quanta--that electromagnetic energy can only be explained in terms of discrete bundles--solved many of the riddles of nineteenth century science, but it also opened a philosophical Pandora's box.

Planck's theory set the stage for modern physics and, in turn, created a crisis among the scientists. Many of the classical physicists could not accept the new theories. Even Planck is said to have spent the remaining 47 years of his life trying to disprove his own theory! The New Physics introduced by Planck and a score of other physicists over the first three decades of this century was a turning point in modern thought.

Quantum mechanics represented a giant, disquieting leap in human understanding in a very short period of time. Physicist George Gamow, a participant in the establishment of the New Physics, titled his popular book Thirty Years That Shook Physics (Dover Press, New York) in recognition of that fact.

Early Scientific Philosophies

Until the New Physics arose early in the twentieth century, man held what might be called a "common sense world view." Certain humorists tell us that "'common sense' is that undefinable property that I possess, and you lack, anytime we disagree." Or, according to Einstein, "common sense is the sum total of all prejudices laid down before the age of eighteen." The common sense world view called into question by the New Physics is based on certain basic premises.

First, the physical universe is real, and independent of man, or man's observation of it. Second, the physical world is not only real, but it is also orderly and rational. The physical world obeys certain immutable laws of nature that are also independent of man. This same idea led directly to the concept of causality, or the "Law of Cause and Effect," which claims that for every effect there is a preceding cause. For example, the baseball flies because the bat hit the ball with a certain force.

Third, the physical world is knowable. The laws of nature can be understood by man once his ignorance is overcome. Given adequate doses of knowledge blended with reason, the unknown becomes known. While something might be presently unknown, it is never unknowable.

There are certain consequences of the basic classical premises that led man to the scientific method of research. Under the classical world view we believe that it is possible to formulate laws of nature from experiment and orderly observation. If they repeated an experiment a thousand times and always got the same result (and never saw a contradictory result), then they generalized statements regarding the results of the experiment. That is, they formulated laws regarding the experiment. This view was so strongly entrenched that traditional scientists--and by extension all rational men--would not make statements of supposed fact that were neither supportable by, nor derived from, experiment or other repeatable empirical evidence. The very language of scientific reason was structured such that concepts which are not verifiable in the laboratory were inherently suspect.

Another assumption derived from these premises was that, for any statement of fact either the statement itself is true or its negation is true. For example, either "that object is a widget" or "that object is not a widget" must be true, but they cannot both be true. In the common sense of the classical world view something cannot be both a widget and a not-widget. In the New Physics, on which modern electronics is based, some experts claim that it is possible for something to be a widget and a not-widget at the same time! In a famous thought experiment called the Paradox of Schrödinger's Cat, it is claimed that a certain apparently confused cat is simultaneously alive and dead. Obviously, minds schooled in the classical world view found such concepts utterly frightening, and they rebelled against them.

A profound problem with Newton's Old Physics is the question of time. According to the Newtonians, every particle in the universe is on a reversible trajectory. There is no reason to suspect that events, indeed time itself, cannot be reversed like a movie running backwards. But something is clearly amiss with this view because certain processes are clearly irreversible outside of divine miracles. For example, reconstituting a burned sheet of paper or a blown light emitting diode (LED) is not possible.

Unfortunately for the comfortable classical scientific world view, by the end of the nineteenth century it became painfully apparent that there were cracks in the Newtonian edifice. Inexplicable phenomena were being observed in experiments that were beyond criticism as to method. Scientists of the day found the experimental method faultless, even though the results destroyed their own world view. The path of Newtonian physics began a torturous descent from the peak, until at the dawn of our present century that path ended at the edge of a precipice. Scientists stood perplexed and bewildered at the edge of that precipice and then--under the teachings of Planck, Einstein, Bohr, and others--took a giant leap forward.

Twentieth Century Developments

The long and successful reign of classical physics was threatened in the late nineteenth century by discoveries regarding black body thermal radiation, radioactivity, X-rays, and the nature of light. A major problem with light and X-rays was that various experiments seemed to contradict each other. Some seemed to show conclusively that light has a wave nature. For example, light refracts when passing from air into water, and passing a light beam through a pair of closely spaced slits forms a characteristic light-dark interference pattern (**Figure 1-3**). Both experiments show purely wave-like actions. Paradoxically, other experiments show equally conclusively that light has a particle nature. Consider the photoelectric effect. When a light beam falls on certain types of metallic plates, electrons are emitted from the plate to form

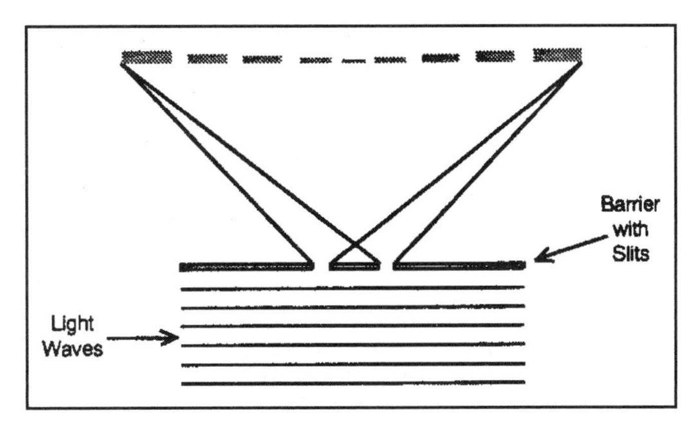

Figure 1-3. Wave model is verified by the double slit experiment.

an electric current (**Figure 1-4**). If light has a wave nature, then one would expect the energy of the electrons to increase when the light intensity increases. In reality, the light intensity has no effect on the energy level of the emitted electrons, but it affects the magnitude of the current flow. Albert Einstein won the 1905 Nobel Prize for physics by applying Planck's theory of quantum energy to the photoelectric effect problem. His paper presented an essentially particle view of light to a world that had become accustomed to thinking of light in terms of waves.

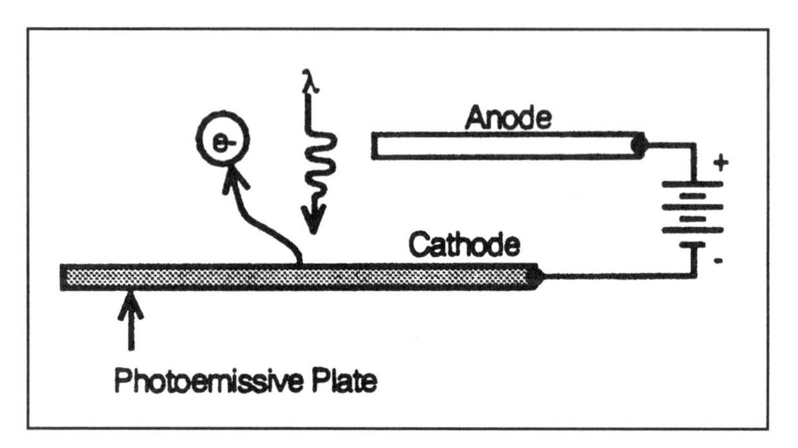

Figure 1-4. The particle model of light is verified by the photoelectric effect.

As scientists delved deeper into atomic phenomena they became increasingly aware that the microworld where these things occur is a very strange place indeed. When Werner Heisenberg worked on these problems at the Bohr Institute he would often take long walks through Copenhagen city parks late at night. As a classically trained physicist, Heisenberg despaired at the schizophrenic weirdness of the quantum world: "Can nature possibly be so absurd as it seemed to us in these atomic experiments?" he lamented.

During the first three decades of the twentieth century, Heisenberg, Bohr and a relatively small band of scientists worked out a system that seems to explain atomic phenomena. As we have seen, that system is called quantum mechanics (QM). It displaced the comfortable cause-and-effect definitions of classical physics, and put in their place a new set of definitions that are based on probabilities and "tendencies to exist." The seeming paradox is that unpredictability and uncertainty appear intrinsic to the universe at the deepest levels. Eventually quantum mechanics was universally accepted as the mathematical construct that best predicts the behavior of matter and describes its nature at the subatomic level. Even though problems persisted, QM became a standard for scientists.

Modeling the quantum world presented new problems. In high school physics we learned the Bohr model of the atom (**Figure 1-5**). It consists of a nucleus of positively charged protons and electrically neutral neutrons surrounded by concentric spheres of negatively charged electrons orbiting the nucleus in fixed paths. Our inadequate model, graphically reinforced by pictures in textbooks and ping-pong ball toys in the classroom, would lead us to view subatomic particles as balls of differing sizes and weights. The problems with the solar system model become apparent when electrons and protons show some very peculiar behavior.

Earlier we noted that light seemed to be a wave in some experiments and a particle in others. In the classical mind, such a situation was unacceptable because particle and wave descriptions seem mutually exclusive. In the 1920s, however, physicist Neils Bohr postulated that these two descriptions were merely complementary to each other, not contradictory. Bohr's *Complementarity Principle* holds that neither description alone is entirely sufficient to describe light--both are needed together. This was a great discovery for twentieth century scientists.

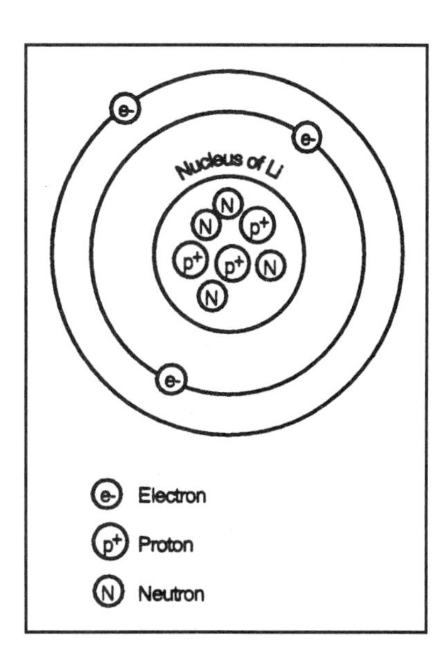

Figure 1-5. Simple model of an Li atom: a nucleus of protons and neutrons, with electrons orbitting.

Waves

Waves are an important part of science, and indeed much of physical reality--including light--is made up of either waves or phenomena that "behave in a wave-like manner." Radio-TV signals, infrared heat, ultraviolet, X-rays, gamma rays, and visible light are all electromagnetic waves.

For the sake of simplicity we will use the sinewave model (**Figure 1-6**) in order to discuss wave behavior. The sinewave is easily understood and easily generated in experiments.

Frequency and Wavelength

Any sinewave exhibits amplitude, velocity, frequency, wavelength, period, and phase (see **Figure 1-6**). *Amplitude* is the measure of the intensity of the wave. *Velocity* is the speed of propagation through any given medium. In air or a vacuum, the electromagnetic wave propagates at the speed of light (denoted by "c"): about 300,000,000 meters per second (3×10^8 m/s), or about 186,282 miles per second.

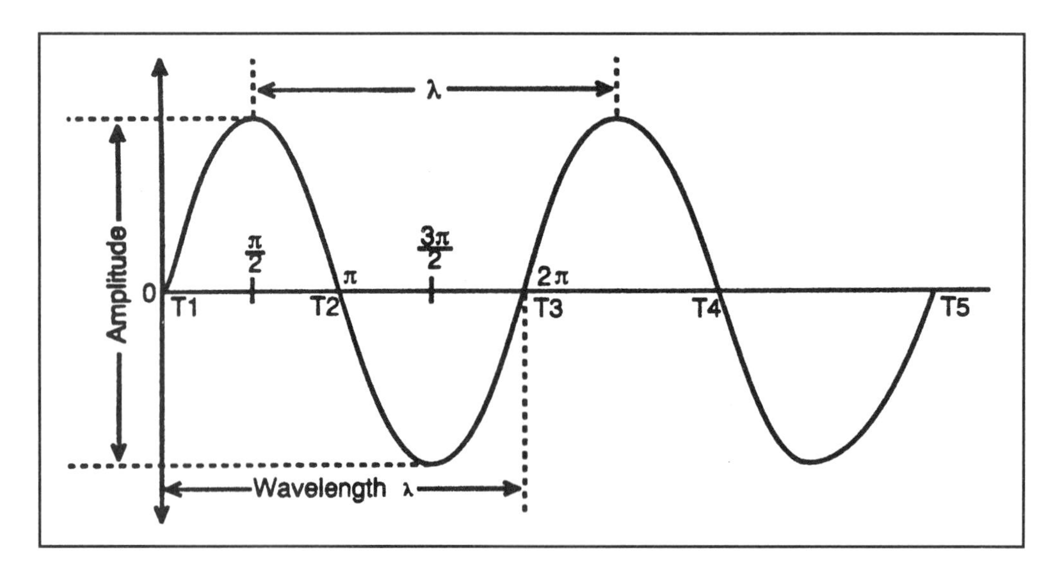

Figure 1-6. A sinusoidal wave.

Frequency. Each wave consists of individual *cycles*. In **Figure 1-6** a cycle is one complete excursion of the waveform, such as T1-T3 or T3–T5. *Frequency* indicates events per unit time, which in the case of electromagnetic waves is measured in cycles per second (cps). In older texts cps was the unit of frequency, but in honor of radio pioneer Heinrich Hertz, the modern unit of frequency is hertz (Hz); one cycle per second equals one hertz (1 cps = 1 Hz).

A single hertz is generally too small for practical work in both sound and radio, so to make calculations easier we adopt superunits by adding a multiplier prefix to the basic unit. Thus, for hertz, the superunit kilohertz (kHz) is 1000 Hz, megahertz (MHz) is 1,000,000 Hz, and gigahertz (GHz) is 1,000,000,000 Hz, or 1000 MHz.

Wavelength. Wavelength is defined as the distance between identical points on successive waves (see **Figure 1-6**). For example, the wavelength can be measured from positive peak to positive peak, or between zero crossings in the same direction.

The wavelength is represented in equations by the Greek letter lambda (λ). For all forms of waves the velocity, wavelength, and frequency are related such that the product of frequency and wavelength is equal to the velocity.

$$\lambda f = V \qquad \text{eq. (1-1)}$$

For electromagnetic waves this relationship can be expressed in the form:

$$c = \lambda f \sqrt{\varepsilon} \qquad \text{eq. (1-2)}$$

Where:

λ is the wavelength in meters.

f is the frequency in hertz (Hz).

ε is a constant of the propagation medium.

V is the velocity in meters per second.

c is the velocity of light (300,000,000 m/s).

The constant, ε, is a characteristic of the medium in which the wave propagates. The value of ε is defined as 1.000 for a perfect vacuum, and very nearly 1.0 for dry air (typically 1.006). In most practical applications, the value of ε in air is taken to be 1.000. For mediums other than air or vacuum, the velocity of propagation is slower, so that the value of ε relative to a vacuum is higher.

Period. The *period* of the wave is the time required to propagate one wavelength, and is the reciprocal of frequency when the medium is consistent. It is represented by *T*. Period is related to frequency and wavelength by:

$$T = \frac{1}{f} \qquad \text{eq. (1-3)}$$

because $\lambda f = c$,

$$\frac{\lambda}{T} = c \qquad \text{eq. (1-4)}$$

The concept of *phase* refers to the time relationships of a wave, and is usually measured in either degrees or radians. **Figure 1-6** depicts the angular measure of the sinewave. In some instances these phase measurements (T1,T2, etc.) are used to specify phase; that is, the instantaneous point being referred to on a single sinewave. At other times, the term phase is used to denote the angular difference between two waves (**Figure 1-7**).

Electromagnetic Waves

Radio signals, infrared radiation, visible light, ultraviolet light, X-rays, and gamma rays are examples of electromagnetic (EM) waves. These waves are similar in structure to each other, except for frequency and wavelength. The EM wave consists of two mutually perpendicular oscillating fields (**Figure 1-8a**) traveling together. One of the fields is an electric field while the other is a magnetic field.

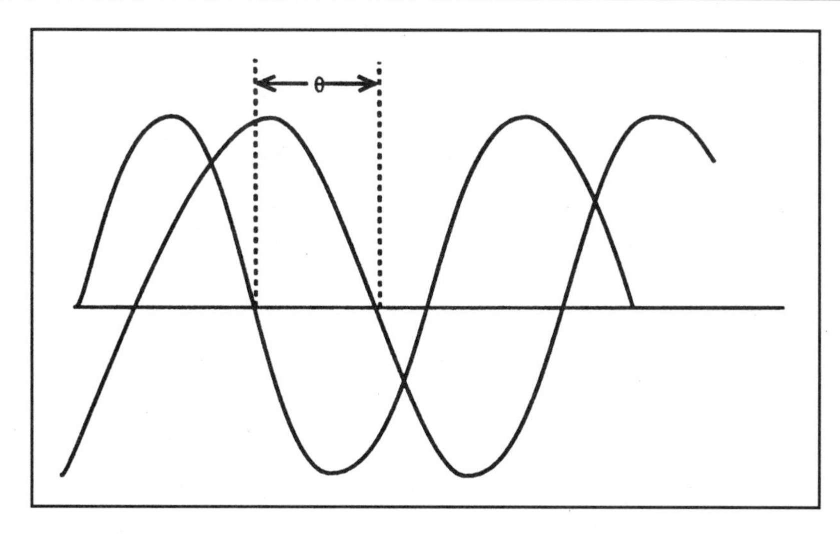

Figure 1-7. "Phase difference" is the difference in timing between two waves.

Electromagnetic wave propagation theory makes use of a construct called an *isotropic source*, which assumes that the radiator, or emitter, is a very tiny spherical source that radiates equally well in all directions. The radiation pattern is thus a sphere with the isotropic source at the center. Because a spherical source is easily calculated mathematically, signal intensities at all points can be determined from basic principles.

The radiated sphere becomes larger as the wave propagates away from the isotropic source. If, at a great distance from the center, we take a look at a small slice of the advancing wavefront we can assume that it is essentially a "flat" plane, as in **Figure 1-8b**. This situation is analogous to the apparent flatness of the prairie, even though the surface of the Earth is a near-sphere. We would be able to "see" the electric and magnetic field vectors at right angles to each other in the flat plane wavefront. *Polarity* in an electromagnetic wave is determined by the nature of the electrical field vector.

An EM wave travels at the speed of light. To put this velocity in perspective, light originating on the Sun's surface reaches Earth in about eight minutes. A terrestrial light beam can travel around the Earth seven times in one second, and such a "signal" can make the round trip between the Earth and the moon in about 2.7 seconds.

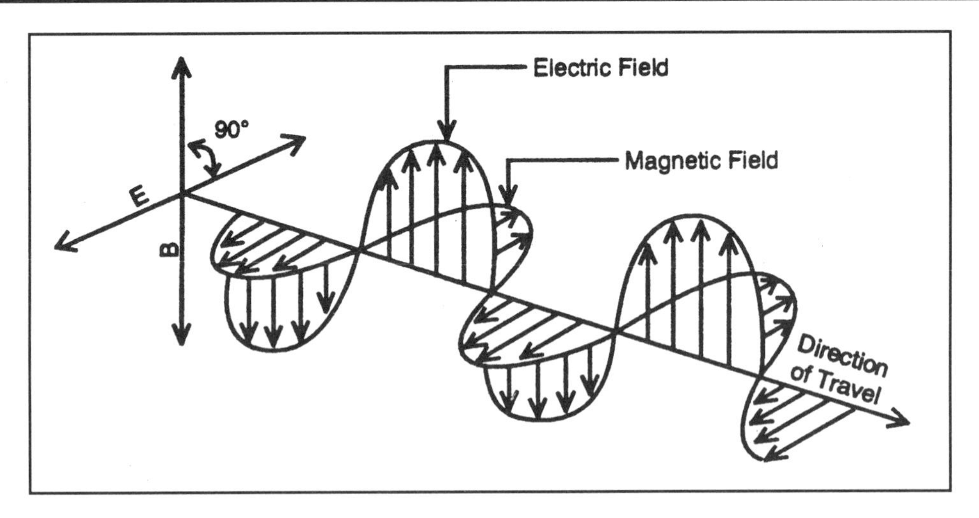

Figure 1-8a. The electromagnetic (E-M) wave consists of mutually perpendicular electric and magnetic fields. This is a linear depiction of an E-M wave propagating in space.

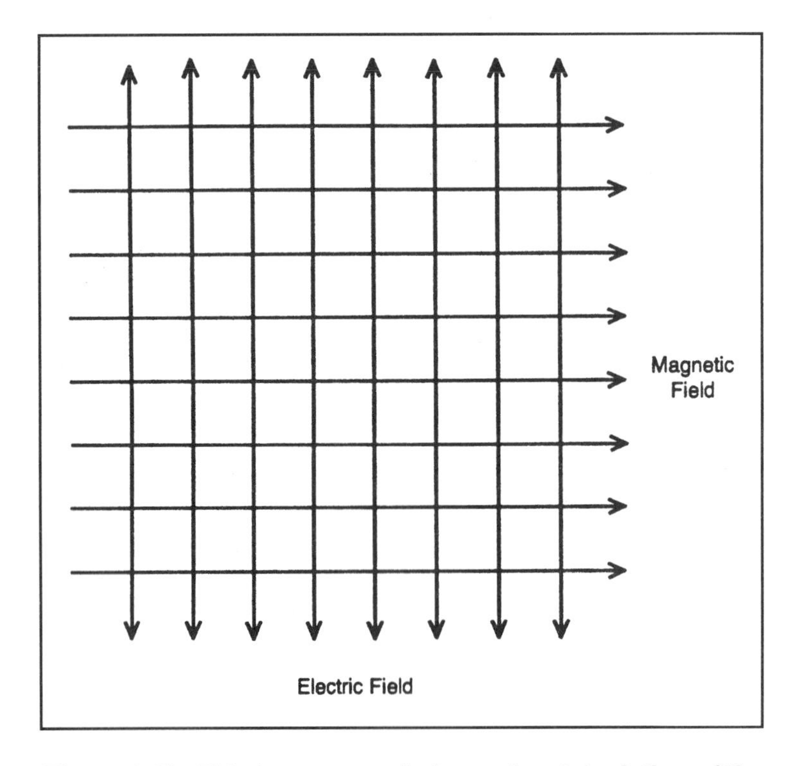

Figure 1-8b. This is an expanded wavefront depiction of the two fields.

The velocity of EM waves slows in denser media, but in air the speed of an EM wave is so close to the free space value of *c* that the same figures are used for both air and vacuum in practical problems. In water, the speed of light is only three quarters of the free space speed.

Wave Propagation

Figure 1-9 illustrates some of the wave behavior associated with light and radio waves: reflection, refraction, and diffraction. All three result from wave propagation. In fact, many situations involve all three in varying combinations.

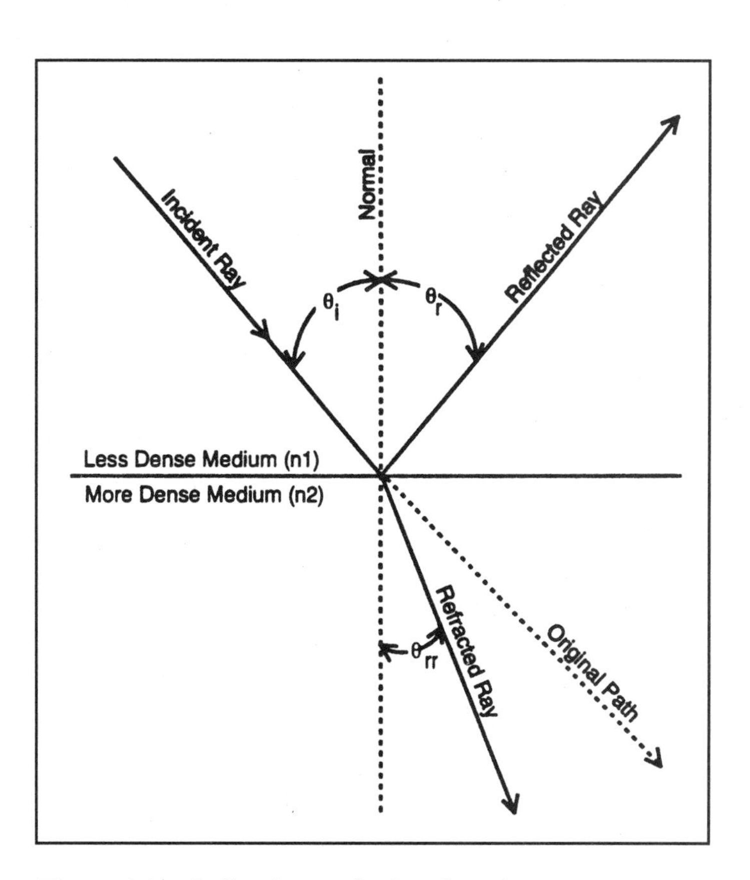

Figure 1-9a. Reflection and refraction phenomena across a boundary between materials of nonequal optical densities.

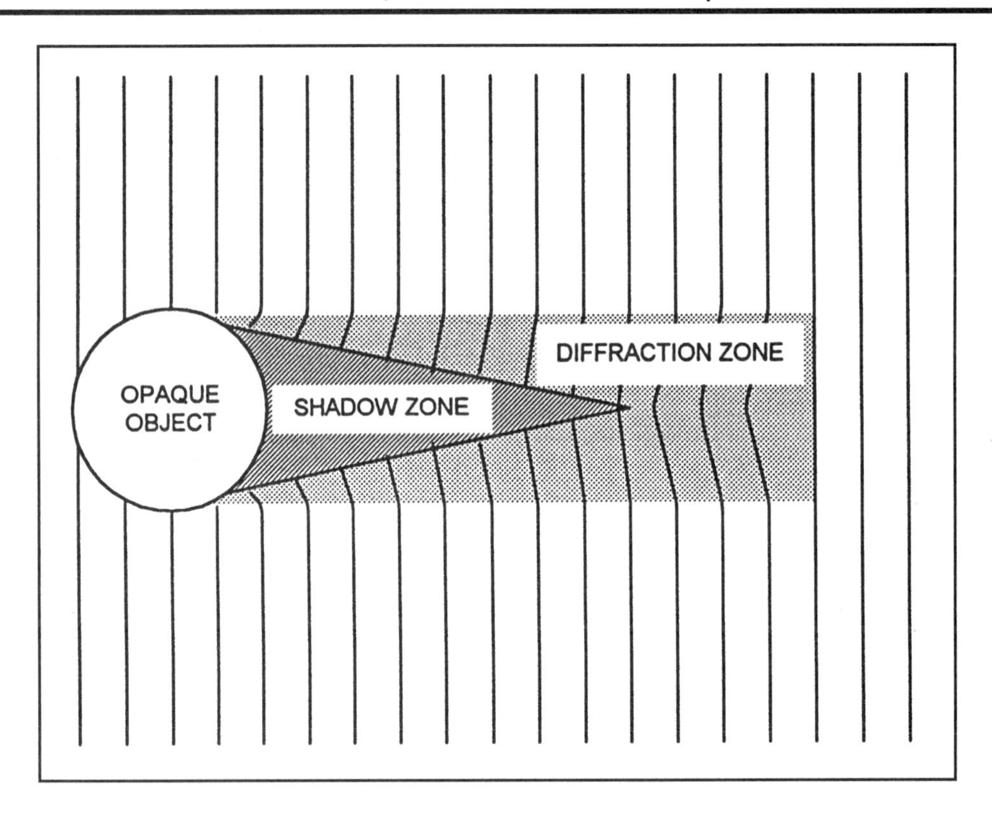

Figure 1-9b. Diffraction around an object.

Reflection occurs when a wave strikes a reflective medium, as when a light wave strikes a glass mirror. The incident wave (shown as a single ray) strikes the interface between less dense and more dense mediums at a certain angle of incidence (θ_i), and is reflected at exactly the same angle (θ_r) (**Equation 1-5**). Because these angles are equal, we can trace a reflected wave back to its origin.

There are two basic forms of reflection: specular (**Figure 1-10a**) and *diffuse* (**Figure 1-10b**). For specular reflections, the medium surface is optically smooth so that a pencil-sized light beam bounced off the surface can be easily traced. In this case, the angle of incidence equals the angle of reflection, or:

$$\theta_i = \theta_r \qquad \textit{eq. (1-5)}$$

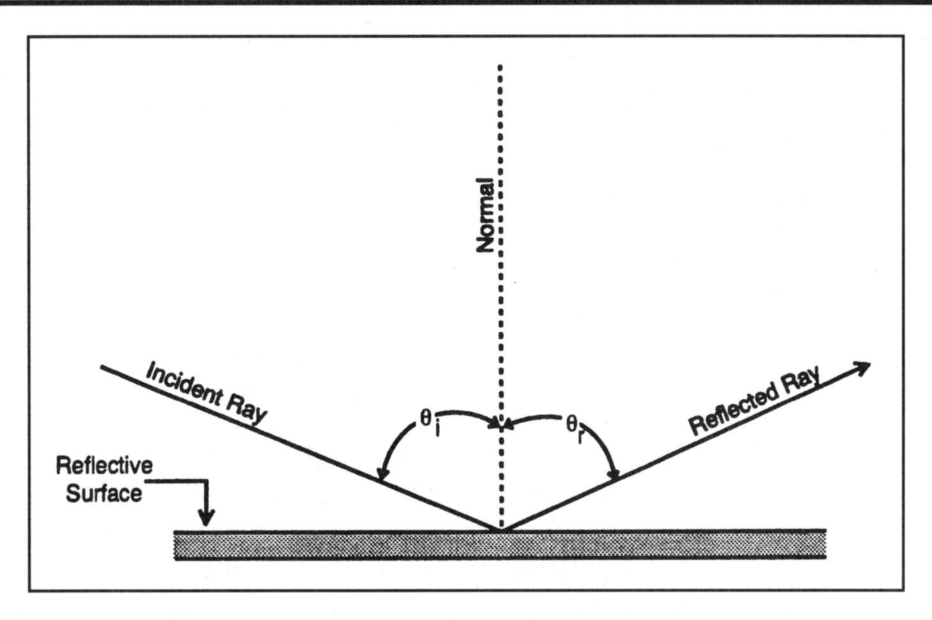

Figure 1-10a. One form of reflection in optical systems, specular reflection, produces a sharp reflected image.

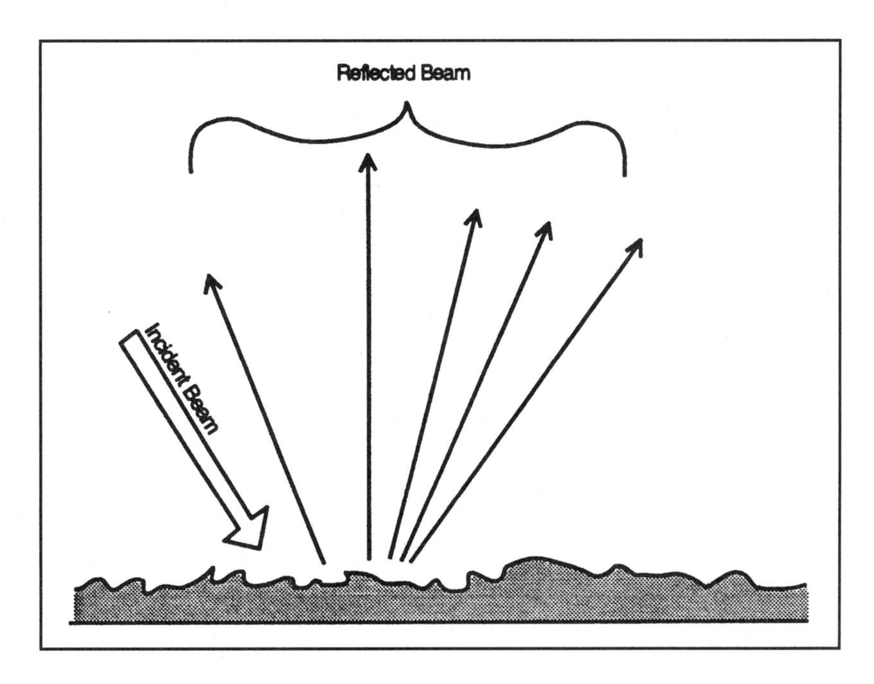

Figure 1-10b. Another form of reflection, diffused reflection, produces a "fuzzy ball" reflected image.

In these measurements the angles are always measured with respect to a "normal" reference line; that is, a line perpendicular to the reflective surface. **Equation 1-5** explains why a person you see in a mirror can also see you if you can see their eyes.

Diffuse reflection occurs from optically rough surfaces. It is difficult to trace the reflected light beam in this situation because one cannot be certain which angle to select. **Equation 1-5** still works, but because each point on the surface may be at a different angle from nearby points, the overall effect is one of a diffused "cloud" rather than a sharp reflected image.

Refraction (see **Figure 1-9a**) occurs when the incident wave enters the region of different density, and thereby undergoes both a velocity and directional change. The amount and direction of the change are determined by the ratio of the densities between the two mediums. If the first medium is markedly different from the second medium, then bending is great. In radio systems, the two media might be layers of air with different densities. This is the phenomena that is responsible for international shortwave "skip" transmissions. It is possible for both reflection and refraction to occur within the same system.

Diffraction is shown in **Figure 1-9b**. In this case, an advancing EM wavefront encounters an object. The shadow zone behind the object is not simply perpendicular to the wavefront, but takes on a cone shape as waves bend around the object. The umbra region (or diffraction zone) between the shadow zone and the direct propagation zone is a region of weak (but not always zero) light strength. In practical situations, light levels in the dark region are raised somewhat by scatter and reflection.

When waves are out of phase and thus interfere with each other in a linear medium, the amplitude of the wave at each point is the algebraic sum of all of the waves present at that point. **Figure 1-11a** shows two waves ("A" and "B") that are exactly superimposed with one another. These two waves are exactly the same frequency and in phase, differing only in amplitude. The resultant wave, shown as a dotted line, is what an observer would see. The resultant is the superposition of the two waves (A + B).

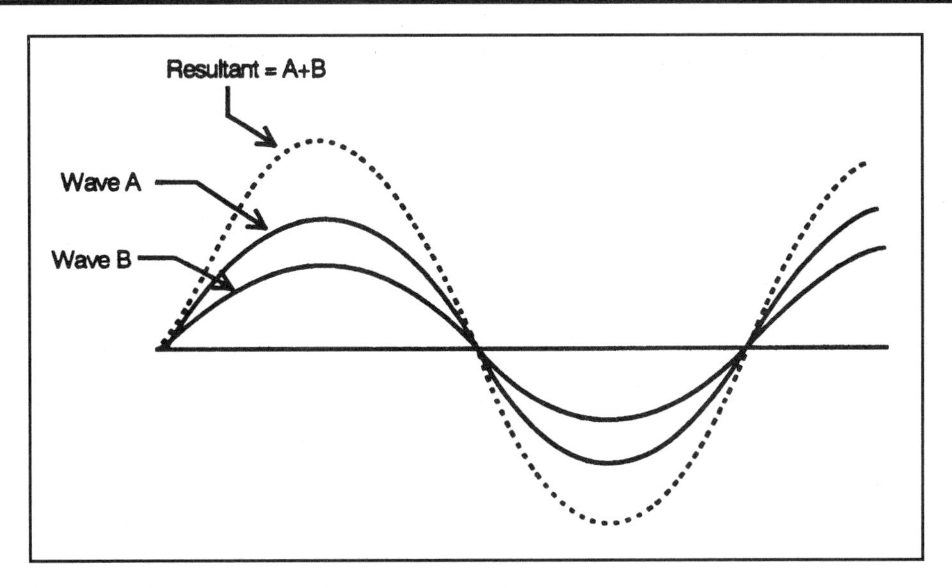

Figure 1-11a. Constructive wave interference.

A different situation is shown in **Figure 1-11b**. In this case, the two waves are exactly out of phase with one another and have equal amplitudes, so the resultant is zero. If the waves were other than equal in amplitude or frequency, or if the phase was neither 0 degrees or 180 degrees, the resultant wave would be some value other than zero or maximum.

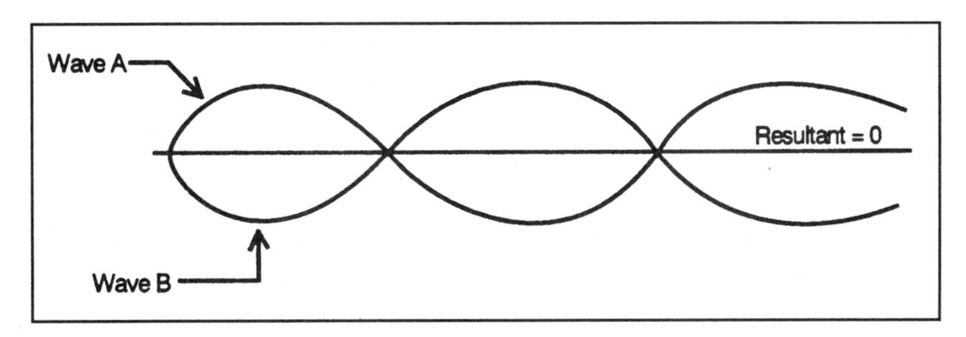

Figure 1-11b. Destructive wave interference.

The Electromagnetic Spectrum

The *electromagnetic spectrum* is a continuum of frequencies and wavelengths from the lowest to the highest. **Figure 1-12** shows the electromagnetic spectrum in slide rule form. The entire radio spectrum, from 40 Hz to 300 GHz, occupies only a tiny portion of the entire spectrum. The chart of **Figure 1-12** exaggerates the low frequency radio bands in order to show them as more than a sliver on the entire range. The visible light spectrum has wavelengths from 400 nanometers (nm) to 800 nm. Longer waves in the adjacent band are called infrared ($\lambda > 880$ nm), and include the radiant energy generated by heat. At the other end of the band is the ultraviolet ($\lambda < 400$ nm).

In electro-optics, the portion of the spectrum that is of greatest interest includes the infrared (IR), visible, and ultraviolet regions. These are not only interesting because we can see and feel them, but also because certain electronic measurement instruments depend on them.

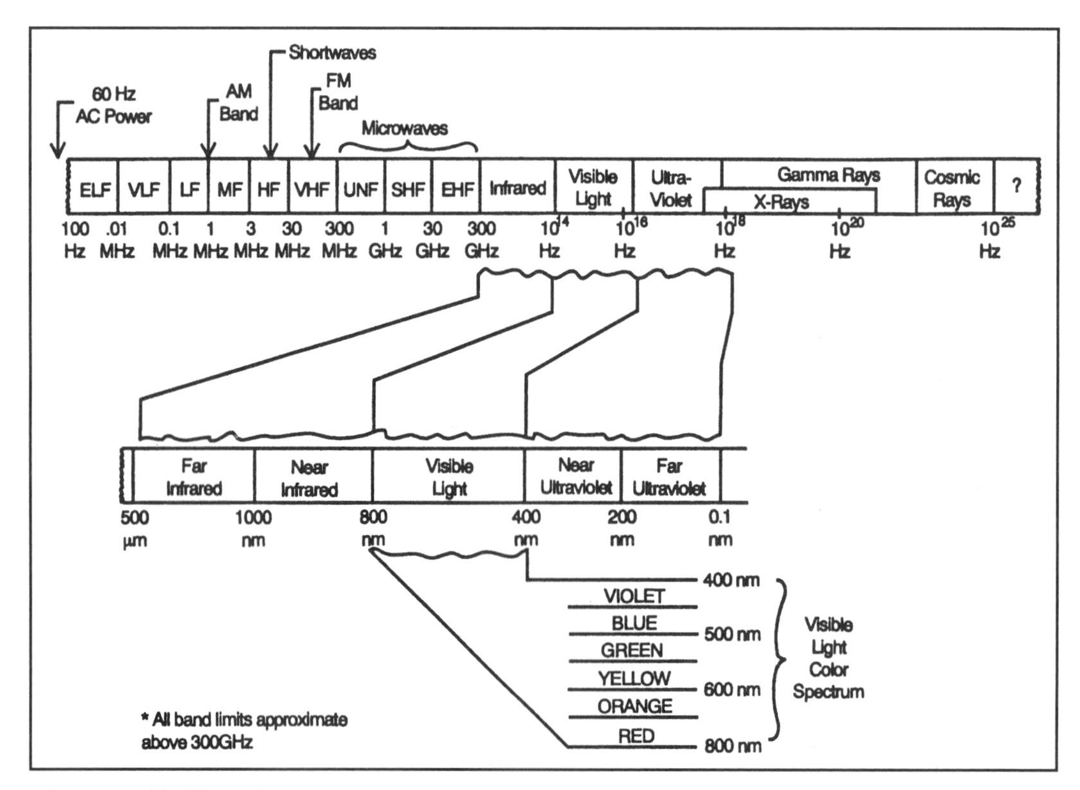

Figure 1-12. The electromagnetic spectrum.

Standing Waves

The *standing wave* is a phenomenon in systems where waves can reflect, thus recombining the incident and reflected waves. We can find this phenomenon in many wave systems, and in radio it becomes critical. When the amateur radio operator talks about standing wave ratio (SWR), or voltage standing wave ratio (VSWR), this combination of waves is what is meant.

Figure 1-13 shows the rope analogy for reflected waves. A taut rope (**Figure 1-13a**) is tied to a rigid wall that does not absorb any of the energy in the pulse propagated down the rope. When the free end of the rope is given a vertical displacement (**Figure 1-13b**) a wave is propagated down the rope at velocity, *v*, (**Figure 1-13c**). When the pulse hits the wall (**Figure 1-13d**) it is reflected (**Figure 1-13e**) and propagates back down the rope towards the free end (**Figure 1-13f**).

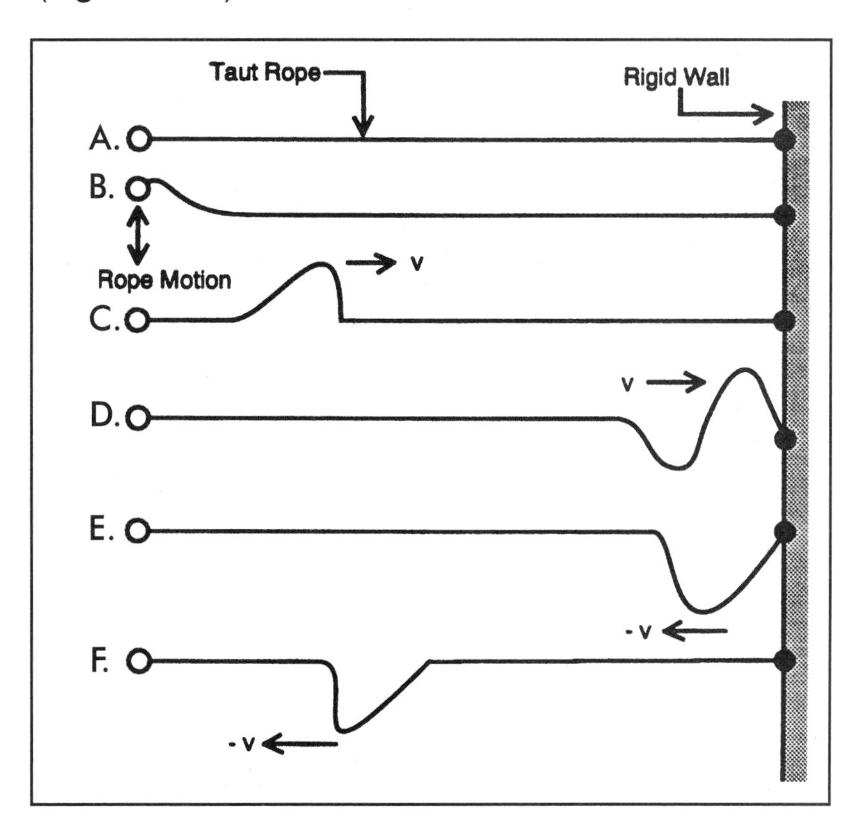

Figure 1-13. Rope analogy for reflected waves.

If a second pulse is propagated down the line before the first pulse dies out, then there will be two pulses on the line at the same time (**Figure 1-14a**). When the two pulses interfere the resultant wave will be the algebraic sum of the two. In the event a pulse train is applied to the line, the interference of incident (or forward) and reflected (or reverse) waves creates standing waves (**Figure 1-14b**).

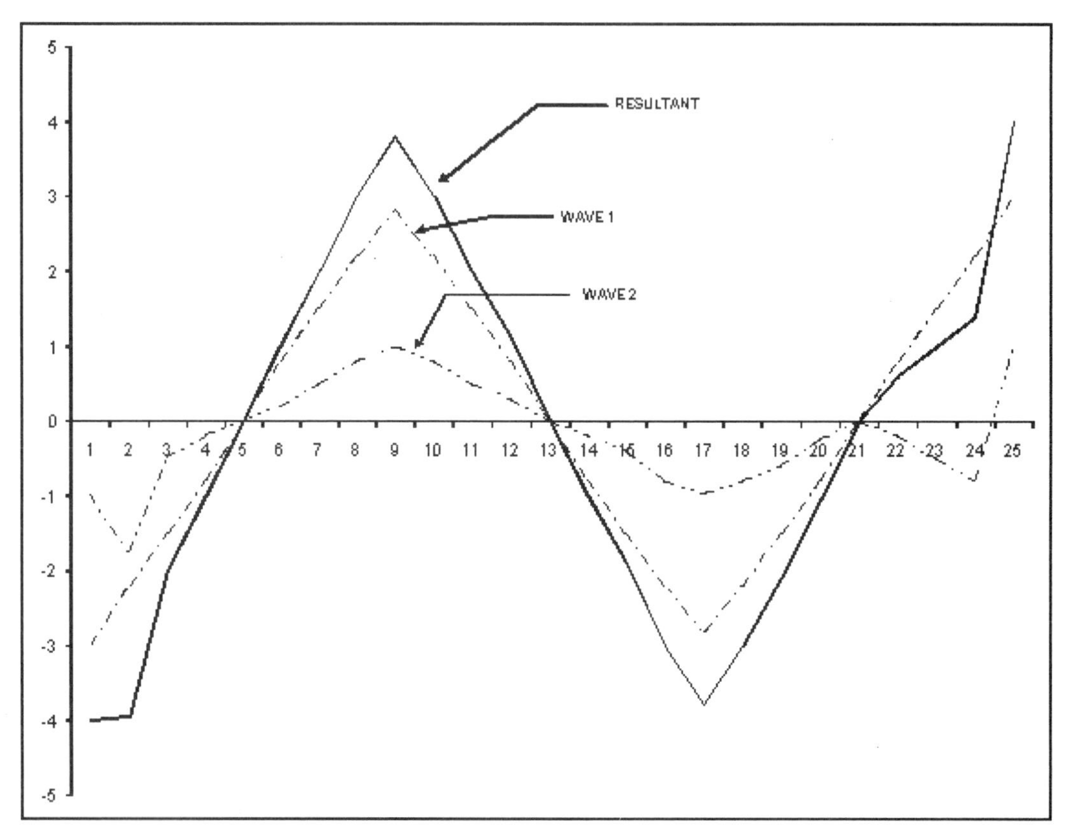

Figure 1-14a. Interference of incident and reflected waves produces a resultant wave that is the algebraic sum of the two original waves.

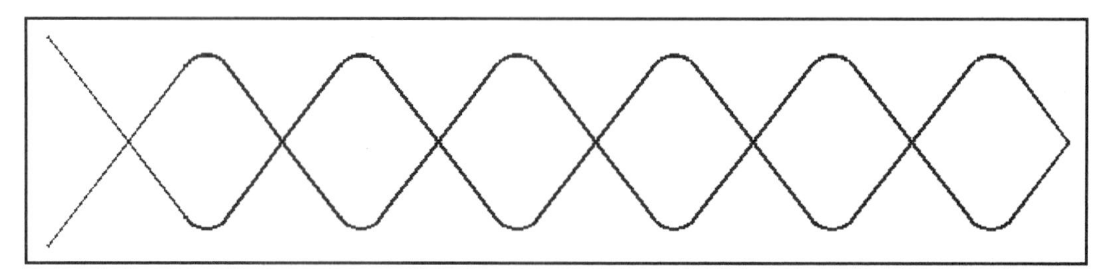

Figure 1-14b. Standing waves result when forward and reflected waves are continuous.

Summary

Thus far we have talked about waves and the nature of light. In the next chapter we will look at more light physics, and the physics of the electron. Electro-optical devices depend heavily on this material, and a good understanding of it is vital to understanding the chapters ahead.

Chapter 2

More Light Physics

This chapter examines two basic areas: *light level measurements* and *electron behavior.* These topics are critical to electro-optics and the operation of the electronic devices that will be discussed in later chapters.

Light Level Measurements

The science of light measurement is called *photometry*, and has traditionally been a difficult area of technology to master. People who are used to the level of precision and accuracy enjoyed by other areas of technology are sometimes struck by the ambiguities of photometry. Photometric measurements are complicated by factors such as stray light artifacts, the spectral content of the light emitted from a given source, nonuniform illumination of the sensor (there is no truly isotropic point source in light measurements), dc (direct current) and thermal drift in the sensors and associated electronics, and the nonlinear spectral response of the sensors typically used to make the measurements.

The earliest practical standard for measuring light level was a comparison standard called the *candlepower* (C.P.), and in modified form this standard, the candela (cd) is still with us today. In the classical candlepower measurement, a specially constructed candle was burned so that unknown light sources could be compared with it. Because of the difficulty inherent in making uniformly burning candles that show little variance from one to another, and the fact that the sensor was the human eye (which exhibits variance of its own, even in the same individual), and that the comparison was essentially a subjective judgment, the early candlepower measurements tended to show a very wide variance. It was frequently the case that the same emitting source would be measured differently by different observers, or at different times by the same observer.

Modern candlepower measurements are made with specially calibrated incandescent lamps as the light source. The United States National Institute for Standards and Technology (NIST), formerly called the National Bureau of Standards (NBS), defines a standard tungsten lamp that has a color temperature of 2,854 K as the source for these measurements. The sensors used in the measurement are thermopiles (two or more thermocouples in a series-adding circuit), and colored filters simulate the response of the normal human eye (**Figure 2-1**).

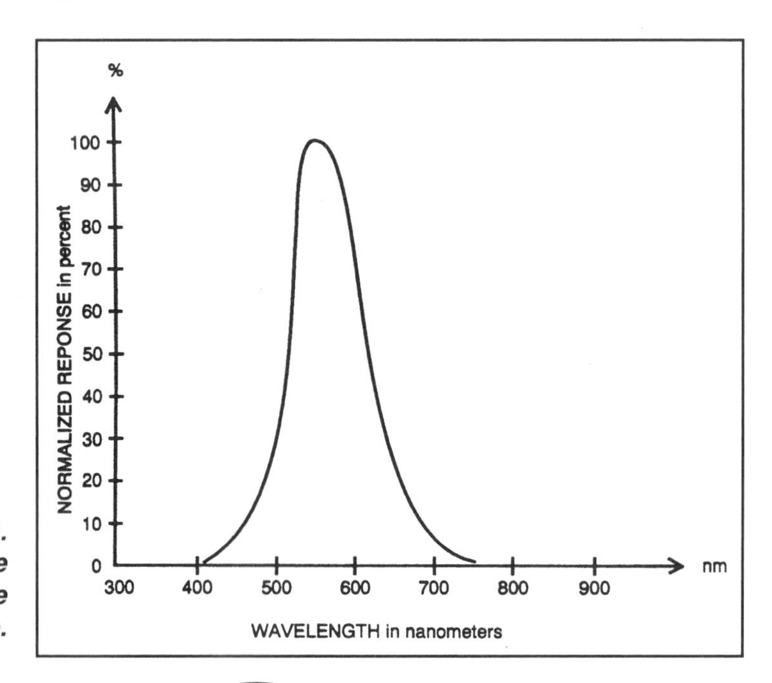

Figure 2-1. Approximate spectral response of the human eye.

Secondary standards are those used in local standards and calibration laboratories to make comparison measurements. Both NIST and private standards laboratories offer calibrated tungsten lamps that produce a specified candlepower output, with a specified spectrum, at standardized voltage and current settings. These lamps are used to make comparison measurements, and when the lamps are supplied or calibrated by NIST the measurements are considered *NIST traceable*.

The output of a tungsten secondary standard lamp is expressed in terms of *horizontal candlepower* (HCP). The light output level "...is normally measured in a plane perpendicular to the lamp's axis at a height in line with the filament."[1] Because the light output level varies with the viewing angle with respect to the lamp filament, the front and back of the lamp are marked. This information allows consistency in measurements between the calibration and subsequent measurements made with the standard lamp.

The measurement procedure calls for varying the distance (D) between the calibrated lamp and the sensor. The distance required to produce any given level of candlepower is:

$$D = \sqrt{\frac{HCP}{Foot\text{-}Candles}} \qquad \text{eq. (2-1)}$$

A potential problem in making footcandle measurements arises from the fact that the incandescent lamp used for the source standard is not a true isotropic point source of light. In order to compensate for that problem, the sensor and lamp should be separated by a long distance (D) relative to the lamp dimensions. The "rule of thumb" is to make the minimum value of D at least six times the largest dimension (L_{max}) of the lamp:

$$D_{min} \geq 6\, L_{max} \qquad \text{eq. (2-2)}$$

At that distance the lamp will approach being an quasi-isotropic point source, and will be more competent for making measurements.

Brightness is measured in units called *footlamberts*. The brightness is usually most appropriate for measuring light reflected from highly diffuse surfaces. By definition, one footlambert (Fl) is the light reflected from a 100 percent reflective diffuse surface that is illuminated with a one footcandle (fc) source. For example, if an 18 percent reflective grey card is illuminated with 1 fc, the brightness of the reflected light is 0.18 Fl.

Brightness measurements are sometimes made through a simulation technique in which an opalized glass screen is interposed between a light source and a sensor. There is a calibration factor published for the glass that relates the brightness in footlamberts on the distal side of the glass (**Figure 2-2**) to the footcandle illumination on the proximal side.

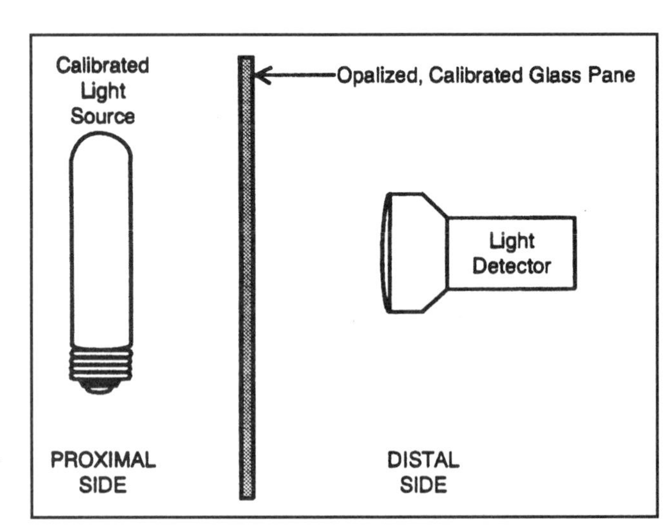

Figure 2-2. Brightness measurement scheme.

Electron Behavior in Electric and Magnetic Fields

Understanding the behavior of electrons in the presence of electrical and magnetic fields is necessary to understand how many of the optoelectronic devices actually work. For readers who just want a view of circuit operation, this material can be omitted.

There is a great deal of mathematics in this chapter. The intention is not so much to provide practical formulas for you to use, but to show the development of the various equations. Several examples are provided, but only the most important points use the most developed form of each equation.

The "Billiard Ball" Model of Electron Dynamics

The classical view of the electron taught in elementary electronics textbooks models this subatomic particle as a point source mass (9.11×10^{-28} gram) with a unit negative electrical charge of e, which equals 1.6×10^{-19} coulombs. The other common charged subatomic particle is the proton, and it has a unit positive charge which balances the electron's charge, producing electroneutrality within the atom. Consider the action of the electron under various conditions.

Electron Motion in Linear Electrostatic Fields

Assume that an electric field E is impressed across a space, s, between two conductive plates (**Figure 2-3**). Assume that the space is a perfect vacuum, so that no inadvertent collisions take place between the electron and gas molecules. The strength of the electric field is E/s (volts/cm). An electron injected into the space interacts with the field in a specific manner. The electron carries a negative charge, so it is repelled from the negative plate while being attracted to the positive plate.

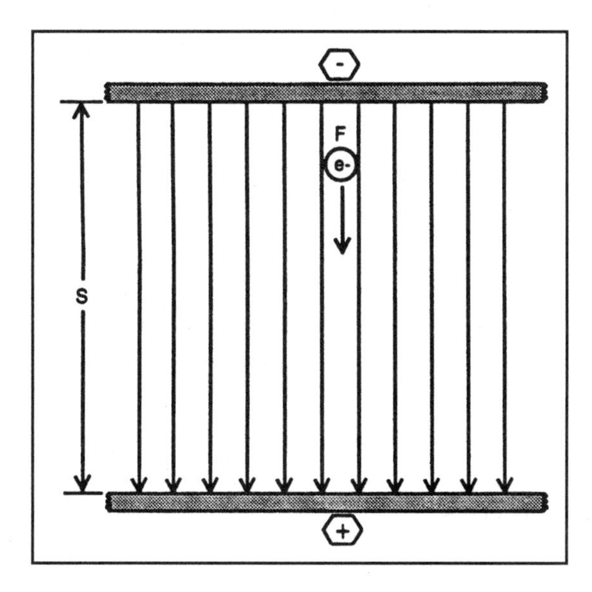

Figure 2-3. Action of a negatively charged electron in a linear electrical field.

The electrostatic field exerts a force on the electron, as proven by the fact that the electron moves. The mechanical work, W_m, done on the electron is:

$$W_m = F\,s \text{ ergs} \qquad \text{eq. (2-3)}$$

and the electrical work, W_e:

$$W_e = E\,e \text{ joules} \qquad \text{eq. (2-4)}$$

The value of the force, F, is:

$$F = \frac{10^7\,E\,e}{s} \text{ dynes} \qquad \text{eq. (2-5)}$$

When the electron is in an electrostatic field, the force is constant if space, s, is constant and the electric field potential, E, is held constant. A particle of mass, m, experiencing a constant force will exhibit a constant acceleration, a, of:

$$a = \frac{10^7\,E\,e}{s\,m} \qquad \text{eq. (2-6)}$$

Where,

a is the acceleration in cm/s^2.

s is the distance in cm.

m is the mass in grams.

E is the electric field potential in volts.

e is the electron charge (1.6×10^{-19} coulombs).

In a perfect vacuum the electron velocity is a function of the applied field strength. Because the acceleration is constant, the average velocity, v_{av}, will be one-half the difference between final velocity, v_f, and initial velocity, v_i:

$$v_{av} = \frac{v_f - v_i}{2} \qquad \text{eq. (2-7)}$$

or, if we assume that initial velocity $v_i = 0$, then **Equation 2-7** reduces to:

$$v_{av} = \frac{v_f}{2} \qquad \text{eq. (2-8)}$$

The time, t, required to traverse the field is the distance divided by average velocity:

$$t = \frac{s}{v_{av}} \qquad \text{eq. (2-9)}$$

Substituting **Equation 2-8** into **Equation 2-9**:

$$t = \frac{s}{\left(\frac{v_f}{2}\right)} \qquad \text{eq. (2-10)}$$

$$t = \frac{2s}{v_f} \qquad \text{eq. (2-11)}$$

Final velocity, v_f, is the product of acceleration and time:

$$v_f = a\,t \qquad \text{eq. (2-12)}$$

Or, by substituting **Equation 2-6** and **Equation 2-11** into **Equation 2-12**:

$$v_f = \left(\frac{10^7\,E\,e}{s\,m}\right)\left(\frac{2\,s}{v_f}\right) \qquad \text{eq. (2-13)}$$

and by combining terms:

$$v_f^2 = \frac{2 \times 10^7\,E\,e}{m} \qquad \text{eq. (2-14)}$$

or,

$$v_f = \sqrt{\frac{2 \times 10^7\,E\,e}{m}} \qquad \text{eq. (2-15)}$$

Note that the terms in **Equation 2-15** are all constants except for field strength, E. This equation therefore reduces to:

$$v_f = \sqrt{\frac{(2 \times 10^7)(1.6 \times 10^{-19}\,coulombs)}{9.11 \times 10^{-28}\,gm}} \times \sqrt{E} \qquad \text{eq. (2-16)}$$

$$v_f = \sqrt{\frac{3.2 \times 10^{-12}}{9.11 \times 10^{-28}}} \times \sqrt{E} \qquad \text{eq. (2-17)}$$

- $v_f = 5.93 \times 10^7 \sqrt{E}\ cm/s$ eq. (2-18)

Example 2-1

An electron is injected into an electric field of 1,200 volts. Calculate the final velocity assuming that the initial velocity is zero.

Solution:

- $v_f = 5.93 \times 10^7 \sqrt{E}\ cm/s$
- $v_f = 5.93 \times 10^7 \sqrt{1{,}200\ volts}\ cm/s$
- $v_f = (5.93 \times 10^7)(34.6)\ cm/\sec = 2.05 \times 10^9\ cm/s$

A moving electron of mass, m, has a kinetic energy, U_k of:

- $U_k = \frac{1}{2} m v^2$ eq. (2-19)

or,

- $U_k = 10^7 E e$ eq. (2-20)

Example 2-2

Find the kinetic energy of a moving electron that has a velocity of 2.05×10^9 cm/s.

Solution:

- $U_k = \frac{1}{2} m v^2$

- $U_k = \frac{1}{2}(9.11 \times 10^{-28}\ g)(2.05 \times 10^9\ cm/s)^2$

- $U_k = \frac{1}{2}(9.11 \times 10^{-28}\ gm)(4.2 \times 10^{18}\ (cm/s)^2) = 1.9 \times 10^9\ ergs$

In a later section we will use these relationships to describe other physical phenomena.

Why Not a Perfect Vacuum?

In the preceding material we assumed that the electron travels in a perfect vacuum. Such an electron can theoretically accelerate until its velocity reaches the speed of light. While this assumption is reasonable for analysis purposes, it fails in real devices. Because of residual internal gas pressure, all real vacuum electronic devices contain atoms or molecules that can collide with individual electrons.

Even when the initial evacuation is nearly perfect, there are at least two mechanisms that introduce gasses into the vacuum device. First, leaks allow air inside, if only a molecule at a time. Second, a certain amount of "out-gassing" from the materials that make up the internal structures of the device adds molecules. Thus, in a stream of electrons flowing in a vacuum device the average velocities reach an upper limit; this limit is called the *saturation velocity*, v_{sat}. Saturation velocity is also found in conductors because of electron interaction with the atoms of the conductor or semiconductor material.

Saturation velocity becomes important in microwave and optoelectronic devices because it limits the speed of travel across devices (this is called "transit time"), and therefore limits the operating frequency. For example, in a vacuum tube the electron must traverse a vacuum space between a cathode and anode. As the period of a waveform approaches the device transit time, the ability of the device to amplify breaks down. Similarly, in a semiconductor device, the saturation velocity is on the order of 10^7 m/s, so there is an upper frequency limit for any given path length through the semiconductor material. If this limit did not exist, there is no reason why light—an electromagnetic wave—could not be amplified similarly to radio waves.

Motion in an Orthogonal Electric Field

In the previous section we discussed the case of an electron injected into an E-field parallel to the electric lines of force (see **Figure 2-3**). Now consider the case of the electron injected into the E-field orthogonal to the electrical lines of force (**Figure 2-4**).

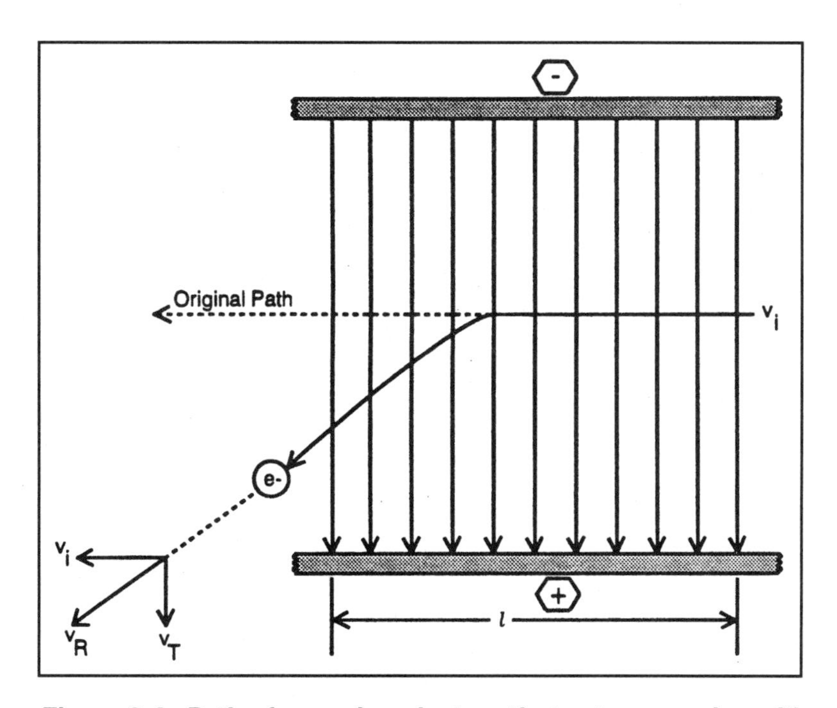

Figure 2-4. Path of a moving electron that enters a region with an electrical field.

In **Figure 2-4**, the electron enters the E-field from the right with an initial velocity, v_i. The negatively charged electron is repelled from the negative end of the field, and is attracted to the positive end. It deflects from its original path to a new curved path in the direction of the positive end of the field. The degree of deflection, measured by angle θ, is a function of the strength of the E-field and the length of the interaction region, L.

During its travel through the E-field, the electron retains its initial velocity, but also picks up a translational velocity, v_t, in the direction of the positive end of the field. The resultant velocity, v_r, along the line of travel is a vector sum of the two components, or in trigonometric terms:

- $$\theta = \arctan\left(\frac{v_t}{v_i}\right)$$ *eq. (2-21)*

and,

- $$v_r = v_i \cos\theta$$ *eq. (2-22)*

Example 2-3

An electron has an initial velocity of 3.2×10^9 cm/s, and a translational velocity of 2.9×10^9 cm/s in an electrical field. Find the angle of deflection of the electron away from its original track in the electric field.

Solution:

- $$\theta = \arctan\left(\frac{v_t}{v_i}\right)$$

- $$\theta = \arctan\left(\frac{2.9 X 10^9}{3.2 X 10^9}\right) = 42.2 \text{ degrees}$$

Example 2-4

Calculate the resultant velocity of the electron in Example 2-3.

Solution:

- $v_r = v_i \cos\theta$
- $v_r = (3.2 \times 10^9\ cm/\sec)\cos(42.2\ degrees)$
- $v_r = (3.2 \times 10^9\ cm/\sec)(0.74) = 2.4 \times 10^9\ cm/s$

Relativistic Effects on Electrons

The electronic mass quoted in most textbooks (9.11×10^{-28} gm) is the *rest mass* of the particle, and its value is usually symbolized by m_o. For velocities $v << c$, where c is the velocity of light (3×10^{10} cm/s), the rest mass is a good value to use. As the velocity increases above about 0.1 c, the electron becomes heavy enough to begin to show errors in calculations. The new mass, according to the Theory of Relativity, is:

- $$m = \frac{m_o}{\sqrt{1 - \frac{v^2}{c^2}}}$$ *eq. (2-23)*

Where:

m is the relativistic mass at velocity v.

m_o is the rest mass (9.11×10^{-28} gm).

v is the electron velocity in cm/s.

c is the speed of light (3×10^{10} cm/s).

Example 2-5

An electron at rest, or traveling at very low velocities, has a mass, m_o, of 9.11 $\times 10^{-28}$ gm. Calculate its relativistic mass at a) the electron saturation velocity for silicon (~5 $\times 10^{9}$ cm/s), and b) a velocity of 2.8 $\times 10^{10}$ cm/s.

Solution A:

- $$m = \frac{m_o}{\sqrt{1 - \frac{v^2}{c^2}}}$$
- $$m = \frac{9.11 \times 10^{-28}\ gm}{\sqrt{1 - \frac{(5 \times 10^{9}\ cm/\sec)^2}{(3 \times 10^{10}\ cm/\sec)^2}}}$$
- $$m = 9.24 \times 10^{-28}\ gm$$

Solution B:

- $$m = \frac{m_o}{\sqrt{1 - \frac{v^2}{c^2}}}$$
- $$m = \frac{9.11 \times 10^{-28}\ gm}{\sqrt{1 - \frac{(2.8 \times 10^{10}\ cm/s)^2}{(3 \times 10^{10}\ cm/s)^2}}}$$
- $$m = 2.54 \times 10^{-27}\ gm$$

The voltage potentials at which these relativistic effects become important are surprisingly low. The general rule is that relativistic effects begin to be important at potentials above 5,000 volts. Therefore, designers sometimes use relativistic mass in their calculations.

Electron Motion in Magnetic Fields

An electron in motion constitutes an "electrical current," so it will react when it encounters a magnetic field. Consider **Figure 2-5**, where an electron is injected into a magnetic field of H, measured on oersteds. (The oersted, symbolized by Oe, is the cgs-system unit of magnetizing force such that one oersted equals 0.796 ampere-turns per centimeter.) In this figure the arrows pointing into the page indicate that the magnetic lines of force are *away* from the observer. The electron is deflected into a circular path of radius r by the magnetic field. The electron does not alter either tangential velocity, kinetic energy or magnitude of momentum.

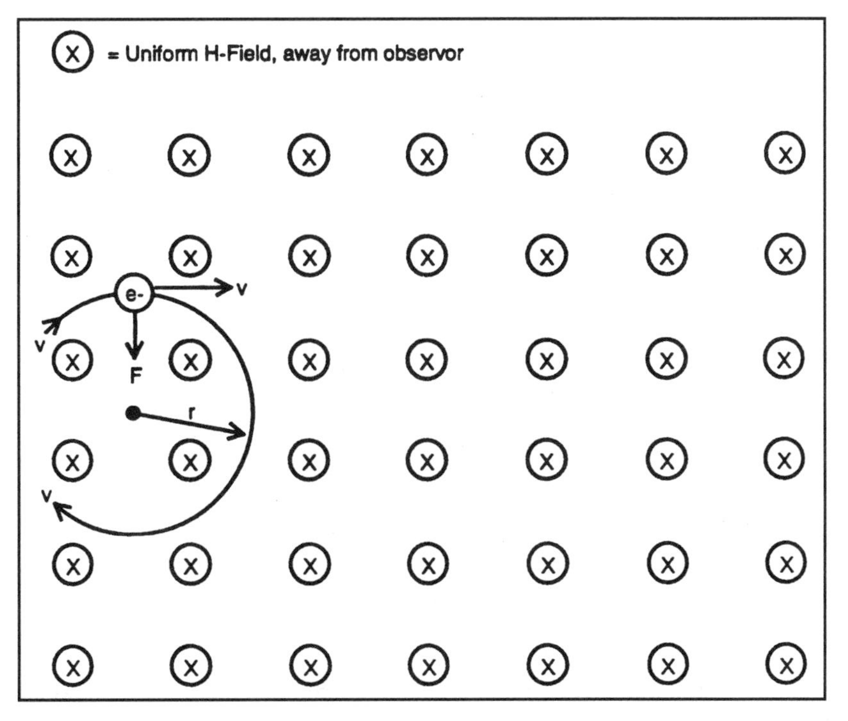

Figure 2-5. Path of an electron entering an orthogonal magnetic field.

The deflecting force, F, is derived from the force exerted on a current-carrying conductor by a magnetic field:

$$F = \frac{H\,v\,e}{10} \text{ dynes} \qquad \text{eq. (2-24)}$$

The *angular velocity*, ω, of the electron is given by v/r, while its acceleration, a, is vw. The acceleration is directed toward the center of the circle along the same vector line as force, F. We know from Newton's laws that F = ma, so we may conclude that F = ma = mvω. Therefore:

$$m\,v\,\omega = \frac{H\,v\,e}{10} \qquad \text{eq. (2-25)}$$

or,

$$\omega = \frac{H\,e}{10\,m} \qquad \text{eq. (2-26)}$$

But, because e and m are constants:

$$\omega = \frac{H\,(1.6 \times 10^{-19} \text{ coulombs})}{(10)(9.11 \times 10^{-28} \text{ gm})} \qquad \text{eq. (2-27)}$$

$$\omega = 1.76 \times 10^{7}\,H \text{ radians/second} \qquad \text{eq. (2-28)}$$

An implication of **Equation 2-27** is that angular angular, ω, is invariant under changes in velocity, v. This fact becomes important when we study certain "vacuum tube" microwave power generators.

The path radius, r, is given by r = v/ω. Thus, by substituting the expressions for v and ω in **Equation 2-26** and **Equation 2-27**, respectively, we arrive at:

- $$r = \frac{v}{\omega}$$ *eq. (2-29)*

Equation 2-29 can also be used to calculate the cyclotron frequency, as is seen in the following example.

Example 2-6

Find the cyclotron frequency of an electron that rotates about an axis in an orbit with a radius, r = 2 cm, and a velocity, v = 8×10^9 cm/s.

Solution:

- $$\omega = \frac{v}{r}$$

- $$\omega = \frac{8 \; x \; 10^9 \; cm/s}{2 \; cm/cycle} = 4 \; x \; 10^9 \; Hz$$

or, about 4,000 MHz. (Insightful readers may recognize this principle as the basis for the microwave radio power generator device called the *magnetron*.)

Further,

- $$r = \frac{5.93 \; x \; 10^7 \sqrt{E}}{1.76 \; x \; 10^7 \; H}$$ *eq. (2-30)*

- $$r = \frac{3.37 \sqrt{E}}{H} \; cm$$ *eq. (2-31)*

Example 2-7

Calculate the radius of rotation of an electron in a combined electric field of 4,000 volts and a magnetic field of 2500 oersteds.

Solution:

- $$r = \frac{3.37\sqrt{E}}{H}$$

- $$r = \frac{3.37\sqrt{4{,}000\ volts}}{2{,}500\ Oe} = 0.085\ cm$$

Electron Motion in Simultaneous E and H Fields

Thus far we have studied the effects of either an electric or a magnetic field on the trajectory of a moving electron. Many electronic devices can be understood with only this information, but some apply both electric and magnetic fields simultaneously. It is necessary to examine the path of an electron in simultaneous E/H fields. Two cases are considered: parallel fields and *crossed fields*. It is not surprising that two different classes of microwave vacuum tube depend on these cases.

Motion in Parallel Fields

Consider the case where two fields, E and H, are parallel to each other in the same space, as in **Figure 2-6a**. The trajectory of the electron in this field depends on the velocity vector at the site of injection. If this vector is parallel to the E and H fields (v1 in **Figure 2-6a**), then the electron possesses no tangential velocity and the magnetic field therefore cannot act on it. If the injection velocity vector is <u>not</u> parallel to the fields (v2 in **Figure 2-6a**), then there is a tangential component to the velocity and the electron interacts with the H-field. The path will be helical as shown in side aspect in **Figure 2-6a**, and end aspect in **Figure 2-6b**.

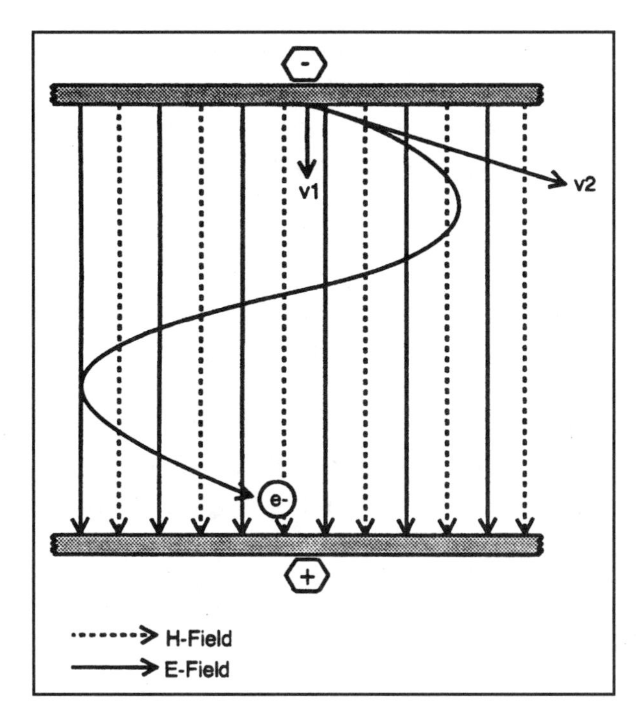

Figure 2-6a. Electron in parallel magnetic and electric fields along the path.

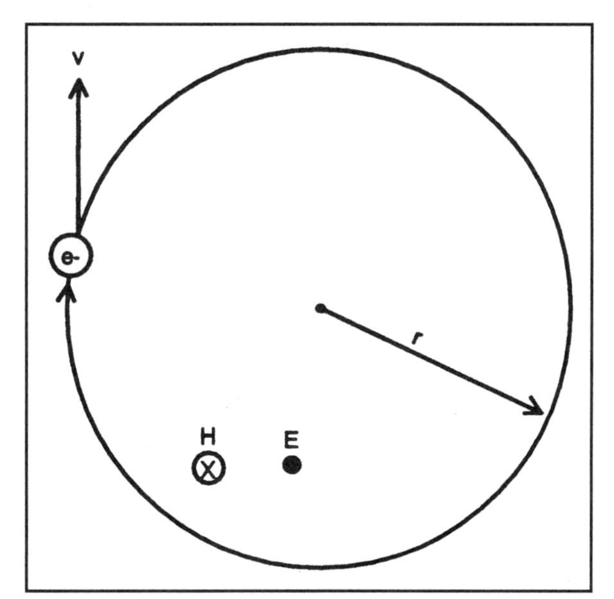

Figure 2-6b. Path of electron when parallel E- and H-fields are orthogonal to the plane of the page.

Quantum Mechanical Description Of Electrons

The "billiard ball" description of electrons works well in some areas of electronics technology, such as low velocity vacuum tube, certain transistor devices, or magnetrons. It does not, however, reasonably explain the "tunneling" phenomenon found in the Esaki ("tunnel") diode. For some devices and phenomena, we must appeal to the quantum mechanical description of matter, including electrons.

The quantum mechanics revolution, which brought us the many benefits of electronics technology, began, as we have seen, on December 14, 1900 at a meeting of the German Physical Society in Berlin. For decades scientists had been working on the problem of black body radiation; that is, the radiation given off by a black body, a theoretical body that absorbs 100 percent of the electromagnetic radiation that falls on it, when it is heated. Planck demonstrated that the equations for predicting radiation levels fit the experimental data if he assumed only certain discrete levels of energy. These discrete energy levels are defined by:

$$U = n h \qquad \text{eq. (2-32)}$$

Where:

U is the energy in ergs.

h is Planck's constant (6.63×10^{-27} erg-s).

n is an integer (1, 2, 3 ...).

From these calculations came Planck's constant, a key concept in quantum mechanics. In 1905, Albert Einstein demonstrated that Planck's theory applied to the photoelectric effect. In this phenomenon, the energy emitted when special photosensitive metallic plates are exposed to light is not linear with light intensity (as expected), but rather a function of light frequency. This discovery raised a question for scientists who had, since Maxwell, viewed light as a wave: this photoelectric phenomenon was more easily explained if light were considered to be a particle. It is now posited that light has a dual nature: it behaves as a particle in some experiments and as a wave in others; the two properties are complementary, not contradictory. In the 1920s Danish physicist Bohr formalized this proposition in his *Complementarity Principle.*

The relationship between the energy of the radiation and the frequency (color) of the incident light is:

$$U = n h f \qquad \text{eq. (2-33)}$$

Where:

U is the energy in ergs.

h is Planck's constant (10^{-7} Joules in SI units)(6.63×10^{-27} erg-s).

f is the frequency of the light in hertz.

n is an integer (1, 2, 3 ...).

Figure 2-7 shows the permitted values of n, U and f. The values of U represent energy state levels. The dots represent the states that are allowed. **Figure 2-8** shows an energy diagram. When an electron is in its lowest natural state (U1), it is said to be in the *ground state*. When the system absorbs energy it can increase its energy state. This is the *absorption line*. The energy level of the electron jumps from U1 to U5. The excited energy state is unstable, and the electron will soon fall back toward ground state, releasing a photon. This phenomenon is shown as the *emission line.*

It had been noted in the 19th century that certain materials emitted light when excited above the ground state. Applying a certain amount of energy to a system would cause the emission of specific colors of light as the system returned to equilibrium. Several sets of light spectra were noted: the Rydberg series, the Lyman series, the Paschen series, and the Balmer series. Bohr explained the several series by applying Planck's theory of quantized energy to the electrons orbiting the nucleus in an atom. When an electron at rest

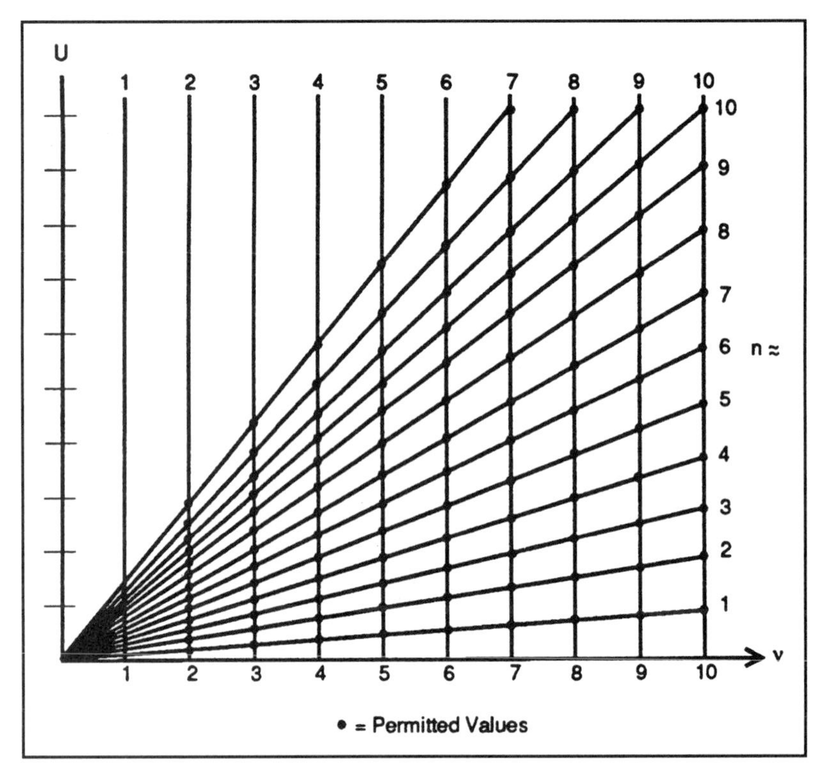

Figure 2-7. Permitted energy states of an electron.

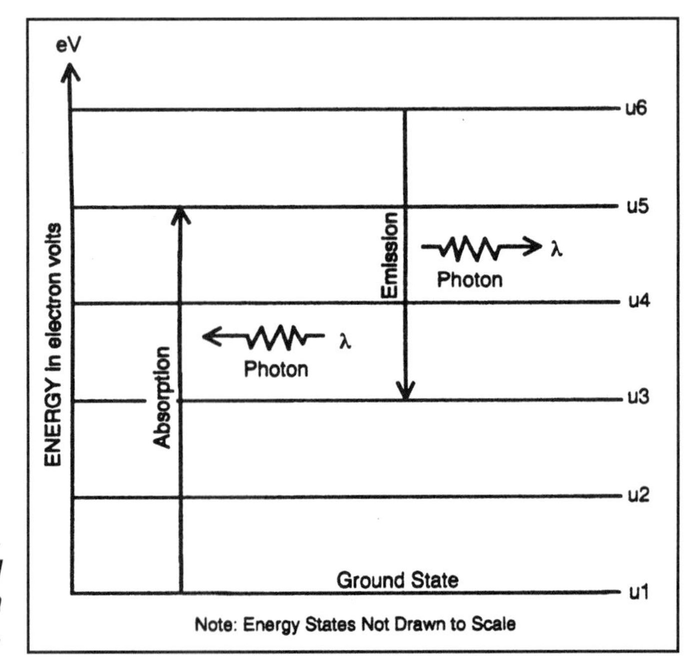

Figure 2-8. Absorption and emission phenomena energy diagrams.

receives energy from an external source, its energy level increases only in certain allowable discrete levels defined by Planck's constant (U2, U3, etc. in **Figure 2-9**). The energetic state is unstable, and the electron soon returns toward ground state. Because of the Law of Conservation of Matter and Energy, that energy must be emitted; this emission takes the form of a photon of light with a frequency according to **Equation 2-33**.

The expectation was that an electron would fall from the excited state back to a lower state in one movement. This motion would cause a single color of light to be emitted corresponding to $U_{6,3}$ (**Figure 2-9**), the difference in energy levels between U_6 and U_3. But in the experiments cited, there was more than one color in the spectra, indicating a different process was taking place. Bohr explained the existence of multiple colors by postulating that the electrons dropped back to the original state in more than one step. In **Figure 2-9** there are two steps ($U_{6,4}$ and $U_{4,3}$); that is, functionally equivalent to $U_{6,3}$ from a conservation point of view, but it allowing for the existence of the multiple color emissions.

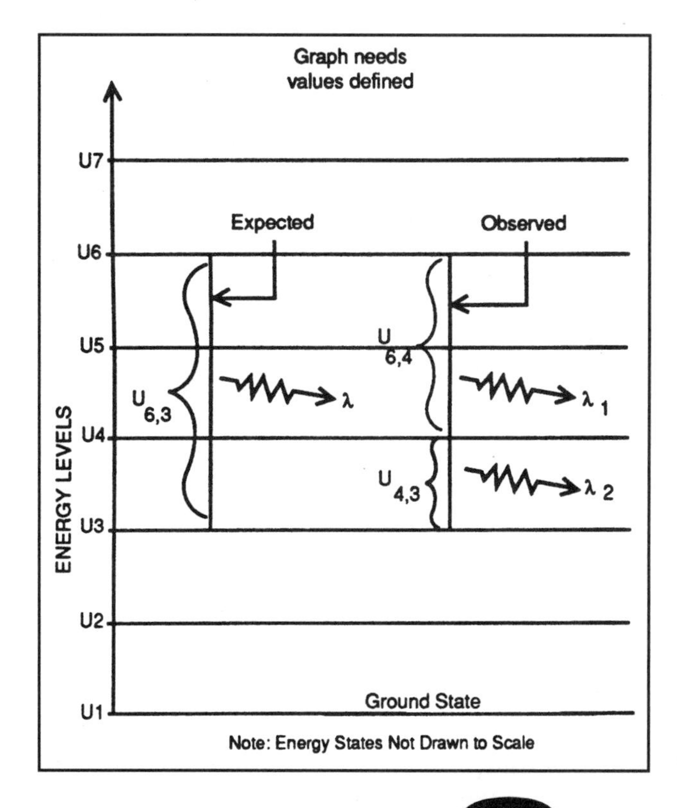

Figure 2-9. Expected one-step and observed two-step emission phenomena.

The phenomenon explained by Bohr is the basis of a number of devices today. It is also the operant phenomenon in such mundane optoelectronic devices as neon glow lamps, fluorescent lamps, and certain semiconductor devices. In the laser (Light Amplification by Stimulated Emission of Radiation), the simultaneous emission of energy from an extremely large number of electrons is coordinated by an external energy source (such as a xenon flash tube) so that the emissions occur in-phase with each other, emitting so-called coherent light, the hallmark of the laser.

The Bohr model of the atom assigned electrons to orbits around the nucleus at specific quantized distances prescribed by their respective energies. Bohr's solar system model of the atom is still accepted today for elementary explanations, even though physicists know that the real situation is more complex.

During the 1920s quantum mechanics became formalized into the modified theory that is used today. In 1924 Duke Louis de Broglie proposed that not only light but matter also has a dualistic complementary wave-particle nature. The wave function of matter is usually symbolized by Ψ. According to de Broglie, the wavelength of a particle such as an electron is given by:

- $$\lambda = \frac{h}{m v}$$ *eq. (2-34)*

or,

- $$\lambda = \frac{h}{\sqrt{2 U m}}$$ *eq. (2-35)*

Where:

λ is the wavelength.

h is Planck's constant (6.63×10^{-27} erg-s).

U is the energy of the particle (ergs).

m is the mass of the particle (gm).

v is the velocity of the particle (cm/s).

Example 2-8

Calculate the de Broglie wavelength of an electron that has a velocity of 10^9 cm/s.

Solution:

- $$\lambda = \frac{h}{m\,v}$$

- $$\lambda = \frac{6.63 \times 10^{-27}\ erg - s}{(9.11 \times 10^{-28}\ gm)(\ 10^{9}\ cm/s)}$$

- $$\lambda = 7.3 \times 10^{-9}\ cm$$

Combining the Bohr and de Broglie theories produced a model in which the simple orbital electron actually became a standing wave. An integer number of de Broglie waves fits into the space allocated by the Bohr theory for an electron of a given energy level (**Figure 2-10**). According to de Broglie's theorem, we can deduce the following relationship in which the circumferences of the orbits are integer multiples of the de Broglie wavelength:

- $$2\pi\, r = \frac{n\,h}{m\,v} \qquad eq.\ (2\text{-}36)$$

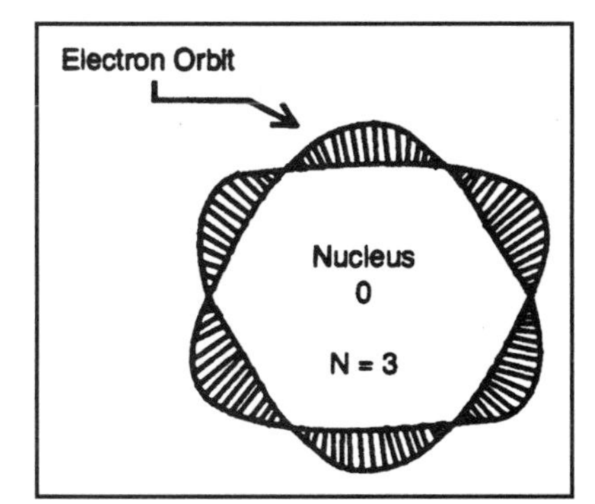

Figure 2-10. Wave model of electron cloud.

Erwin Schrödinger disputed Bohr's solar system model of the atom. Instead of a billiard ball nucleus surrounded by billiard ball electrons, Schrödinger proposed (1933) an entirely new model based on a probabilistic interpretation of the wave function. Like light waves, elementary subatomic particles sometimes behave like particles and other times like waves. Again we have a complementary system, but in this case the waves are the "matter waves" postulated by de Broglie.

According to Schrödinger's view, the atom consists of a matter wave nucleus surrounded by matter wave electrons. Schrödinger's wave equation describes the matter waves, Ψ, in terms of probability, Ψ^2. It is important to realize that matter wave equations do not describe a real chain of events the way water wave equations describe real movement by real water particles. The equations describe only the *probabilities* of finding a real particle at a specific place at a given time.

To these factors we must now add another facet: the *Uncertainty Principle*. In 1927 Werner Heisenberg proposed that certain pairs of properties of atomic particles can not both be measured with accuracy simultaneously. For example, it is impossible to precisely measure both the position and momentum of an electron. Momentum is the product of mass and velocity, so it follows that we cannot know both where the electron is located and how fast it is traveling. The product of the accuracies with which each can be known cannot, Heisenberg found, be less than Planck's constant. Stated mathematically, Heisenberg's uncertainty principle is:

$$\Delta P_x \Delta X \geq h \qquad \text{eq. (2-37)}$$

Where:

ΔP is uncertainty in momentum.

ΔX is uncertainty in position.

Do not confuse the uncertainty principle with the inability to measure some parameters due to a disturbance. Many physical measurements are inaccurate because the act of measurement (or the nature of the instruments) disturbs the system and thereby changes the value of the measurement enough to introduce errors. For example, a low impedance voltmeter disturbs a high impedance circuit enough to introduce serious errors. What the QM scientist is telling us, however, is that the electron actually does not possess both a precise location and a precise momentum.

Schrödinger's probability interpretation leads to anomalies, such as the tunnel diode (**Figure 2-11**). In an ordinary PN junction the transition region between N-type and P-type semiconductor material consists of a dipole layer created by relatively immobile charged electrons and holes. This dipole layer has an associated electric field of up to 10 kV/cm in an unbiased junction, and even more in a reverse biased junction. Although the band gap energy is the same on both sides of the junction, a difference in the potential energy on the two sides results in an *electrical potential barrier*, E_B. The potential barrier blocks electrons trying to pass over the junction. Unless an electron has sufficient energy, it cannot leap over the potential barrier.

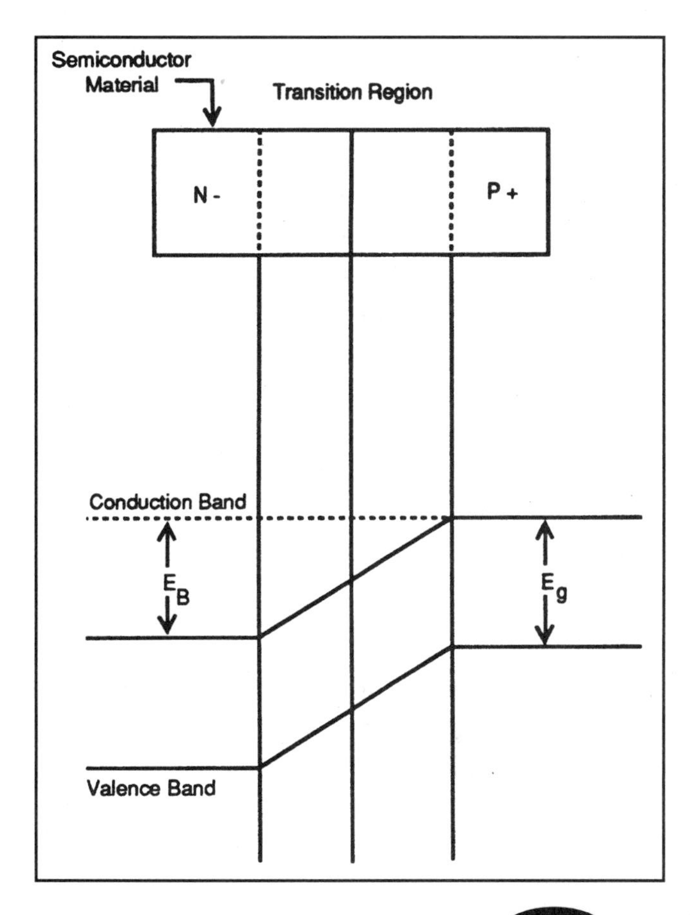

Figure 2-11. Energy diagram of a PN junction.

The Esaki tunnel diode has a transition region between P-type and N-type semiconductor material on the order of 100 angstroms (Å) rather than the 10,000 Å that is typical in ordinary PN junction diodes. In addition, very high doping levels result in much greater charge carrier concentrations. These two conditions taken together alter the situation so that quantum effects can take place. As an electron approaches the junction there is a small, but finite, probability that it will somehow pass through the junction even though it lacks sufficient energy. Furthermore, the electron has the same energy on the other side of the barrier as it had to begin with.

Let's consider a macro-world analogy to the tunnel diode to see what the scientists are actually claiming. Suppose there is a player on a racquetball court. The "particle" is not an electron, but rather it is the racquetball. The ball receives its energy when it is smacked by the player. Assume that the walls of the court are so high that the ball can not fly high enough to get over it no matter how hard the player whacks it. According to the standard wisdom that ball should spend eternity bouncing around the one racquetball court.

But now suppose that the court is a special quantum court. According to QM there is a small but finite chance that the ball will approach the wall, seem to pass right through it, and strike your neighbor on the head. Furthermore, the ball will have the same energy on both sides of the wall. This "tunneling" phenomenon does not really exist in the way we normally could understand it. What QM is really saying is that an electron disappears on one side of the barrier, and another identical electron appears on the other side. It's not that the electron passed through, but that something magic-like happened. In electronics we tend to glibly toss around electrons as if they were micro-ping-pong balls. While it works in our circuit descriptions, it isn't what the physicists are telling us is real.

Summary

During the first three decades of the twentieth century Bohr and a small band of scientists working in Copenhagen and elsewhere devised a system that seems to explain atomic phenomena. Physicists developed a world where matter (including the electron) has a complementary wave-particle nature. Instead of a mass of tiny balls, electrons and protons are a solid motion of probability waves. These theories displaced the cause-and-effect definitions of classical physics, and put in their place a new set of definitions that are based on probabilities and "tendencies to exist." The seeming paradox is that unpredictability and uncertainty appear intrinsic to the universe at the deepest levels. Quantum Mechanics is accepted by scientists as the mathematical construct that best predicts the behavior of matter at the subatomic level. Even though problems with the theory persist, QM has become the standard for scientists.

In the next chapter we will continue our discussion of physics by looking at optics; that is, the study of how light interacts with devices such as lenses and mirrors.

References

1. Clairex Data Book, Clairex Electronics, Mount Vernon NY.

Chapter 3

Mirrors, Prisms and Lenses

Electro-optical instruments cannot be discussed properly without reference to the optical components used in those instruments: mirrors, prisms, and lenses. These devices, used singly or in combination with each other, form the basis of sophisticated E-O instruments. In this chapter we will take a look at some of these components.

The Keys to Electro-Optical Instruments

Understanding optical instruments involves a knowledge of reflection, refraction, and Y. In order to simplify discussions of optical instruments we resort to ray tracing methods. Light beams are represented by pencil-thin rays (lines) that permit us to use simple arithmetic to describe the phenomena.

Refraction and Total Internal Reflection

The phenomenon of refraction takes place when a light beam travels from a less dense medium (such as air) to a more dense medium (such as water.) In this system, the incident light ray is bent towards the *normal* (**Figure 3-1a**), which is orthogonal (90 degrees) to the surface boundary between the two media. The amount of bending in refractive systems is proportional to the ratio of the indices of refraction of the two materials.

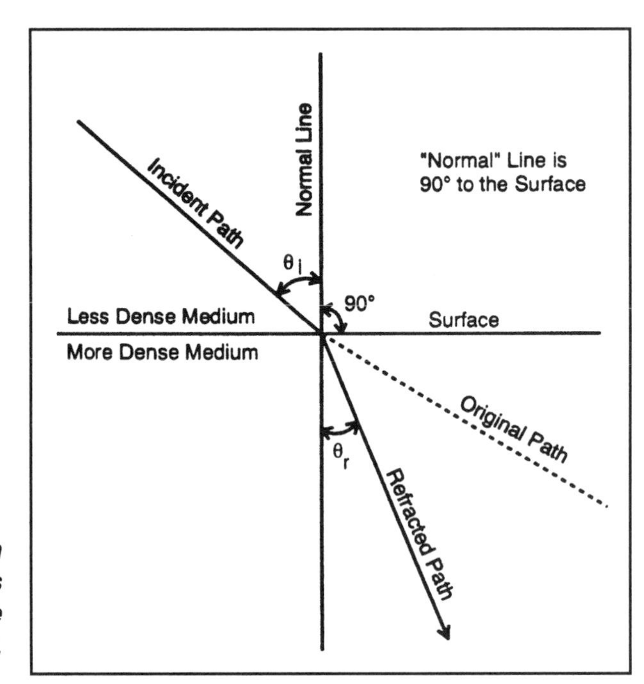

Figure 3-1a. Refraction occurs when light travels from less dense to more dense material.

The opposite reaction is seen when the light source propagates from the denser to the less dense medium; for example, a light source in water that has a surface boundary with air. In **Figure 3-1b**, light beam AO approaches the boundary between the media with angle of incidence, θ_i. When the beam crosses the boundary into the less dense region it is refracted away from the normal line (OB)—exactly the opposite of the situation in the less-dense-to-more-dense case.

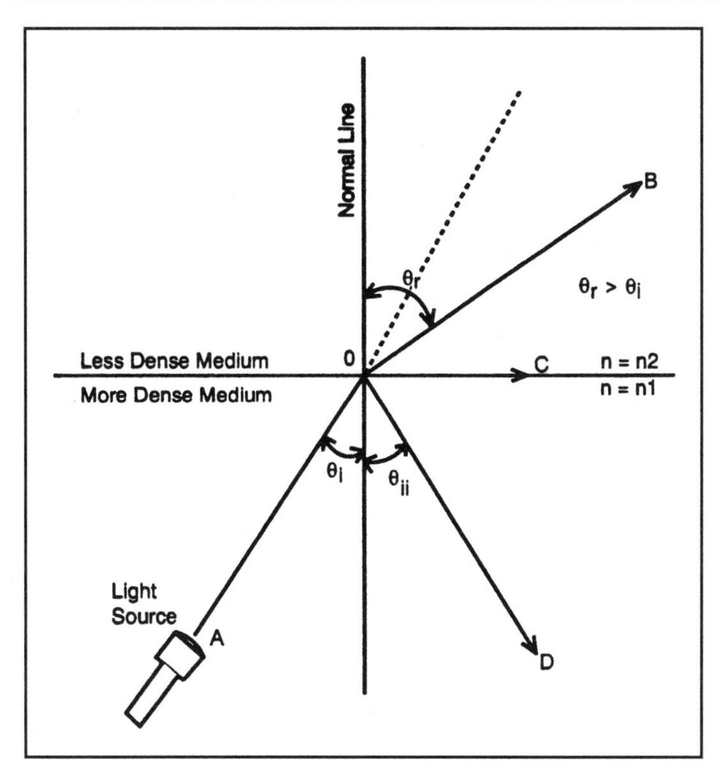

Figure 3-1b. Refraction when light travels from more-to-less optically dense materials.

As the angle of incidence is increased, a certain critical angle, θ_{ic}, is reached at the point where the angle of refraction equals 90 degrees. In this case, the exiting beam travels along the boundary surface (line OC in **Figure 3-1b**). The critical angle is found from the Snell's law of refraction:

- $$n_1 \sin(\theta_i) = n_2 \sin(\theta_r)$$ *eq. (3-1)*

- $$n_1 \sin(\theta_{ic}) = n_2 \sin(90)$$ *eq. (3-2)*

Where:

θ_{ic} is the critical angle.

n_1 is the index of refraction of the first material.

n_2 is the index of refraction of the second material.

or, rewritten:

- $$\theta_{ic} = \arcsin\left(\frac{n_2}{n_1}\right)$$ *eq. (3-3)*

Example 3-1

A light source originates in water (n = 1.33) and propagates into dry air (n = 1). Calculate the critical angle for this system.

Solution:

- $$\theta_{ic} = \arcsin\left(\frac{n_1}{n_2}\right)$$

- $$\theta_{ic} = \arcsin\left(\frac{1.00}{1.33}\right)$$

- $$\theta_{ic} = \arcsin(0.7519) = 48.76 \ degrees$$

At angles greater than the critical angle, the angle of refraction increases above 90 degrees to the normal, so the light beam reflects from the surface boundary back into the denser medium (line OD in **Figure 3-1b**). This phenomenon is called *total internal reflection* (TIR), and as we shall see later is an important factor in light propagation through optical fibers.

An instrumentation application of TIR is seen in **Figure 3-1c**. Prisms are used in periscopes, microscopes, binoculars, and other instruments to change the path of an incident light beam. An orthogonal incident light beam (OA) is totally internally reflected inside the 45 degree isosceles prism (OCD) emerging as beam DE on the other face—with its direction translated 90 degrees.

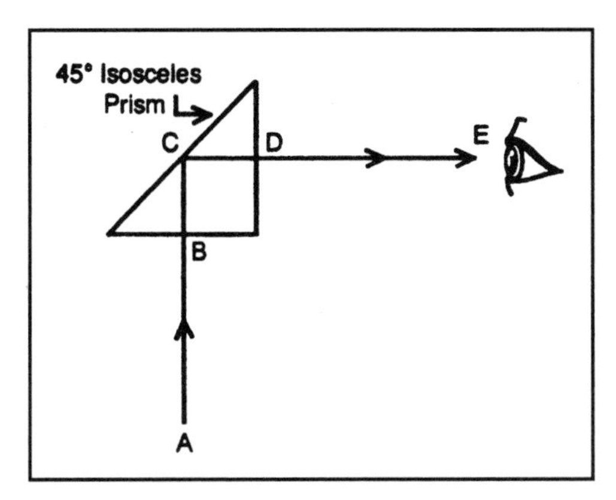

Figure 3-1c. When the critical angle is exceeded, total internal reflection occurs; light direction change in prism.

The change from external refraction to total internal reflection does not occur abruptly at the critical angle, but rather there is a transition region about the critical angle where the percentage reflected and percentage refracted varies. When the light is normal to the surface, a portion of the light is reflected back towards the source and a portion is refracted in the other medium. The fraction of light that is reflected is specified by a *reflection coefficient*, Γ:

$$\Gamma = \left(\frac{n_2 - n_1}{n_2 + n_1}\right)^2 \qquad \text{eq. (3-4)}$$

This fraction changes with both angle of incidence and the ratio of the indices of refraction of the two materials, n_1 and n_2.

Example 3-2

Calculate the reflection coefficient of the case where the incident material (n_1) has an index of refraction of 1.33, and the second material has an index of refraction (n_2) of 2.2.

Solution:

- $$\Gamma = \left(\frac{n_2 - n_1}{n_2 + n_1}\right)^2$$
- $$\Gamma = \left(\frac{2.2 - 1.33}{2.2 + 1.33}\right)^2$$
- $$\Gamma = \left(\frac{0.87}{3.53}\right)^2$$
- $$\Gamma = (0.247)^2 = 0.061$$

As can be seen from Example 3-2, the percentage of reflection is small (about 6.1 percent).

Refraction/Reflection from Slabs with Parallel Faces

A *slab* is a piece of optical material that has a thickness, *d*, and two or more parallel faces (**Figure 3-2**). Assuming that a light beam originates in the less dense region, n_1, there will be both specular reflection (AOB) and refraction (AOC) taking place. All specular reflections obey Snell's law of reflection: the angle of incidence equals the angle of reflection. The refraction inside the slab obeys Snell's law of refraction, already stated:

$$n_1 \sin(\theta_i) = n_2 \sin(\theta_2) \qquad \text{eq. (3-5)}$$

When the light beam reaches the rear surface at point C (**Figure 3-2**), reflection takes place and the beam heads back towards the surface boundary. At point D it again encounters the boundary and refraction occurs. The emerging beam (DE) is an offset reflection that is parallel to the original, or main, reflection ($OB \parallel DE$). The beam is seen at the same angle of reflection, but is displaced along the surface, X.

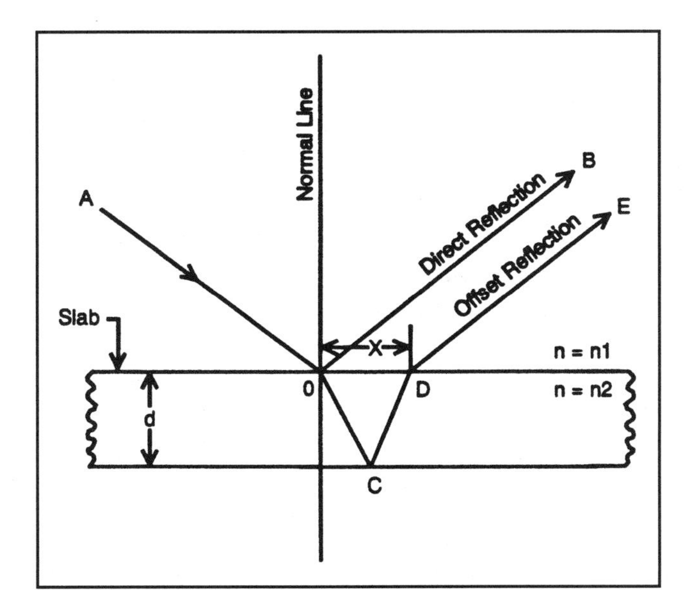

Figure 3-2. Double reflection from parallel face slab.

Shimmering, Mirages, and Looming

While the born skeptic might say: "I'll believe it when I see it...", we know that our eyes can and do play tricks on us. Optical illusions cause us to see things that are not there. We also see things differently than they really are because of optical phenomena. For example, stars don't really twinkle, except when viewed from close to the Earth's surface. The twinkling is caused in part by the light passing through regions of the atmosphere where significant changes of index of refraction occur.

A similar phenomenon, called *shimmering*, occurs when light passes from cold air through hot air, to cold air again. Shimmering is seen in deserts, above a space heater, over heated pavement, or across the hood of an overheated automobile. Shimmering causes objects on the other side of the turbulent heated air to appear wavy, or as though they were moving.

Mirages are familiar phenomena. You can see water mirages on highways during the summer months. The road ahead will appear wet, but as you approach the "wetness" it evaporates in an instant. Mirages are due to differential refractive indices between hot air near the surface and cold air at higher points.

Figure 3-3 shows how mirages can occur. A mountain peak in the distance is in cooler air, while the observer is on the hot surface. If the direct light path was through air of uniform temperature, then it would not bend appreciably, and the observer would see a mountain top. But if a temperature gradient changes the index of refraction as little as 0.03 percent, then the light path becomes bent. If the angle of incidence is sufficiently large, then total "internal" reflection occurs. The light arrives at the observer as if it had originated at point "A," and appears like light that had reflected off a lake or river (hence it looks "watery").

Looming is the opposite of mirages, and occurs when the surface is substantially colder than air a few meters above the surface. Looming occurs over snow-covered terrain, extensive ice packs, and cold bodies of water such as the Great Lakes or oceans. In looming, the rays bend toward the surface rather than away from it, but the perceived line of sight is straight. In **Figure 3-4**, the ship appears in the air above itself. Looming accounts for islands, lighthouses, and other objects being sighted further away than the curvature of the Earth would normally permit; that is, before they are actually at or above the horizon.

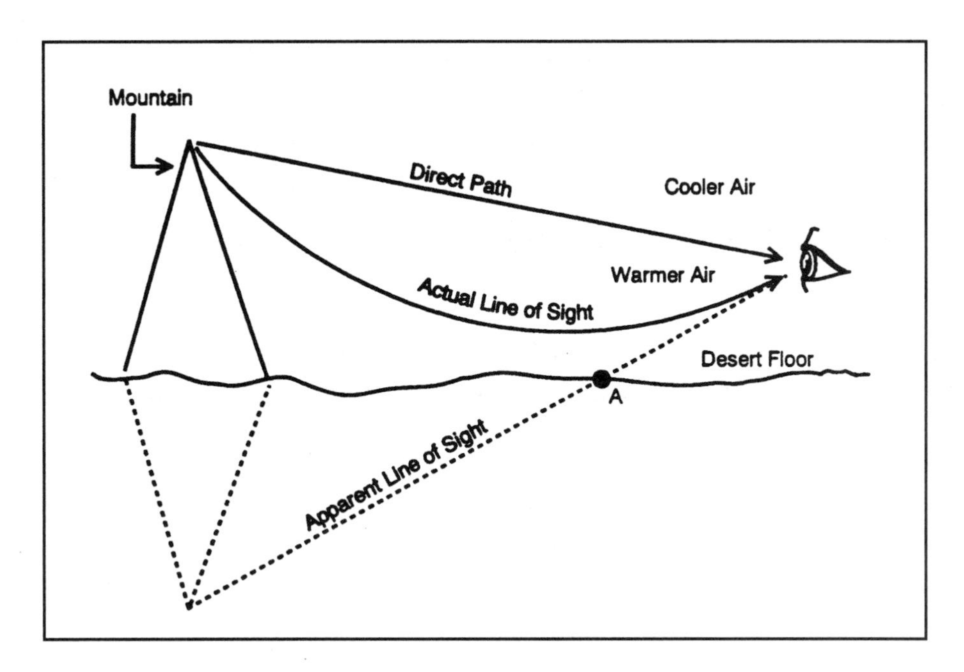

Figure 3-3. Mirage effect.

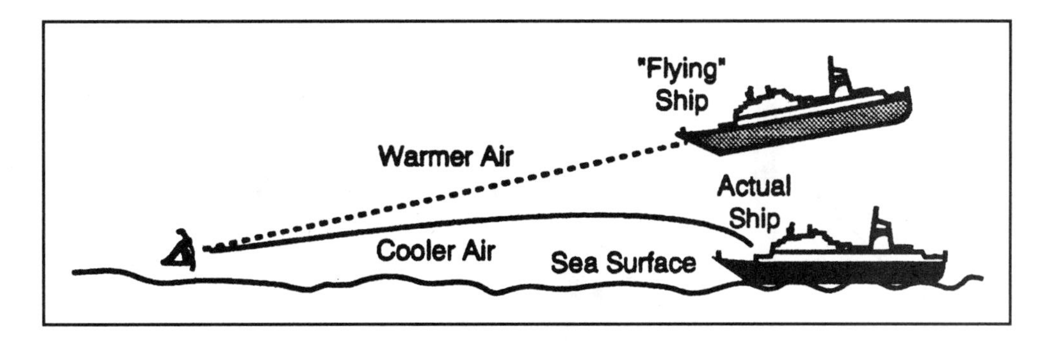

Figure 3-4. Looming effect.

Mirrors

A *mirror* is a surface so well polished that the largest irregularities on its surface are less than one-quarter wavelength of the impinging light wave. Such a surface provides specular reflection, rather than diffuse reflection, and will obey the following rules:

1. The incident light beam, reflected light beam, and a normal light beam to the surface at the point of reflection, all lay on the same plane, and
2. The angle of incidence and the angle of reflection, relative to the normal, are equal.

The earliest mirrors were made of highly polished metals that were readily available (even though expensive) such as copper, silver, or gold. Metal mirrors can only be polished to a certain point and are easily deformed by ordinary handling. Better mirrors can be made by depositing a silver or aluminum layer onto a polished glass surface. Although metal-coated glass plates are optically superior, they are brittle compared with metal mirrors. Glass mirrors can be either made sufficiently robust for normal use, or protected for rigorous uses.

Mirrors take three common shapes: planar, concave, and convex (**Figure 3-5**). The planar mirror consists of a flat surface, the concave mirror has a curved inside surface, and the convex mirror has a curved outside surface.

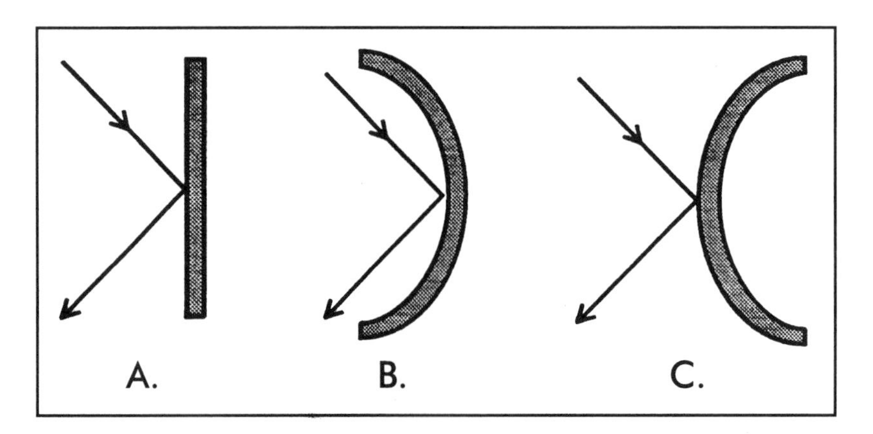

Figure 3-5. Reflection from different shape surfaces: a) Flat; b) Concave; c) Convex.

Ray Tracing and Sign Conventions

All three mirror types follow similar laws and are described by similar equations. Careful and consistent use of a sign convention for the differing characteristics of the three mirror types is imperative.

The sign and symbol convention used herein is based on ray tracing. The point on the object selected for ray tracing is usually a prominent point, such as the base or the peak. There is no reason a random point cannot be selected, but analysis of the situation becomes more difficult.

Labels are assigned to points and distances in a ray diagram. Uppercase letters *C* and *F* are used for the center of curvature and focal point, respectively. A point on an object is labeled with an uppercase letter, and its conjugate point on the image is labeled with the same letter, primed (E on the object and E′ on the image). The exception is that point Q on the principal axis has a conjugate point P on the image, and not Q′. The distance between the vertex of the mirror and the focal point is labeled with a lowercase f; the distance from the vertex to the base of object Q is labeled q; the distance from the vertex to P is labeled p.

For each situation in **Figure 3-6**, there are at least three rays used:

1. In **Figure 3-6a**, a ray is drawn from a point on the object, E, through the center of curvature. This ray is reflected back on itself (line ECD).
2. A second ray is drawn from the same point on the object to the mirror along a path parallel to the principal axis. This ray, EG, is reflected through the focal point, F, (path EGFH).
3. A third ray is drawn from the point on the object to the mirror through the focal point. This ray is reflected parallel to the principal axis (path EFBI).
4. In some cases, a fourth ray is drawn from the point on the object to the vertex of the mirror. It will be reflected at the same angle away from the principal axis (path EAJ in **Figure 3-6a**), such that ∠EAF = ∠FAJ.

Note that all four standard rays cross at the conjugate point, E′. If they don't, then the object is too large for the diameter of the mirror.

The sign conventions to determine whether the distances are positive (+) or negative (-) are taken by considering a line from a point on the object to the mirror along a path that is parallel to the principal axis (line EG in **Figure 3-6a**), for example). The following rules apply:

1. If parallel ray EG goes from the object to the mirror, and then reflects through the focal point, F, then p is positive (see **Figure 3-6a**); if the ray does not pass through the focal point after reflection (**Figure 3-6b**), then p is negative.
2. If the ray passes through the focal point, F, after reflection (**Figure 3-6a**), then f is positive; if the ray does not pass through the focal point then f is negative.
3. If the ray passes through the center of curvature, C, before being reflected, then it will be reflected back on itself, along the same path.

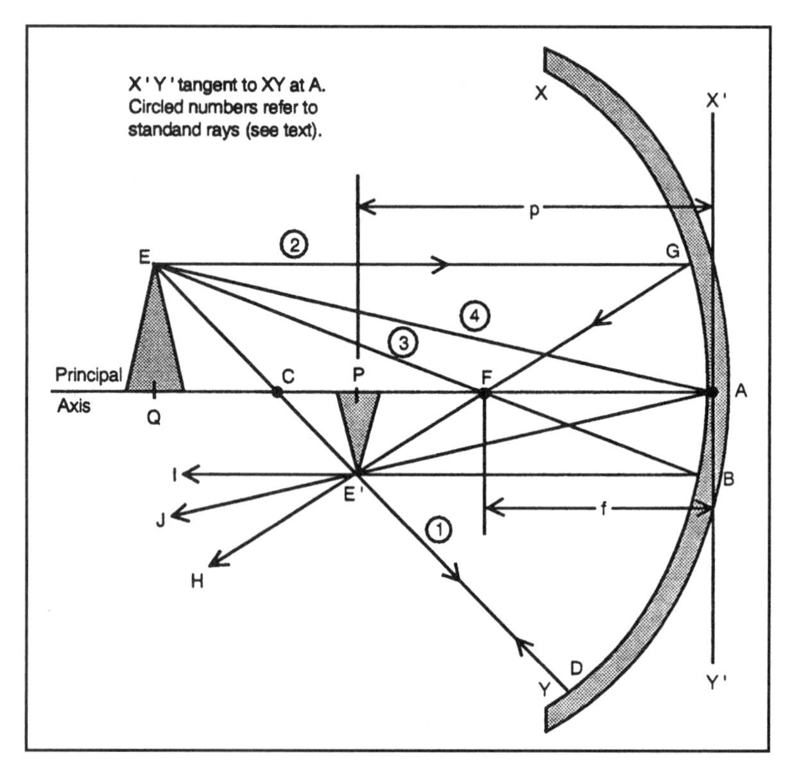

Figure 3-6a. Reflection from the internal surface of a spherical mirror.

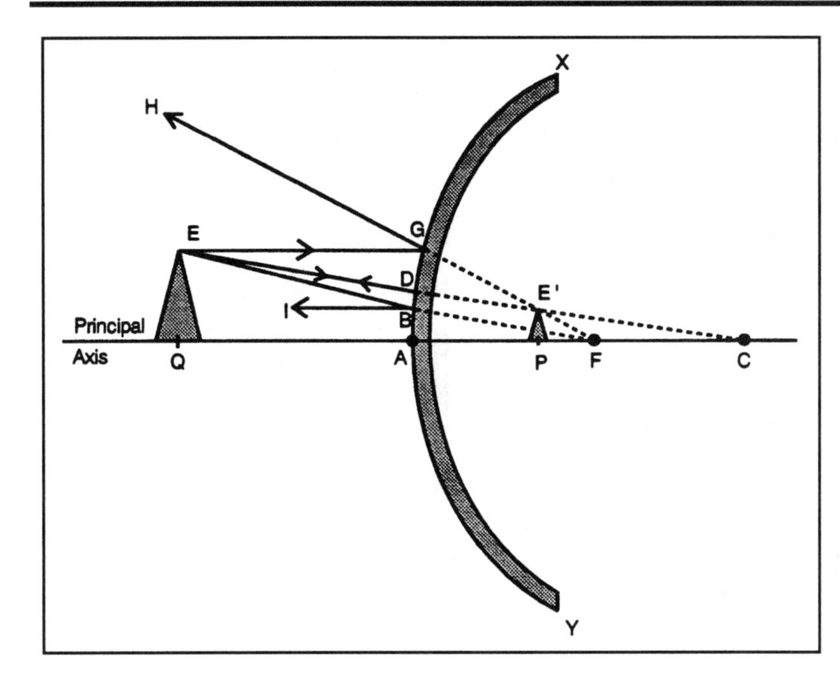

Figure 3-6b. Reflection from the external surface of a spherical mirror.

Planar Mirrors

A planar mirror consists of a flat sheet of reflective material; that is, the entire surface lies in one plane. **Figure 3-7** shows a planar mirror, MP, consisting of a reflective slab. If the light emitting source, O, is placed in front of the mirror, then it will send out rays in all directions. Some rays (OCA, ODB, and ON in **Figure 3-7**) strike the mirror and are reflected according to the rules previously stated.

An observer looking at the mirror will see an image of the light source that appears to be behind the mirror. If the actual source is at a distance X from the mirror surface, then the image of the source will appear at the same distance X′, where X′ = -X, behind the mirror. Because the light rays from the image only appear to come from the image, and do not actually pass through point O′, this type of image is called a virtual image. A true image, on the other hand, is one in which the light beams actually pass through the point.

The virtual image is not totally identical with the object, but rather it is reversed with respect to the reflecting plane (**Figure 3-8**). In other words, if a person standing in front of the mirror has a watchband on the right arm, then

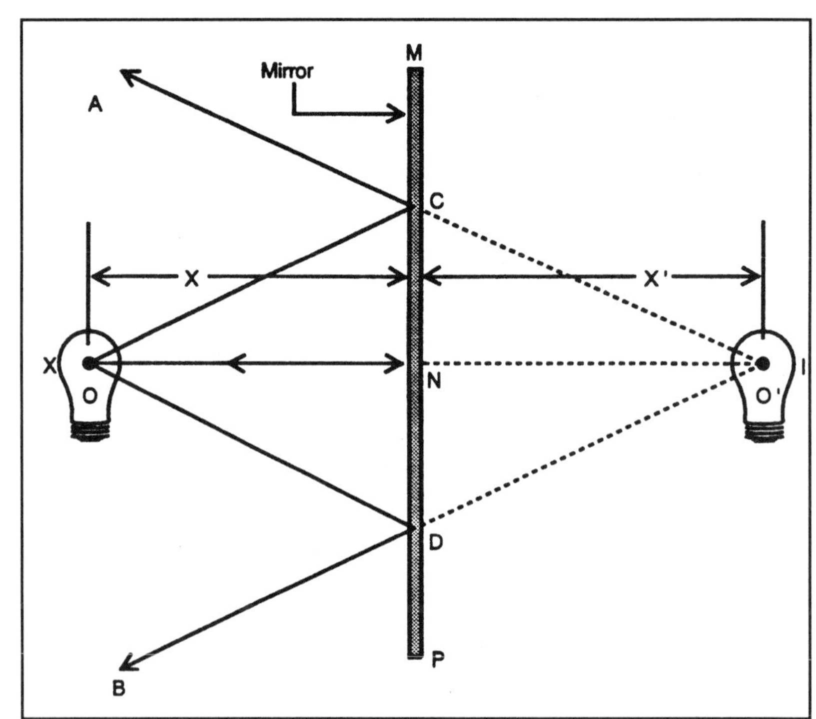

Figure 3-7. Reflection from a flat mirror.

the virtual image of that person has a watchband on the left arm. That is, the mirror image is reversed for an observer on the same side of the mirror surface as the object.

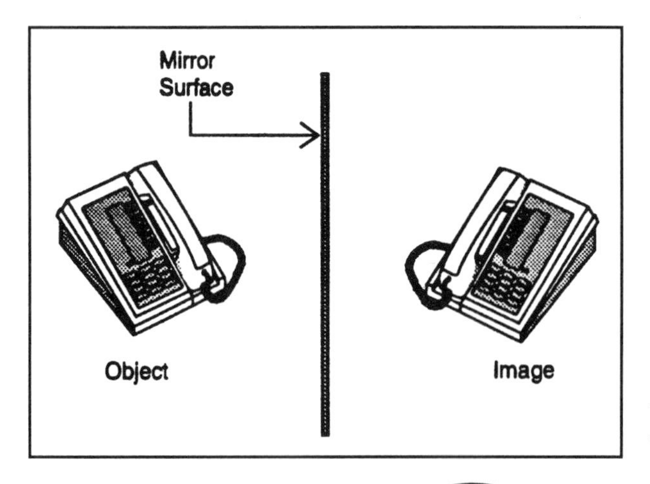

Figure 3-8. Image reversal in flat mirror.

If a mirror reverses left and right, then how come it doesn't reverse up and down? Because the mirror surface reflects point-for-point and doesn't really "reverse" anything. An XYZ coordinate system of the object will show this phenomenon clearly.

Spherical and Parabolic Concave Mirrors

The spherical and parabolic concave mirrors are related types, so can be considered together. We will, however, consider the spherical mirror first. A spherical concave mirror consists of a segment of a hollow sphere that has been polished on the inside surface (**Figure 3-9**). A parabolic concave mirror is similar except that the shape of the surface is parabolic rather than spherical.

Figure 3-9 is a spherical concave mirror, XY, in which the center of the sphere is point C. This point is called the center of curvature of the mirror. A straight line between point C and any point on the surface of the mirror forms a radius of curvature. One particular radius (CA in **Figure 3-9**) passes through the midpoint, A, between X and Y. This point, A, is called the vertex of the mirror, and its radius CA is called the principal axis of the mirror.

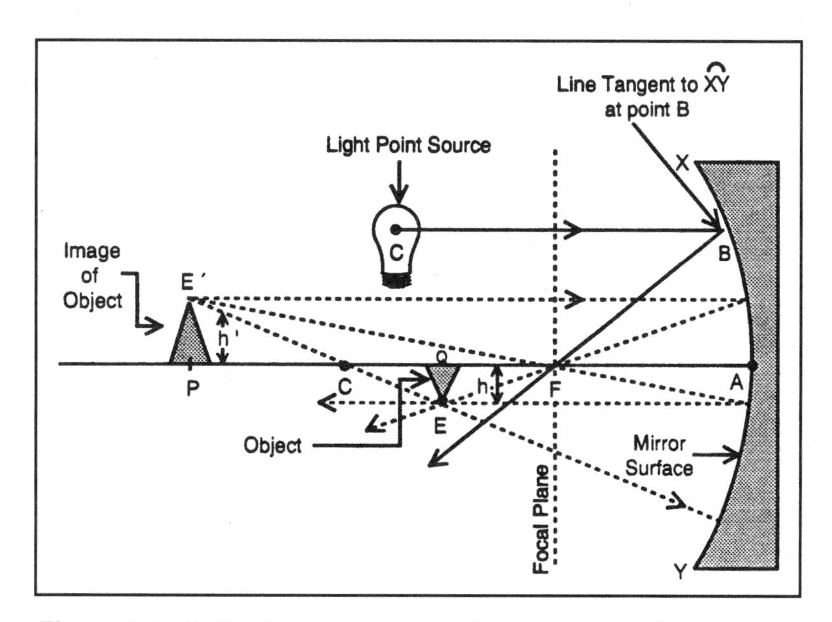

Figure 3-9. Reflection from a spherical concave mirror.

A point source of light at point C will emanate rays in all directions from the source. Some of them are parallel to the principal axis (CA). Rays parallel to CA are reflected from the mirror surface according to this modified rule of reflection:

> On a curved surface, the angle of reflection will be equal to the angle of reflection as measured with respect to a line normal to a line that is tangent to the point of reflection (**Figure 3-10**).

All rays impinging the mirror along paths parallel to the principal axis will be reflected through a point on the principal axis, F, that is halfway between the vertex and the center of curvature, C; (CF = FA, or FA = ½CA). This point, F, is called the focal point of the mirror. A spherical concave mirror will reflect light beams parallel to the principal axis into the focal point. A line drawn perpendicular to the principal axis at the focal point is called the focal plane.

Spherical and parabolic concave mirrors reflect rays toward the principal axis, making such mirrors converging mirrors.

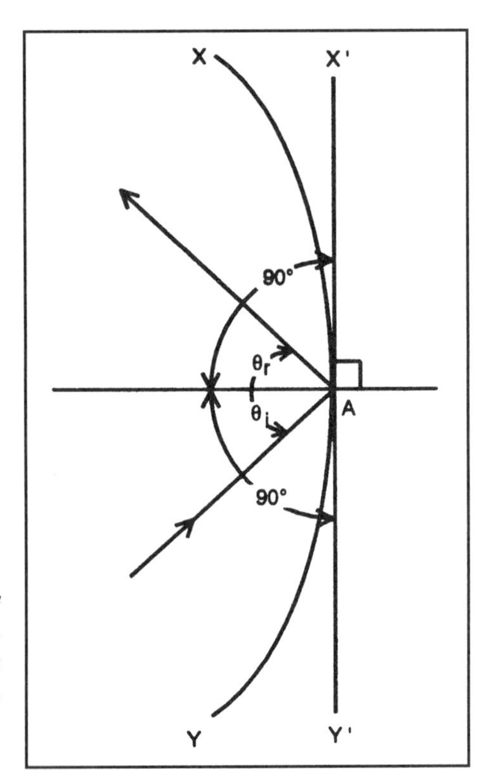

Figure 3-10. Angles of incidence and reflection on a curved surface are measured from a line tangent to the point of reflection.

This feature of both spherical and parabolic concave mirrors has certain practical applications. If an observer is placed at the focal point, then they can see the total collection of parallel rays. Modern astronomical telescopes are built in this manner.

Another application is a solar furnace or oven. An object that is placed at the focal point of a concave mirror that is aimed at the sun will be heated. Backpackers and boaters have been known to cook small meals without fire in this manner. Solar stills for desalinating salt water have been made this way (**Figure 3-11**). A water vessel placed at the focal point of a solar furnace mirror will collect enough heat to boil the water, producing steam. If the steam is passed through a condensing tube to cool it, then it will convert back to water, leaving the salt behind.

Still another application is a search or spotlight. If an optical system is bidirectional, (containing no devices that transmit light in only one direction), then light traveling backward in the system will follow the same path as light entering the system. A light source at the focal point will send rays out in all

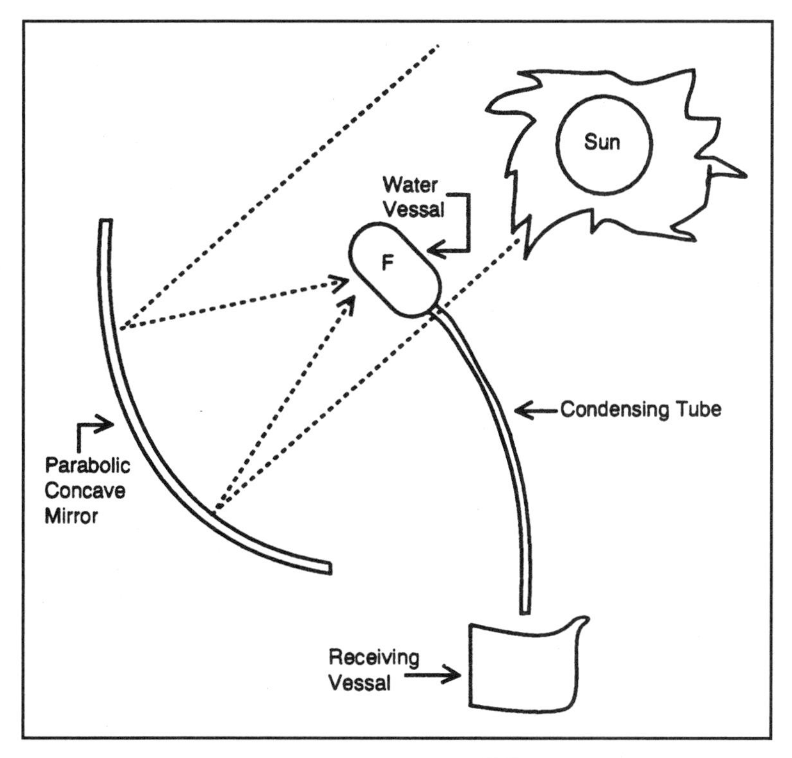

Figure 3-11. Solar still.

directions, and those that strike the mirror will be reflected to the outside world along paths that are parallel to the principal axis. The mirror search light can produce a collimated beam of greater intensity than a bare light source alone. Modern automobile headlights work on this principal.

Now let's consider how light from other objects is reflected as images from the concave mirror. There are two general cases: (1) an object placed inside the focal point (between F and A), and (2) an object placed outside the focal point.

Figure 3-9 contains object, QE, at point Q, which lies outside of the focal point. A true, but inverted, image of the object is formed at PE′. In this case, the following relationships exist:

[Let p = PA, q = QA, and f = FA]

- $$\frac{p}{q} = \frac{P'E}{QE}$$ *eq. (3-6)*

- $$\frac{p}{q} = \frac{PA}{QA}$$ *eq. (3-7)*

- $$\frac{p - f}{f} = \frac{p}{f} - 1$$ *eq. (3-8)*

- $$\frac{p}{q} = \frac{p}{f} - 1$$ *eq. (3-9)*

- $$\frac{1}{p} + \frac{1}{q} = \frac{1}{f}$$ *eq. (3-10)*

Example 3-3

A concave mirror is fashioned such that a 100 mm high object produces an image that is 300 mm high. Calculate the focal length, f.

Solution:

- $$\frac{1}{p} + \frac{1}{q} = \frac{1}{f}$$

- $$\frac{1}{300\ mm} + \frac{1}{100\ mm} = \frac{1}{f}$$

- $$0.0033 + 0.01 = \frac{1}{f} = 0.013$$

- $$f = 75\ mm$$

If q is the object distance from the mirror, and p is the image distance, then the height of the image, h′, with respect to the height of the object, h, is:

$$\frac{h'}{h} = \frac{p}{q} \qquad \text{eq. (3-11)}$$

The quantity h′/h is called the *Transverse Linear Magnification* (TLM) of the mirror.

Example 3-4

Find the TLM and height of image, h′, if the height of the object, h, is 3.7 cm, and it is placed 60 cm in front of a concave spherical mirror that has a 20 cm focal length.

Solution:

1. From h′/h = p/q:

$$h' = \frac{hp}{q}$$

2. Calculate h′:

$$h' = \frac{(3.7\ cm)(60\ cm)}{20\ cm} = 11.1\ cm$$

3. Calculate TLM:

- $$TLM = \frac{h'}{h} = \frac{11.1}{3.7} = 3$$

For an object that is placed outside of the focal plane, the image is inverted, magnified and further from the focal plane than the object.

For an object that is placed inside of the focal plane, the image is an erect, magnified, virtual image "behind" the mirror (**Figure 3-12**).

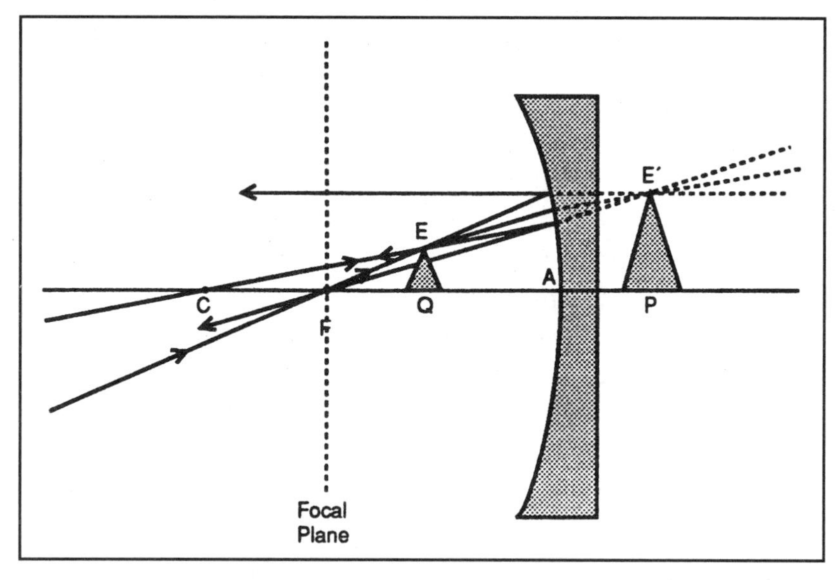

Figure 3-12. Reflection from a mirror when the object is inside the focal plane.

Spherical Aberration

The concave spherical mirror is sharply focused only when the width of the mirror is small compared with its radius of curvature. When the size of the mirror approaches the radius of curvature, however, the rays parallel to the principal axis are reflected through a small region around the focal point, rather than the focal point itself. Such a mirror will produce blurred images. This effect is called *spherical aberration*. It can be eliminated by using a mirror surface that is part of a parabola, rather than a sphere. Such mirrors are called *parabolic reflectors* rather than spherical reflectors, and are used much more extensively than are spherical reflectors in applications requiring a concave mirror.

Convex Mirrors

Convex mirrors are *diverging* mirrors. Light rays impinging the surface of the convex mirror are reflected away from the principal axis. The convex mirror can be modeled as a sphere segment, as in **Figure 3-13**, with center of curvature C, focal point F, and radius of curvature CA, and that is polished on the outside surface.

Ray EBC in **Figure 3-13** impinges the mirror at B, and is reflected back into space at an angle away from the principal axis of the mirror, towards point C. An observer will see ray BC and make the assumption that it originated behind the mirror at focal point F. The top of the image pyramid is seen on line FBC, rather than EBC. The image in a convex mirror is erect, reduced in size relative to the object, and is a virtual image. In working problems involving convex mirrors, **Equation 3-10** still applies, but the quantities q and f are negative.

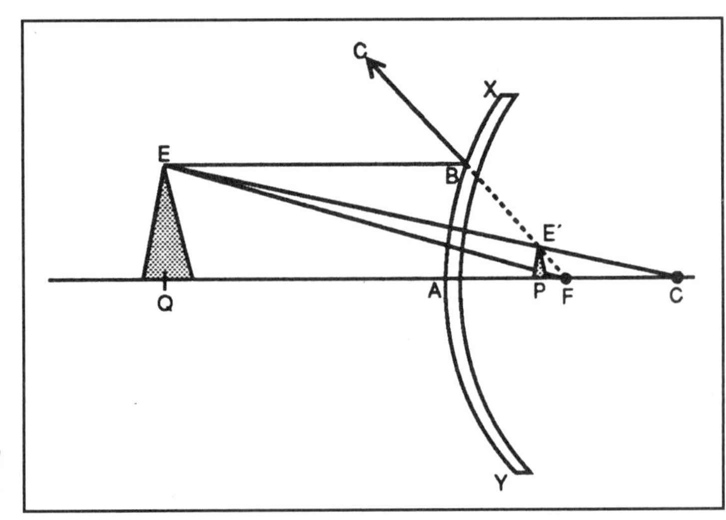

Figure 3-13. Reflection from a convex mirror.

Lenses

Lenses are optical elements that transmit light and redirect parallel light rays: if the rays are redirected towards a focal point on a principal axis, then the lens is converging; if the rays are redirected away from the principal axis and the focal point, then the lens is diverging.

Figure 3-14 shows how a lens can be modeled from a prism. Recall that a prism refracts light internally according to Snell's law. If a beam of parallel light rays enters the prism (**Figure 3-14a**), it is then refracted the same amount and exits in parallel with each other but traveling in a different direction.

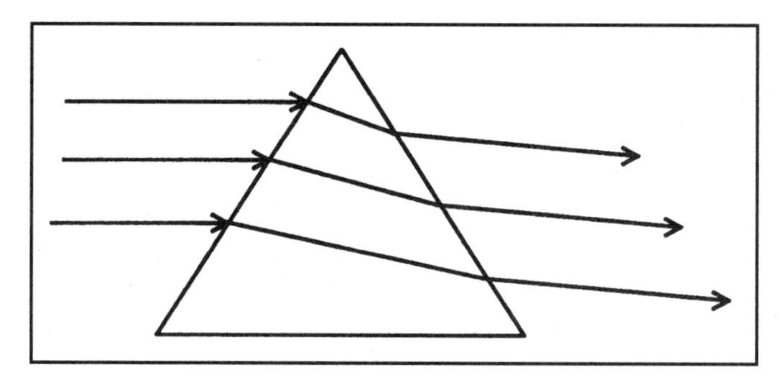

Figure 3-14a. Striaght prism model of a lens.

Now suppose that a genius comes along and constructs a composite prism (**Figure 3-14b**) in which a large number of angled flat surfaces, like a prism, are spaced along the body of this new optical object. The light rays would enter in parallel andbe refracted the same amount, but at different angles because the normals to each surface are different. When these now-nonparallel rays exit the other side, they would either converge together or diverge from one another (depending on prism arrangement), rather than being parallel. In the case shown for **Figure 3-14b** the rays cross at a focal point (F), so the structure is converging.

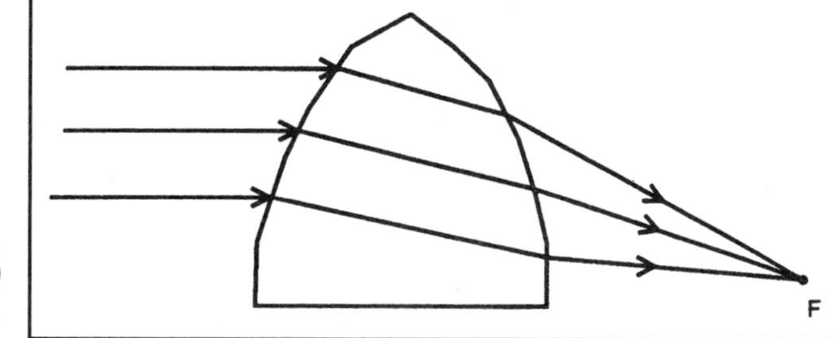

Figure 3-14b. Sectioned (composite) prism model of a lens.

Now further suppose that the new composite prism is remanufactured by the genius' teacher so that there is an infinite number of tiny flat surfaces (**Figure 3-15a**). The surfaces are so small in extent that the overall surface looks to an outside observer like it is perfectly smooth. The same optical action takes place; in the case of **Figure 3-15a** the action is converging, while in **Figure 3-15b** it is diverging.

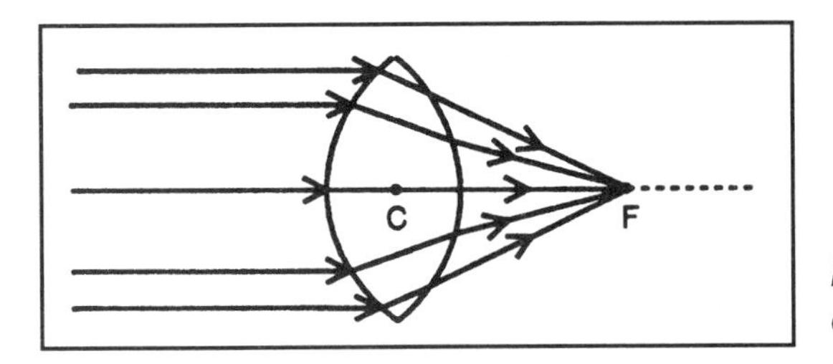

Figure 3-15a. An example of a converging lens.

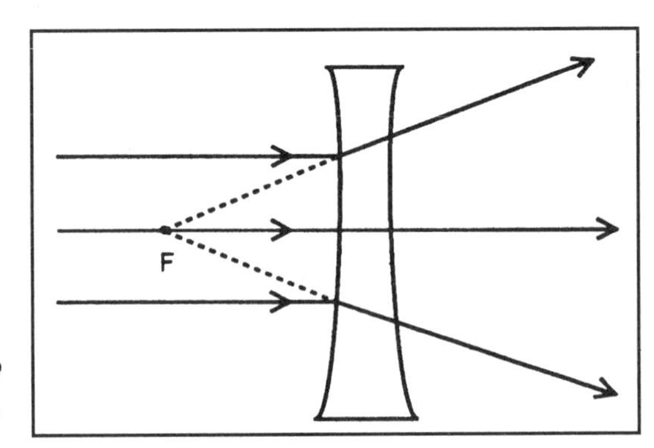

Figure 3-15b. An example of a diverging lens.

Figure 3-16 shows several of the many types of lens shapes available. The lenses that were discussed before are shown in **Figure 3-16a** and **Figure 3-16b**. The converging version is called a double-convex lens, while the diverging version is called a double-concave lens. **Figure 3-16c** shows a plano-convex lens; i.e., one surface is planar (flat). The plano-plano lens in **Figure 3-16d** is not actually a lens, but is an optical flat. Such surfaces are used for filters, protective covers, and other applications in optics. The concave-convex lens is shown in **Figure 3-16e**, while a plano-concave lens is shown in **Figure 3-16f**.

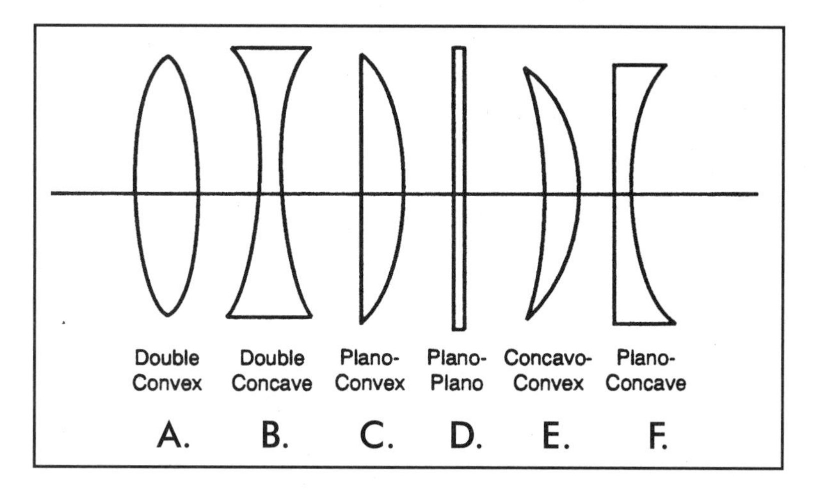

Figure 3-16. Various shapes of lenses; a) Double convex; b) Double concave; c) Plano-convex; d) Plano-plano; e) Concavo-convex; f) Plano-concave.

Standard Rays and Labeling for Lenses

As with mirrors, we can analyze lenses with a standard ray diagram, as shown in **Figure 3-17**. In this case, however, we reverse the meanings of P and Q. The center of the base of the object, not the image, is labeled P. Similarly, the distance between the optical center of the lens, C, and object, P, is labeled p. The distance from C, to the center of the base of the image, Q, is labeled q. The double-convex lens has two focal points, F and F′, called *principal foci*, which can be made equal to each other if the radius of curvature of each surface is the same. The focal length of the lens is the distance from the optical center of the lens to the focal point, F′.

Four standard rays are used to describe lens characteristics:

1. A ray from a point P on an object that lies on the optical axis passes through the lens undeviated to the conjugate point, Q.
2. A ray from the same point, E, on the object passing through the near-side principal focus, F, is deviated by the lens, becoming parallel to the optical axis on the other side. It passes through the conjugate point, E′, on the image.

3 A ray is drawn from a point E on the object along the optical axis to the optical center, C, passes through the lens undeviated to the conjugate point, E′ (see line ECE′ in **Figure 3-17a**).

4. A ray from point E on the object parallel to the optical axis is deviated by the lens so that is passes through the principal focus, F, on the other side of the lens.

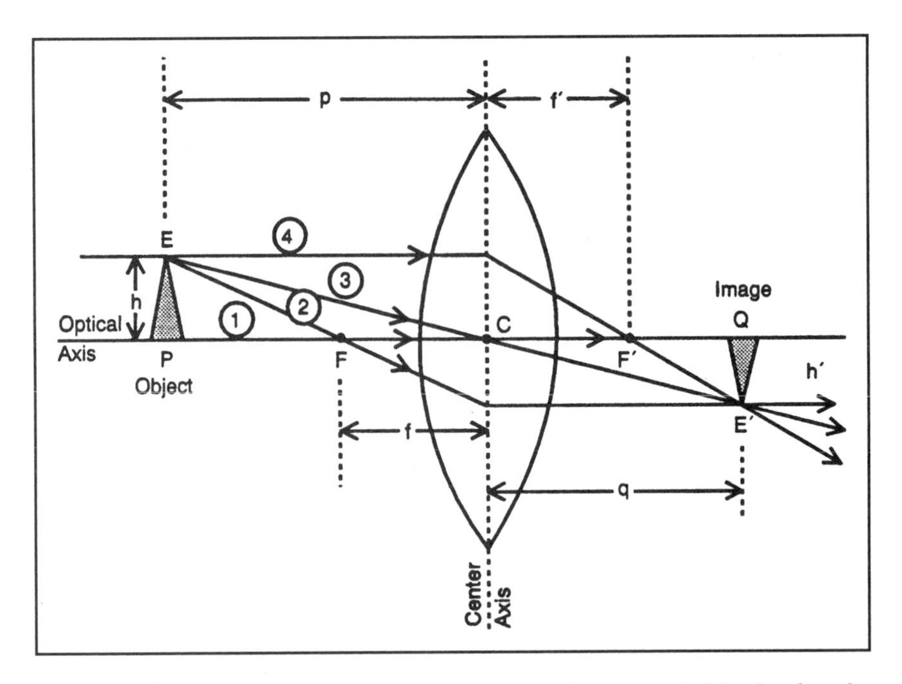

Figure 3-17a. Concave lens action when object is outside the focal plane.

The relationship of the object/lens/image distances in a lens obeys the same rule as in a mirror:

$$\frac{1}{p} + \frac{1}{q} = \frac{1}{f} \qquad \text{eq. (3-12)}$$

Note that f and f′ can be used interchangeably in **Equation 3-12**.

The magnification of the lens is the ratio of the image height, h′, to the object height, h:

$$Magnification = \frac{h'}{h} \qquad \text{eq. (3-13)}$$

The guidelines for images made by the double-convex lens are:

1. For an object outside the principal focus, F, the image will be real, inverted, and on the other side of the lens (the case shown in **Figure 3-17b**).
2. For an object between the principal focus, F, and the surface of the lens, the image will be erect, enlarged (magnified) and virtual (appears on the same side of the lens as the object). This is the case for magnifying glasses.

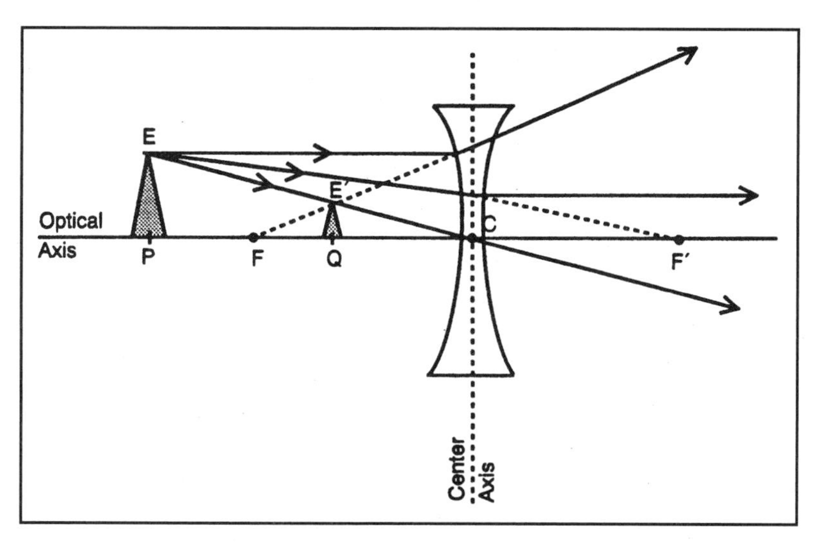

Figure 3-17b. Convex lens operation.

The focal length of the lens is a function of the index of refraction of the lens material and the respective radii of curvature of the two surfaces (**Figure 3-18**):

$$\frac{1}{f} = (n-1)\left(\frac{1}{R_1} - \frac{1}{R_2}\right) \qquad \text{eq. (3-14)}$$

Where:

n is the index of refraction for the lens material.

R_1 is the radius of curvature of the rear surface.

R_2 is the radius of curvature of the incident light surface (front surface).

Equation 3-14 is called the lensmaker's equation. In standard convention, the radius of curvature of the front surface, R_2, is given a positive sign (+), while that of the rear surface, R_1, is given a negative sign (-).

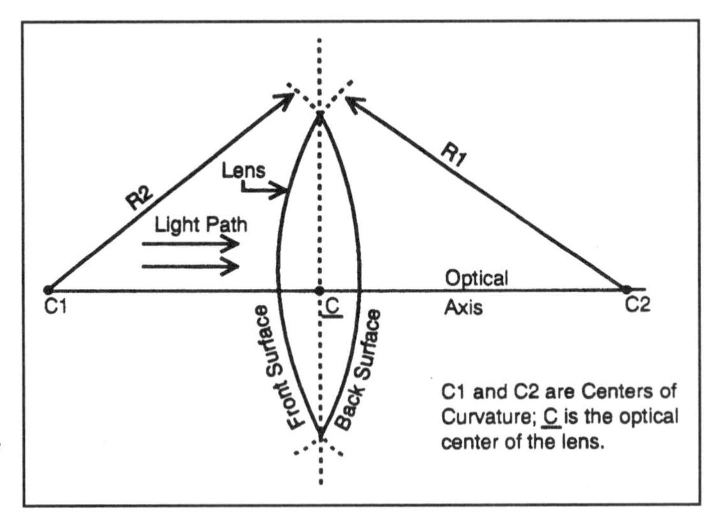

Figure 3-18. Radii of curvature definitions.

Example 3-5

A double-convex lens is made from a material that has an index of refraction of 1.72. Calculate the focal length f if the radius of curvature for the rear surface, R_1, is 15 cm, and for the front surface, R_2, is -25 cm.

Solution:

- $$\frac{1}{f} = (n-1)\left(\frac{1}{R_1} - \frac{1}{R_2}\right)$$

- $$\frac{1}{f} = (1.72-1)\left(\frac{1}{15} - \frac{1}{-25}\right)$$

- $$\frac{1}{f} = (0.72)\left((0.067) - (-0.04)\right)$$

- $$\frac{1}{f} = (0.72)(0.107) = 0.077$$

Taking the reciprocal of $1/f$,

- $$f = \frac{1}{\left(\frac{1}{f}\right)} = \frac{1}{0.077} = 12.98\,cm$$

Put in more general form, the lensmaker's equation can be expressed in the form:

$$\frac{1}{p}+\frac{1}{q}=\left(\frac{n_{lens}}{n_{inc}}-1\right)\left(\frac{1}{R_1}-\frac{1}{R_2}\right) \qquad \text{eq. (3-15)}$$

Where:

n_{lens} is the index of refraction of the lens material.

n_{inc} is the index of refraction of the medium from which the incident light strikes the lens (n_{inc} = 1 for air).

All of the terms previously defined.

When you get glasses, the optometrist will usually prescribe the lens in terms of diopters. The diopter is the power of the lens and is defined as the reciprocal of the focal length expressed in meters (m). Diopter power, P, is:

$$P=\frac{1}{f_{meters}} \qquad \text{eq. (3-16)}$$

Example 3-6

Find the diopter measurement of a lens that has a 55 mm focal length.

Solution:

$$P=\frac{1}{f_{meters}}$$

- $$P = \frac{1}{55\,mm\ x \left(\frac{1\,m}{1000\,mm} \right)}$$

- $$P = \frac{1}{0.055}$$

- $$P = 18.2$$

Many optical instruments use two or more lenses (called elements) in combination, usually sharing a common optical axis. Telescopes, microscopes, and camera telephoto lenses are examples of this concept. The lens elements may be cemented together, or spaced some distance from each other. When thin lenses are cemented together, the total power is:

- $$P_{total} = P_1 + P_2 \qquad \text{eq. (3-17)}$$

and the total focal length:

- $$\frac{1}{f_{total}} = \frac{1}{f_1} + \frac{1}{f_2} \qquad \text{eq. (3-18)}$$

Lens Defects

The perfect lens exists only as an ideal and all lenses may suffer from certain defects. Some of these are as follows:

Spherical Aberration. If the size of a lens is comparable to the size of the object being imaged, then light from the object will pass through both the center and edges of the lens. The rays entering near the edge of a double-convex lens will be refracted more strongly than rays passing through the lens near its optical center. This phenomenon is similar to the spherical aberration of mirrors, and results in a poorly focused image.

Coma. This defect is quite complex, but can be easily detected. Coma defect causes distortion of the object such that the image of a rectangle or square is either barrel-shaped or pincushion-shaped.

Astigmatism. This defect distorts the image when light rays from an object located off the optical axis pass through the lens obliquely, and do not converge at exactly the same point on the other side. Astigmatism can be corrected by using two or more lens elements with proper spacing.

Chromatic Aberration. A prism disperses the colors found in an incident white light beam. This dispersion occurs because the glass bends the different colors (wavelengths) by differing amounts. In a lens, this effect is seen as a set of colored rings, or halos, around the image. Chromatic aberration is prevented by cementing a converging crown glass lens and a diverging flint glass lens together, making an achromatic lens (**Figure 3-19**).

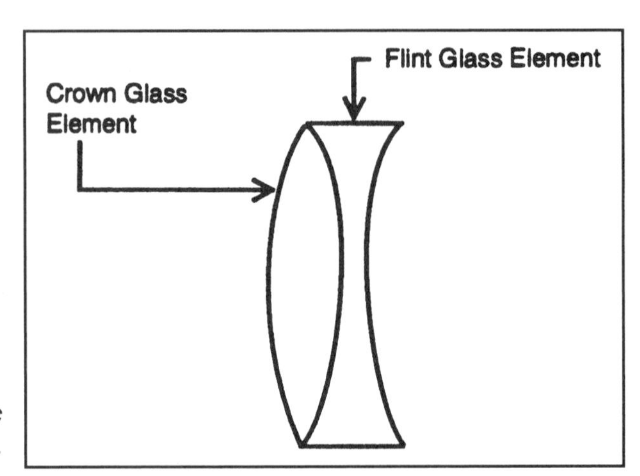

Figure 3-19. Achromatic lens construction.

Chapter 4

Photoemissive Sensors: Phototubes and Photomultiplier Tubes

A *photoemissive sensor* is a specially constructed vacuum tube diode (a two electrode device) that produces a current, I_o, proportional to the intensity of light striking the sensitive surface. Photoemissive sensors fall into two categories, *photoelectric tubes* and *photomultiplier tubes*. Such devices depend on the photoelectric effect for their operation.

The Photoelectric Effect

The *photoelectric effect*—the photoemission of electrons—was discovered in the early nineteenth century. Scientists noticed that certain primitive batteries worked more efficiently when exposed to light. In other experiments, it was found that certain metallic plates in a vacuum were able to emit electrons when exposed to visible or ultraviolet light. As early as 1887, Hertz was using a spark-gap receiving apparatus in his experiments. He noted that the spark-gap receiver was more sensitive when the gap region was illuminated with ultraviolet light. William Hallwachs explained this phenomenon in 1888 by describing the photoelectric effect, also called the *Hallwachs effect*. The cause of the photoelectric effect eluded scientists until early in this century.

Planck's 1900 paper that dealt with blackbody radiation did not specifically address the photoelectric effect, but it was his work that led Albert Einstein to the solution to the photoelectric phenomenon. In 1905, Einstein published three major papers in the German science journal Annalen der Physik: the Theory of Relativity, Brownian motion, and a theory of the photoelectric effect that used Planck's energy quanta.

Prior to the publication of Einstein's paper, scientists studying the photoelectric effect were perplexed by strange behavior observed in the laboratory. The intensity of the impinging light beam affects only the amount of current (the number of electrons) emitted, but not the energy of the emitted electrons. Oddly, however, it was noted that the color of the light affects the energy of the electrons. Electrons emitted from a photoemissive surface under the influence of blue light were more energetic than electrons emitted under red light. The explanation of this behavior can be inferred from **Equation 4-1**.

$$\frac{1}{2}mv^2 = hf - E_\omega 1 \qquad \text{eq. (4-1)}$$

Where:

m is the mass of the electron.

v is the velocity of the fastest emitted electrons.

h is Planck's constant (6.63×10^{-27} erg-s).

f is the frequency of the incident light.

E_ω is the energy in ergs required to permit an electron to escape from the photoemissive surface.

The photoelectric effect is not seen at all wavelengths. For an electron to be emitted, the applied photon energy must be greater than the work function energy (E_ω) of the material. The *work function* is the amount of energy required to dislodge an electron from its associated atom. Therefore, for any given material, there is a critical wavelength below which the photon energy is less than the work function energy. The maximum wavelength can be found from:

$$\lambda \leq \frac{c\,h}{E_\omega} \qquad \textit{eq. (4-2)}$$

Where:

λ is the wavelength in meters.

c is the speed of light ($\approx 3 \times 10^8$ m/s).

h is Planck's constant (6.63×10^{-27} erg-s).

E_ω is the work function energy for the material illuminated.

The construction of a typical photoemissive sensor tube is shown in **Figure 4-1**. The device is a diode because it has two electrodes inside an evacuated glass or metal housing. The photoemissive surface is called the *cathode*, and the electron collector electrode is called the *anode*. These elements are housed inside an envelope, which is either under a high vacuum or filled with an ionizable inert gas at low pressure.

The anode is constructed of a small loop of wire or a thin rod, usually made of either tungsten, platinum or an alloy of these metals. The cathode is made of metal, but the actual surface is enhanced for improved emissivity by a special light sensitive coating. Typical coating materials include antimony, silver, cesium and bismuth, mixed with trace quantities of other elements.

Although photoemission does not require an external voltage source, the electrons must be collected and delivered to an external circuit before the device is useful. In order to collect electrons, an electrical potential is connected across the tube such that the anode is positive with respect to the cathode. Electrons, being negatively charged, are repelled from the negative cathode and attracted to the positive anode. For most common photoemissive sensors, the potential is between 0 and 300 volts.

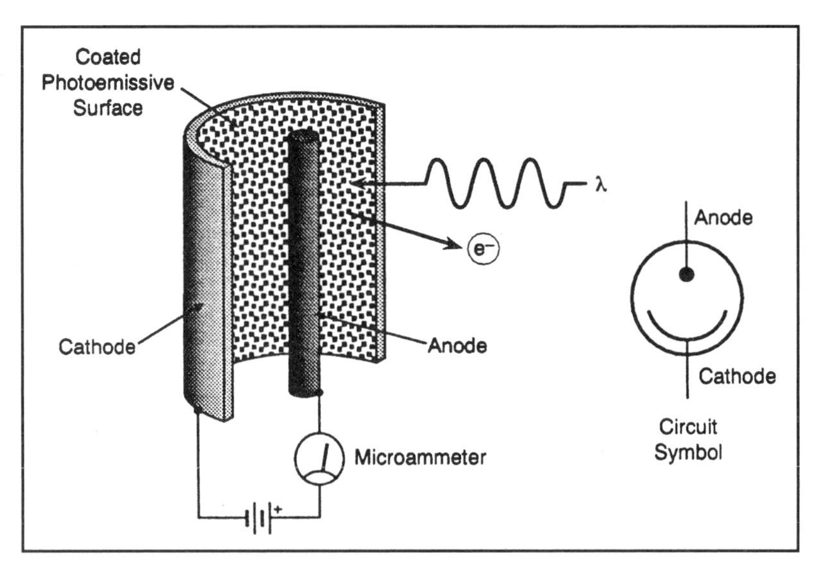

Figure 4-1. Photoelectric tube structure.

There are two general forms of tube construction. In *side-excited* designs (**Figure 4-2a**) the cathode is constructed such that the light must enter the side wall of the tube. The other directions may be blinded by internal blackening or silvering. This prevents stray light from exciting the photoemissive surface. The *end-excited* design is constructed with the photoemissive surface facing the end of the housing (**Figure 4-2b**). Again, all surfaces other than the optical window may be blinded against stray light.

Photoemissive sensors are also categorized according to whether the inside is maintained under a vacuum or is gas-filled. A high vacuum photodetector is evacuated of all air and gases. The device produces a small current, I_o, that is linearly proportional to the intensity of the impinging light.

The output current levels from the high vacuum photodetector are generally too low to be measured directly by a meter's movement, so they are typically converted to an output voltage, V_o, by passing the current through a load resistor, R_L (**Figure 4-3**).

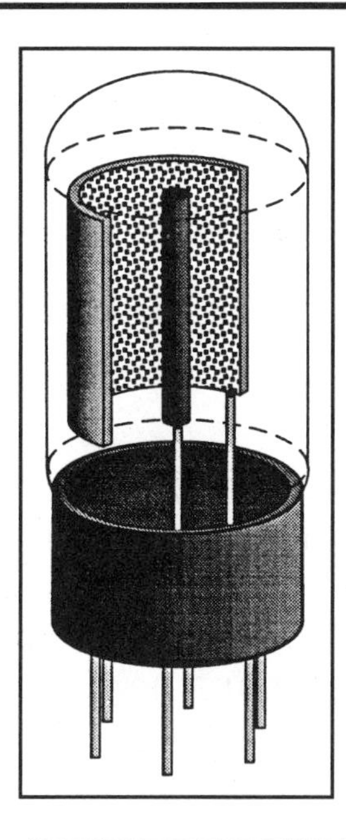

Figure 4-2a. A side-excited photoelectric tube.

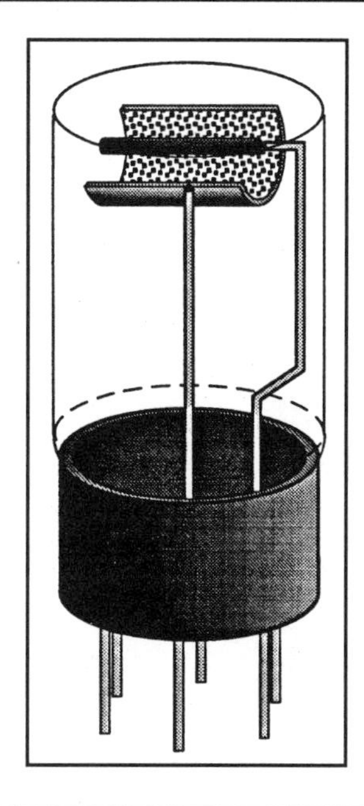

Figure 4-2b. An end-excited photoelectric tube.

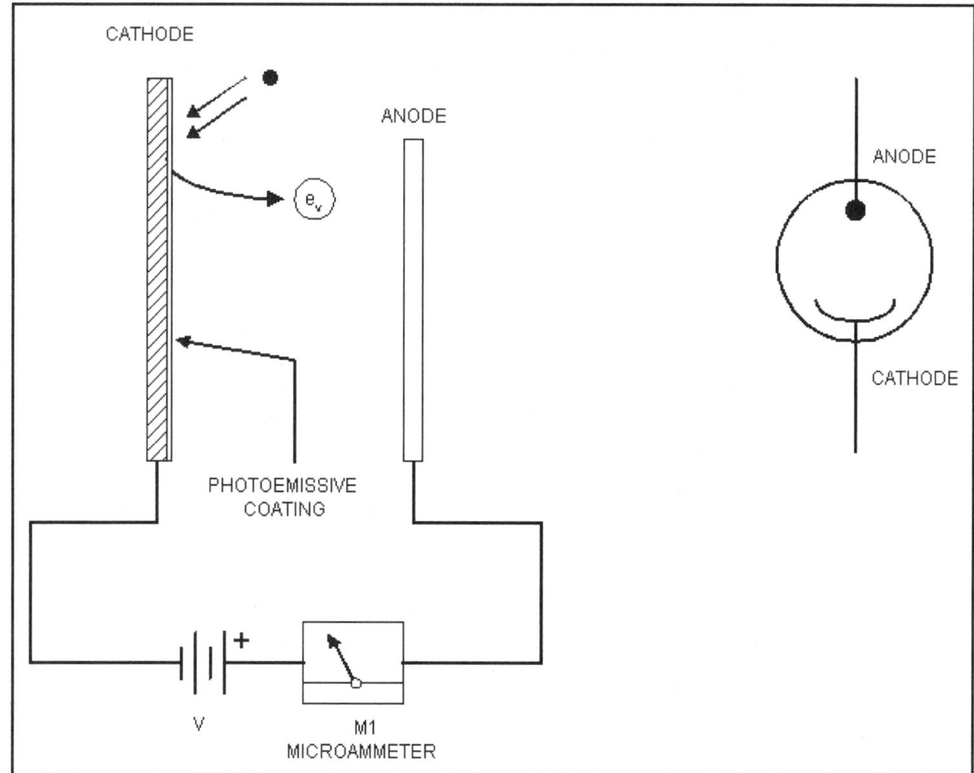

Figure 4-3. Photoelectric tube circuit for light sensing.

The output voltage, V_o, is the difference between the power supply potential, V, and the voltage drop, V1, across R_L caused by the photodetector output current, I_o. By Ohm's law, V1 is equal to the product I_oR_L, so the output voltage is:

$$V_o = V - I_o R_L \quad \text{eq. (4-3)}$$

A gas-filled photoemissive sensor is first evacuated of air, and then refilled with an ionizable, inert gas. Photoelectrons emitted by the cathode collide with the gas molecules, dislodging other electrons and leaving positive ions. This is called the *secondary emission*. Each of the emitted electrons produces a number of secondary electrons, so that the overall current flow is ten to one-thousand times greater than in a similar high vacuum device. The gas pressure is carefully regulated to ensure that this process happens without either dying out or running away. In addition, the cathode to anode potential, V_{ka}, must be kept low enough to prevent imparting sufficient kinetic energy to the electrons to cause complete ionization of the gas. If this were to happen, then the phototube would emit light in the same manner as a glow lamp. The gas filled tube produces higher output current than the high vacuum types, but does not have as linear a light intensity versus output current.

Photoemissive sensors have a characteristic called the *dark current*, I_d; that is, a current that flows from cathode to anode when there is no light impinging on the photoemissive surface. For gas-filled tubes, I_d is on the order of 10^{-8} to 10^{-7} amperes, while for high vacuum devices it is 10^{-9} to 10^{-8} amperes.

The *response time* of a photoemissive sensor is the time required for the device to respond to changes in applied light levels. For high vacuum devices, this time is on the order of one nanosecond, while for gas-filled tubes it is one millisecond.

Photomultiplier Tubes

The photoemission process is less efficient than is needed in many cases, especially under low light-level conditions. The photoelectric tube system can be more efficient by using a photomultiplier tube, or PMT (**Figure 4-4a**). In this photosensor there are positively charged anodes, called *dynodes*, that intercept the electrons. When light impinges on the photoemissive cathode, electrons are emitted. They are accelerated through a positive high voltage potential, V1, to the first dynode, acquiring substantial kinetic energy during this transition. When an electron strikes the metal dynode, it gives up its kinetic energy. Some energy is converted to heat, and some dislodges additional secondary emission electrons from the dynode surface.

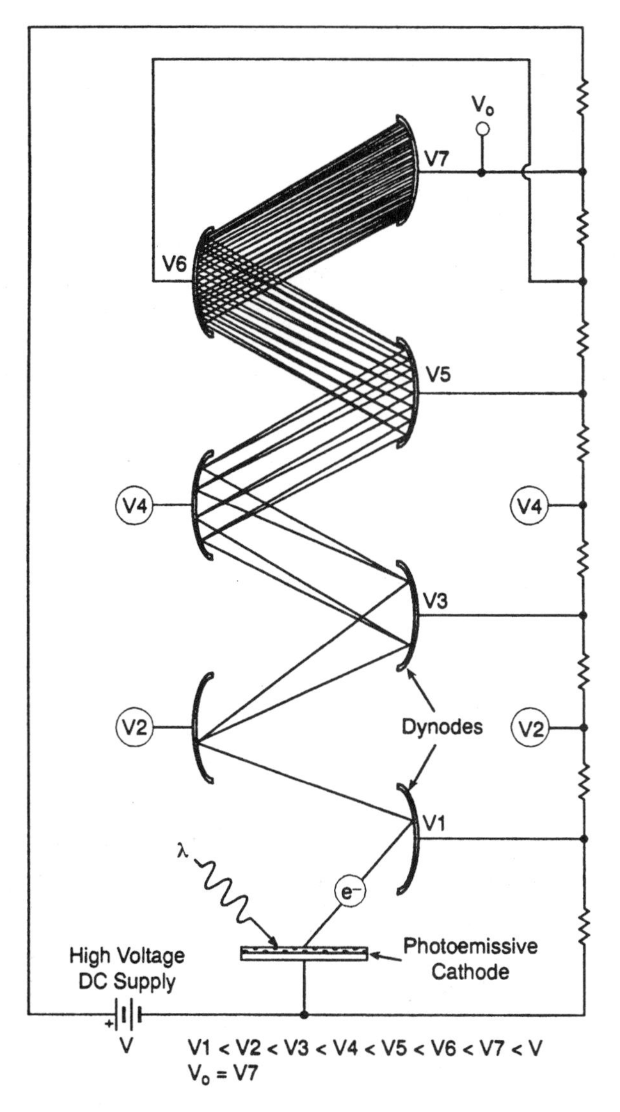

Figure 4-4a. Photomultiplier tube structure.

A single electron can thus cause two or more additional electrons to be dislodged. These electrons are accelerated by the high voltage potential V_2, and produce the same effect at the second dynode. The process is repeated several times with each step increasing the potential by 75 to 100 volts and releasing several more electrons to join the cascade. Finally, the electron stream is collected by either the last dynode or a separate anode, and can be used in an external circuit.

Figure 4-4b shows a typical circuit using a photomultiplier tube. The tube operates from a dc power supply of 200 to 3000 volts, depending on the design of the tube and its intended application. Notice that the photoemissive cathode is connected to the negative terminal of the power supply, so that electrons are forced away from its surface. The anode is connected to the positive side (close to ground), facilitating the collection of electrons. The bias voltage for the series of dynodes is provided by a resistor voltage divider ladder circuit connected across the high voltage power supply. All resistors are equal to each other, and have a value from 25 kΩ to 220 kΩ.

The output current, I_o, is connected to the input of an operational amplifier. Although the specifics of operational amplifier operation will be discussed in Chapter 10, the output voltage is a function of the input current, which in **Figure 4-4b** is also the PMT output current, I_o, and the feedback resistance of R1. **Equation 4-3** applies to this circuit.

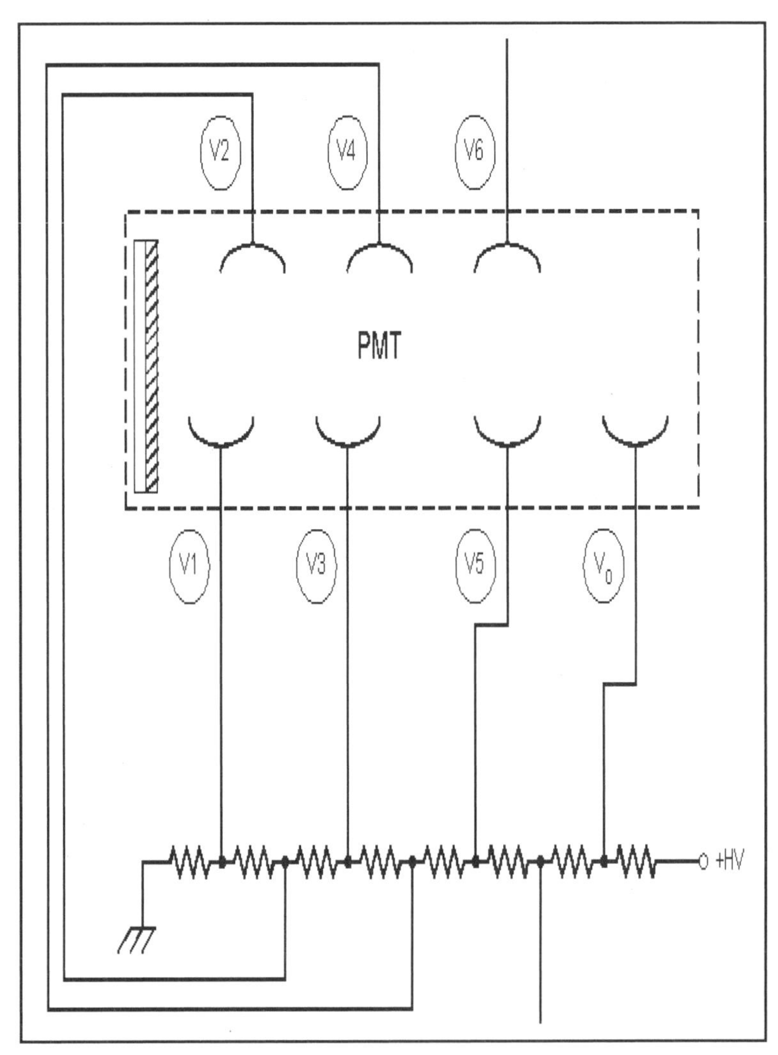

Figure 4-4b. Typical PMT circuit.

The output current of the PMT is a function of both the dc power supply voltage potential between the photocathode and anode, and the applied light level. **Figure 4-4c** shows a typical family of characteristic curves relating the output current to the anode-cathode voltage for various applied light levels. There will always be a small dark current flowing, as long as the dc power supply voltage is applied, even when no light impinges the photomultiplier tube.

At low voltages, **Figure 4-4c** is noticeably nonlinear. In addition, there is a possibility that noise is mixed with the photosignal. Both of these occurrences are reduced by keeping the final dynode voltages stable. In **Figure 4-4b** this is accomplished by connecting capacitors from the last several dynodes to ground. These capacitors stabilize the voltage at each associated dynode, and decouple noise signals to ground.

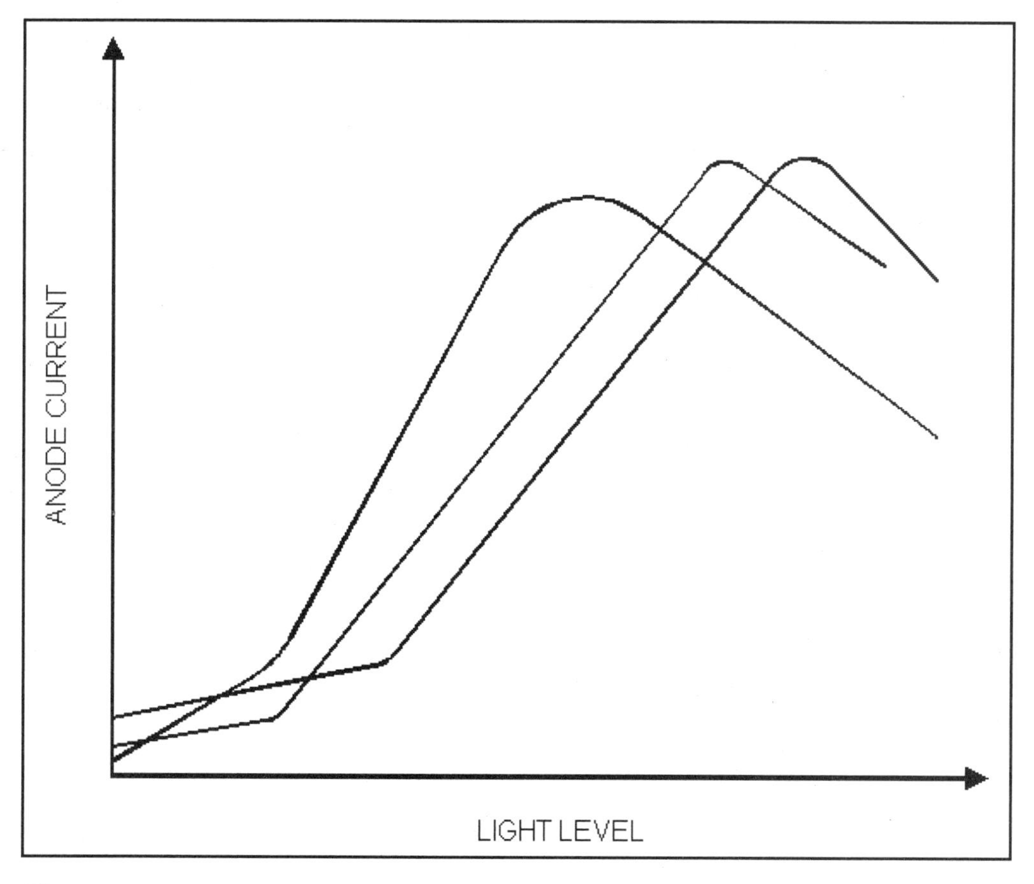

Figure 4-4c. PMT response at various light levels.

The gain (any change of parameters) of a photomultiplier tube is a function of the applied voltage, or more specifically the difference in voltage across the last dynode in the chain:

$$\frac{\Delta A}{A} = \frac{6.3\,\Delta V_d}{V_d} \qquad \text{eq. (4-5)}$$

Where:

A is the gain of the photomultiplier tube.

V_d is the dynode voltage of the last dynode.

Spectral Response of Photoemissive Sensors

The *spectral response* of any photosensor describes its relative sensitivity to various wavelengths of light. The spectral response is stated in graph form as relative output level (usually current) versus wavelength. **Figure 4-5** shows a typical family of spectral response curves. In addition to the graphical data, one can also order photosensors according to a standardized S-number designation. Examples of these S-numbers are shown in **Table 4-1**.

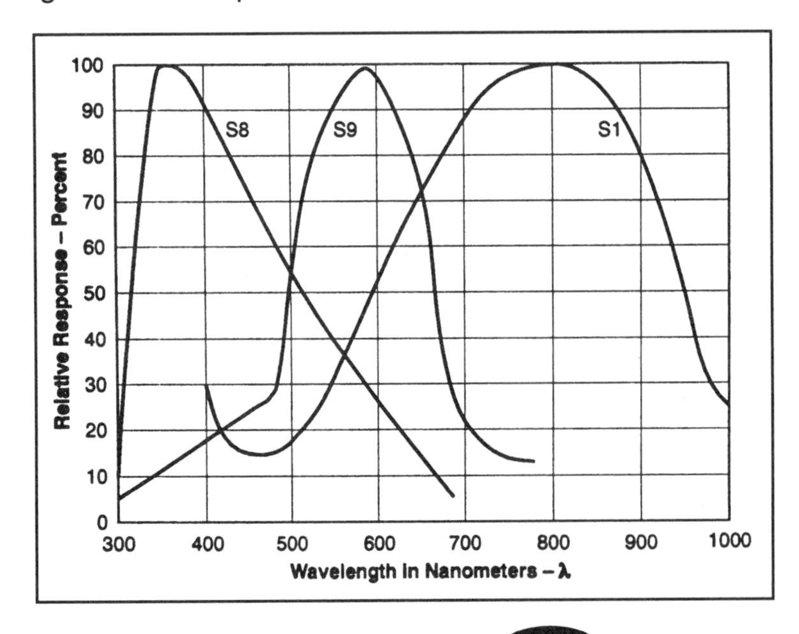

Figure 4-5. Spectral responses of several PM tubes.

Spectral Designator (S-Number)	Wavelength for Peak Response (nm)	Half-Points (nm)
S1	800	620, 950
S3	420	350, 640
S4	400	320, 540
S5	340	230, 510
S8	370	320, 540
S10	450	350, 590
S11	440	350, 560
S12	500	(narrow band)
S13	440	260, 560
S14	150	760, 1730
S20	420	325, 595
S21	450	260, 560

Table 4-1. Photosensor Order by S-number.

Most photoemissive sensors have a spectral response in ultraviolet (near 200 nm) and visible light wavelengths, but a few respond in the near-infrared region (close to 800 nm). Many sensors that respond in the UV region are limited because the sensor tube's glass window does not transmit UV.

Only the S14 sensor in **Table 4-1** and **Figure 4-5** has a narrow band response. When a narrower bandpass characteristic is needed, it is necessary to install the photosensor behind a filter window that passes only the desired wavelengths of light (**Figure 4-6**). The tube must be installed in a light-tight housing that is internally blackened against stray reflections.

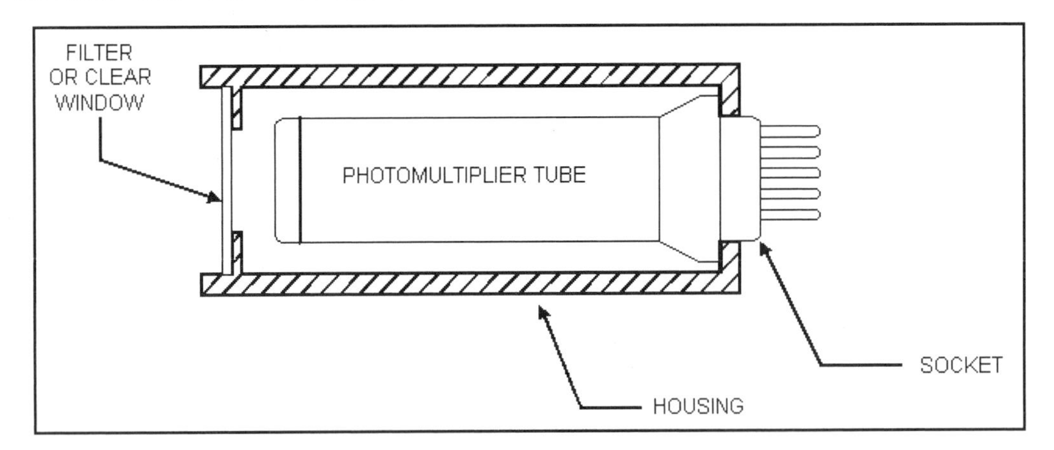

Figure 4-6. Structure of a PM tube sensor.

Phototube and Photomultiplier Tube Applications

Photoemissive sensors are used in a variety of applications ranging from intruder alarms to medical X-ray apparatus. Common phototubes are used where the light levels are relatively high, and low sensitivity sensors are suitable; photomultiplier tubes are used where the light levels are very low, and a high sensitivity sensor is therefore needed.

Intruder alarms follow two basic designs. First, a light source is placed on one side of a doorway, and a phototube on the other side. When a person walks through the light beam, he or she will interrupt the light and cause the output of the phototube to drop. A relay or electronic circuit will announce the intrusion. A second scheme is shown in **Figure 4-7**. In this type of intruder alarm, the phototube and lamp are located inside a common housing. The light beam shines across the protected space to a mirror which reflects the light to the photocell in the lower portion of the housing.

X-rays and gamma rays cannot be detected by most photoemissive sensors. An exception is the *scintillation counter*, shown schematically in **Figure 4-8**. An end-excited photomultiplier tube is fitted with a scintillation crystal, a material that produces flashes of visible light when excited by X-ray photons. The visible photons are picked up by the photomultiplier tube, which amplifies the photoemissive current to a usable level.

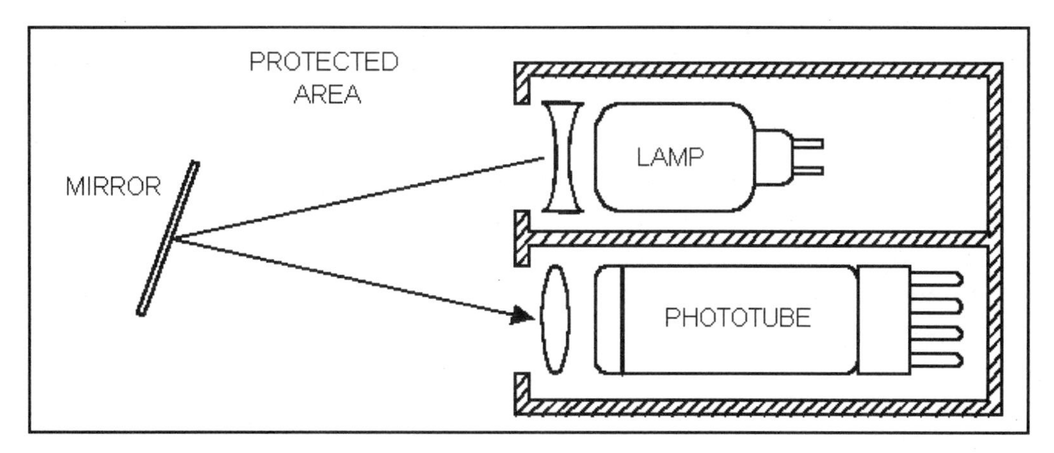

Figure 4-7. Intruder alarm sensor.

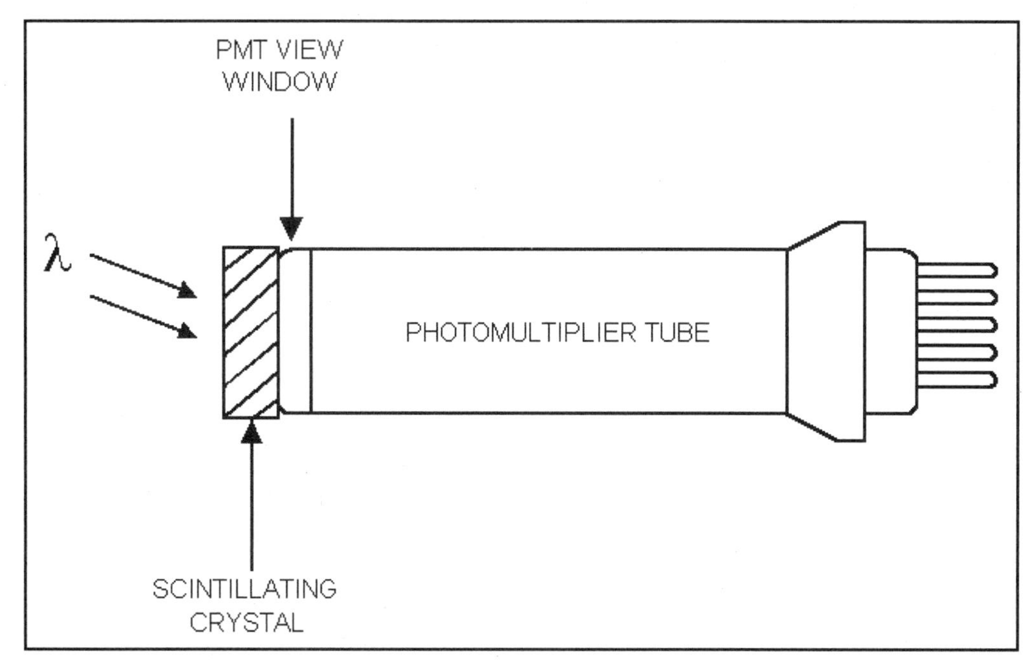

Figure 4-8. Scintillation counter structure.

Another class of applications is shown in **Figure 4-9**. A cathode ray tube (CRT) is a device much like a television picture tube. An electron gun in the rear of the tube produces a beam of electrons, which are accelerated by a high voltage (see Chapter 2) to a high kinetic energy state. These energetic electrons strike a phosphor-coated viewing screen, producing a bright spot at the point where the beam hits. If the beam is swept from side to side, then a horizontal line is produced. Alternatively, if the beam is swept up and down, a vertical line is described on the screen. If both horizontal and vertical sweep are applied to the electron beam (as in a TV set), and the sweep rates are high enough, then the entire screen appears to light up. In reality, the entire screen is not lighted simultaneously, but rather a series of very closely spaced parallel lines are inscribed on the screen.

If the viewing screen is aimed at a photomultiplier tube, then the output current of the PMT is proportional to the brightness of the spot on the CRT screen. Thus, we have a time history of the brightness as the CRT beam sweeps back and forth, up and down.

If a transparency film is interposed between the CRT viewing screen and the PMT, then the brightness of the light reaching the PMT varies as the optical density of the film varies. The varying output current from the PMT represents the image on the film.

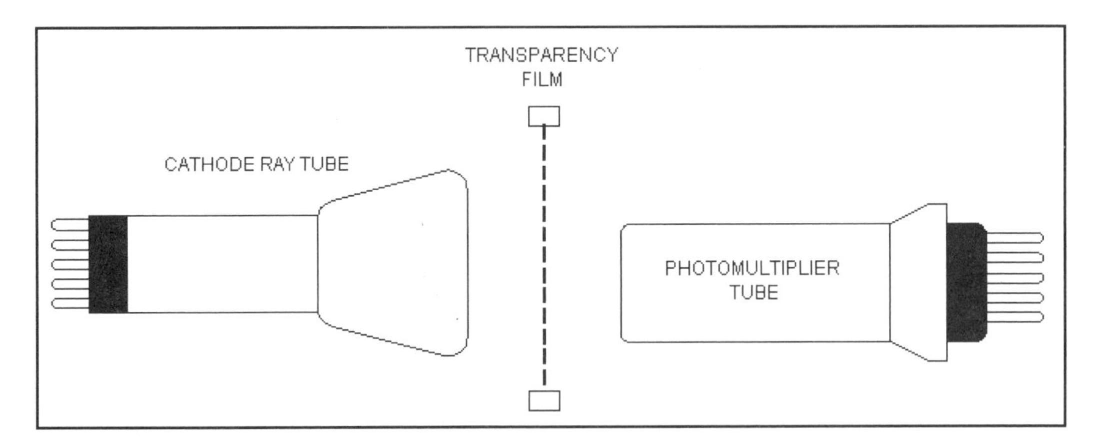

Figure 4-9. Slow-scan television system.

There are two main applications of the system of **Figure 4-9**. The first is slow-scan television (SSTV). Some amateur radio operators, security companies, and industrial inspection systems use SSTV instead of regular fast-scan television (FSTV) because the PMT's output signal can be transmitted over normal audio channels such as telephone circuits and radio communications transmitters.

The other application is densitometry, the art of converting X-ray films to an electronic signal. An X-ray film is interposed between the PMT and CRT. The PMT output current is proportional to the optical density of the film at each point viewed as the electron beam scans. This scheme is used not only to transmit X-ray images, but also to measure radiation exposure of workers with X-ray apparatus. These workers wear badges that contain a small piece of X-ray film, which becomes progressively exposed as it encounters X-rays. When developed, the optical density of the film patch indicates the amount of X-ray exposure experienced by the wearer. These are often scanned by a densitometer, as in **Figure 4-9**, as an occupational safety measure.

Chapter 5

Light Sources

Many applications of the circuits and instruments that use electro-optical sensors depend heavily on the nature of the light source. In many such applications, the *spectral content* (color mix) of the light source is the key factor. **Figure 5-1** shows the spectral content of several different common light sources, ranging from sunlight to the nearly monochromatic spectra of several popular types of light emitting diode.

The light from the sun ranges from the infrared (providing warmth), through the visible spectrum (which allows us to see), to the ultraviolet (which causes skin wrinkling, aging and skin cancer). Fire is another light source. When a material is rapidly oxidized, it will give off a characteristic electromagnetic spectrum that ranges from infrared to bright bluish-white light. Flames are easy to see because of their visible light content.

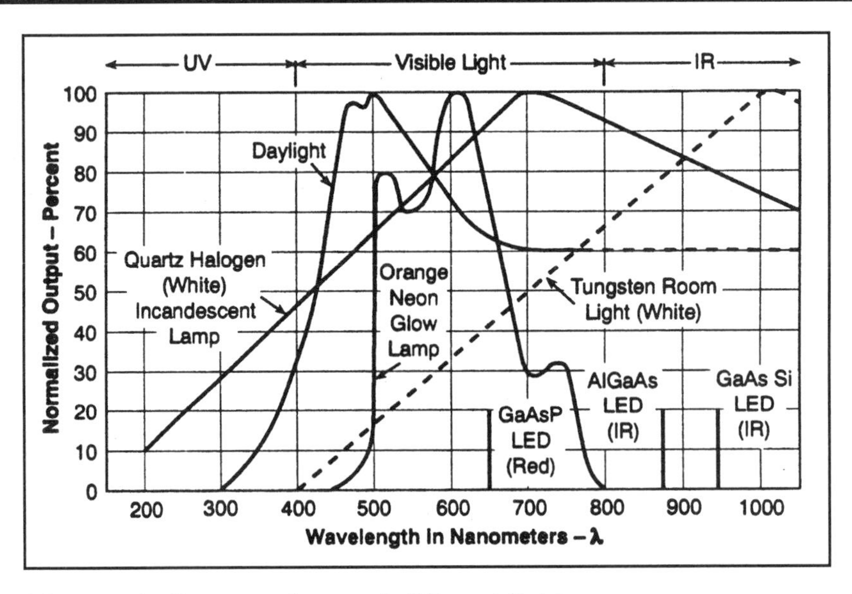

Figure 5-1. Spectra of several different light sources.

The Sun, Fire and Flames

Some electronic instruments depend on the color of the light emitted when a sample substance is burned. A *flame photometer*, for example, measures the sodium and potassium content of blood by burning the sample along with a calibration substance, and noting the relative output levels of different colors of the visible light spectrum. One can also guess the constituents of a fire by noting the mix of colors in the flame. Sodium burns with a yellow flame, while a methane (natural gas) flame is very blue. One furnace repairman told me that the presence of too much yellow (sodium) flame is an indicator that the burner should be cleaned of foreign material and soot.

Incandescent Light Source

For centuries the only source of artificial light was fire, with its practical problems and inherent dangers. In the 1870s, American inventor Thomas Edison found an artificial source that produced light from electricity. Edison's invention was the incandescent lamp (**Figure 5-2**). In this type of light source, a metal wire filament, now usually made of tungsten (Edison used a carbon filament), is mounted to electrodes inside an evacuated glass bulb. The electrodes are connected to an electrical power source. A current flows in the resistive wire at such a level that it will heat the wire to incandescence, the point where it will give off light.

Compare the curves in **Figure 5-1** that show the spectra for the ordinary tungsten room lamp and the quartz halogen lamp. Quartz is used for the envelope of the bulb, because it will survive a lot higher temperature than ordinary glass. The efficiency and spectral content of the incandescent lamp is improved by inserting small amounts of certain inert gases into it.

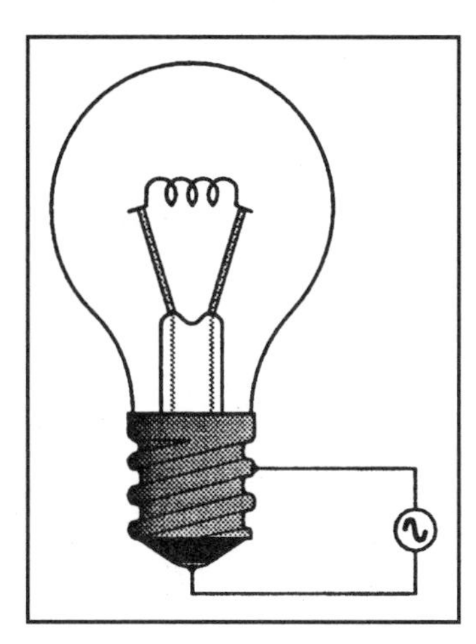

Figure 5-2. The incandescent lamp invented by Thomas Edison still provides most residential illumination in the world.

Glow Lamps

A *glow lamp* (**Figure 5-3a**) consists of a pair of electrodes inside a glass envelope that contains only partial pressure of an inert gas or a mixture of inert gases. When an electrical potential is applied across the electrodes, the gas ionizes, giving off a glow. If direct current is applied, then only one electrode lights (which one depends on the polarity); but if alternating current is applied, both electrodes are lighted.

The spectrum for the glow lamp is considerably narrower than the spectrum for incandescent lamps (see **Figure 5-1**), ranging from about 500 to 800 nanometers. The particular color produced by any given glow lamp is a function of the mixture of gases inside it. The neon glow lamps (NE-2, NE-51, etc.) give off an orange light; other gases or gas mixes produce other colors.

This same phenomenon is the basis for the neon sign industry; not all of the gases used are neon, as evidenced by the wide range of colors found in those signs.

In practical glow lamp circuits (**Figure 5-3b**) a series resistor is used to limit the current flow to a safe value. When the gas ionizes, the electrical resistance between the electrodes breaks down to a low value. At voltages required to produce ionization (~70 volts in neon lamps), the current that will flow when the gas ionizes is immense. A similar gas ionization principle is used to operate common fluorescent lamps.

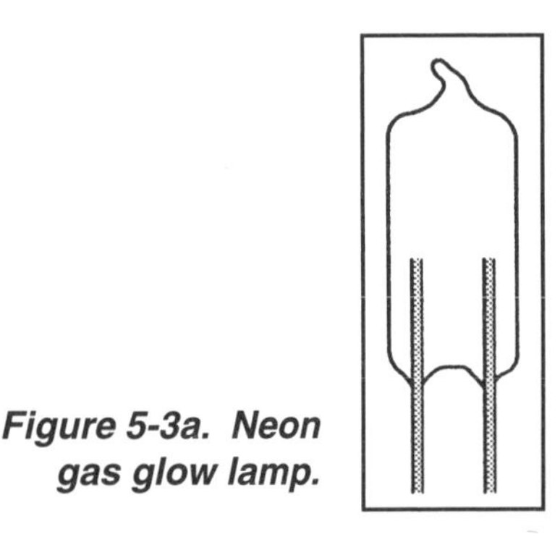

Figure 5-3a. Neon gas glow lamp.

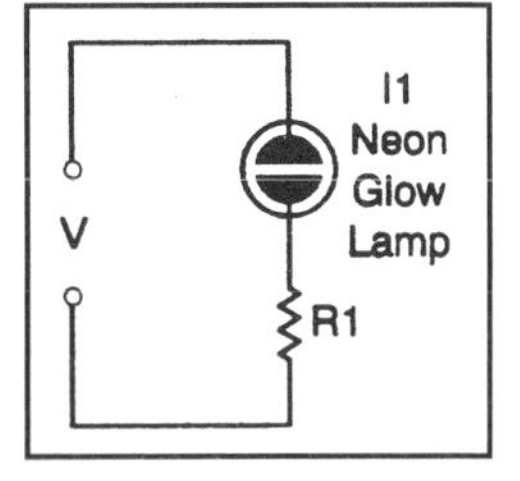

Figure 5-3b. Typical gas glow lamp circuit.

Solid-State Light Sources

In 1907, H.J. Round discovered that semiconductor materials produced light when an electrical potential was applied. After 1960, solid-state light sources became easily available in the form of LEDs.

Figure 5-4a shows a stable atomic configuration. This particular atom (neon) has ten protons and ten orbiting electrons. The first (innermost) shell is completely filled, and therefore stable, with just two electrons.

The difference between hydrogen (H) and helium (He) illustrates what such stability means. Hydrogen has one orbital electron, while helium has two, thereby filling its single shell. Hydrogen easily forms chemical compounds with other elements and is vigorously unstable. Helium is very stable, and will not form compounds. When a shell is completely filled, chemical reactions and electrical current flow become more difficult because sufficient energy must be delivered to the system to strip away an electron from the stable shell.

The second shell in a neon atom contains the remaining eight electrons. All shells other than the first shell are filled when eight electrons are present; the configuration is called a *stable octet.*

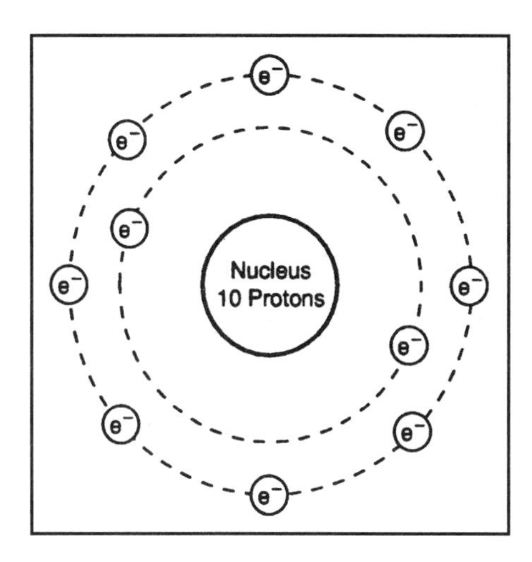

Figure 5-4a. Atom with ten electrons and ten protons provides a model for our discussion. The energy level of the orbital electrons depends on their radial distance from the nucleus: the higher the orbital radius, the higher the energy level.

With the exception of helium, the noble gases (formerly called inert gases: radon, xenon, krypton, argon, neon and helium) exhibit the stable octet outer shell configuration. They are not readily chemically active, nor will they conduct electrical current unless a sufficiently high electric potential is applied to strip away one or more electrons.

As stated in Chapter 2, the electron shell represents an energy state. Lower energy electrons are in lower shells, closer to the nucleus, while more energetic electrons are further away. When energy is absorbed (**Figure 5-4b**), the energy state is changed. The electron jumps to a higher energy shell, leaving a hole behind in the previous shell location. But when the electron gives up this energy (emission line), it falls from a higher energy state back to the lower energy state (filling the hole). The Law of Conservation of Matter and Energy requires that the energy lost when the electrons fall back to a lower state be accounted for. This energy is converted into an electromagnetic wave, either infrared or visible light. Because the photon resulted from an electron recombining with a hole, it is called *recombination radiation.*

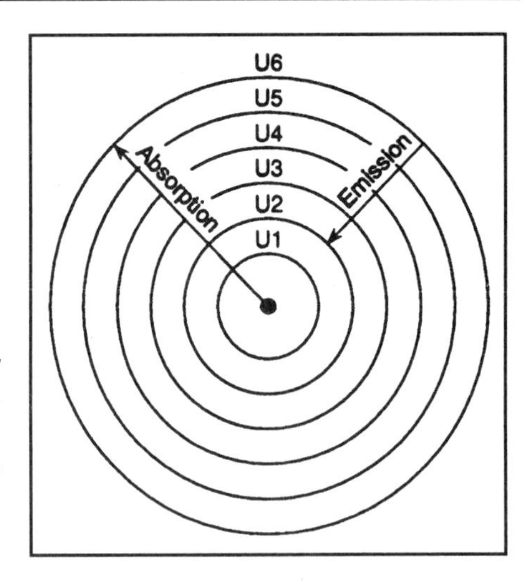

Figure 5-4b. Energy level diagram for absorption and emission phenomena of an atom's orbital electrons. With absorption, electrons move to orbits more distant from the nucleus; with emission of a photon, electrons drop back to closer orbits.

Figure 5-5 shows this phenomenon in more detail. At point A, an electron at ground state (energy level, E_g) is excited by external incident energy, E_i. Under this stimulation the electron jumps to a higher energy level, E_c, leaving a hole in its former energy level. The difference between the conduction and valence bands is called the *forbidden band* or *bandgap energy*, and is usually expressed in electron volts (eV).

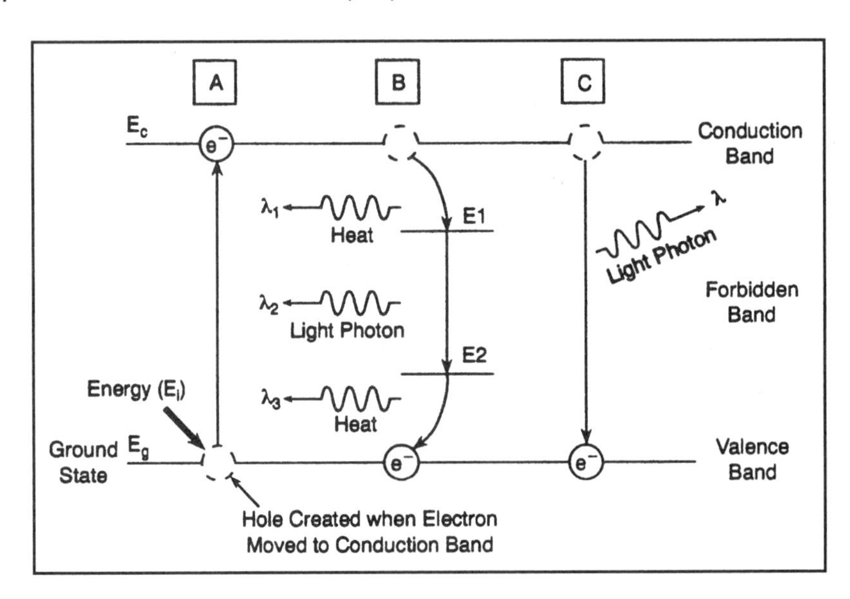

Figure 5-5. Light from a solid-state source is created by emission when an energetic electron falls back to the ground state. Absorption is illustrated at A, B shows two-step emission, and C shows single-step emission.

There are two basic forms of recombination emission: multistep and one step, as shown at points B and C, respectively. In the multistep version, the electron energy drops back to E_g in two or more steps. Different EM wavelengths are emitted at each transition (λ_1 for $E_c \rightarrow E1$, λ_2 for $E1 \rightarrow E2$, and λ_3 for $E2 \rightarrow E_g$). The more efficient one-step version is shown at C. The transition goes directly from E_c to E_g, emitting a single photon in the process.

As stated in **Equation 5-1**, the relationship for wavelengths is:

$$\lambda = \frac{c\,h}{E} \qquad \text{eq. (5-1)}$$

When the constants are combined, and the units are converted to express wavelength λ in nanometers, the equation above reduces to:

$$\lambda_{nm} = \frac{1{,}237}{E_g(eV)} \qquad \text{eq. (5-2)}$$

Where:

λ_{nm} is the wavelength in nanometers (1 nm = 10^{-9} m).

E_g is the bandgap energy between the conduction and valence bands.

Equation 5-2 reveals why LED light sources tend to be nearly monochromatic. The emitted wavelength is a function of the bandgap energy of the material used in the LED. Different material formulations, displaying different bandgap energy levels, produce different wavelengths of monochromatic emission spectra. For example, in gallium arsenide (GaAs), the material used for many light emitting diodes, the bandgap energy ranges from 1.32 eV to 1.36 eV, 1.34 eV being the nominal value. For these levels the emitted light wavelength is 937 nm, 910 nm and 923 nm, respectively.

Because they have a very fast response time, LEDs can be switched on and off rapidly. For a GaAs or GaAsP (gallium arsenide phosphide) LED, the switching time can be from one to ten nanoseconds, depending on structure. Although silicon (Si) LEDs are more efficient than GaAs or GaAsP devices, they are slower. The Si LED can be switched in 300 nanoseconds. Yet silicon LED switches are faster than any other solid-state or most other light sources except the GaAs/GaAsP devices.

Because of their fast switching times LEDs are used in chopped or modulated circuits. The Si LED can be modulated at frequencies or pulse repetition rates up to 1 MHz, while GaAs/GaAsP devices can be modulated at frequencies up to 100 MHz.

The life expectancy of light sources can be calculated in terms of *mean time between failure* (MTBF), measured in hours. This statistical measurement reflects the length of time the average device operates before failure. The MTBF of LEDs can be 10,000 hours or more (as opposed to 750 to 1500 hours for incandescent lamps), but circumstances such as ambient temperature, operating current and duty cycle (on/off cycles) may affect individual applications.

PN Junction Light Radiators

Light emission from semiconductors can occur in bulk materials such as gallium arsenide, but the phenomenon is most effective when the material is formed into a PN junction. A *PN junction* consists of P-type and N-type semiconductor material in intimate contact with each other. N-type semiconductor material is doped with impurities that give it a surplus of electrons, while a P-type semiconductor is doped to give a deficiency of electrons in the crystalline matter.

How are these materials formed? Semiconductor materials tend to be *tetravalent*; that is, they have four electrons in the outer electron shell (called the valence shell). In order to achieve stability similar to the stable octet configuration, these atoms share electrons with each other in covalent bonds to simulate a situation where eight electrons exist in the valence shell. In doing so, the atoms form up into a crystal lattice array (**Figure 5-6a**). Because of the stability of the covalent bonds, the semiconductors are not good conductors of electrical current - thus their name.

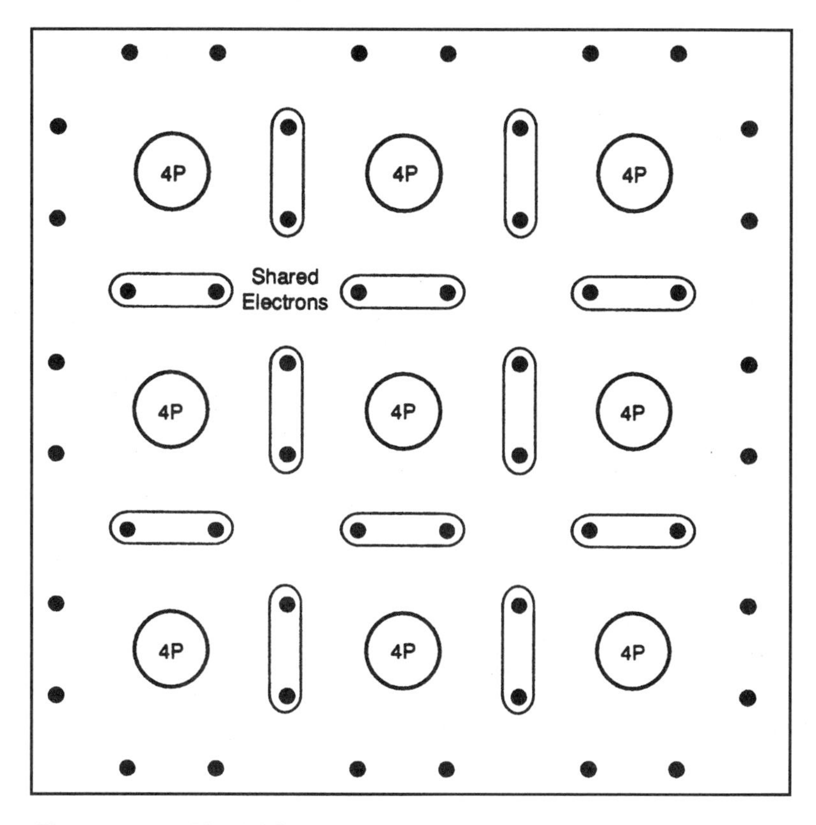

Figure 5-6a. Model for tetravalent atoms forming covalent bonds to simulate a stable octet.

Free electrons are needed to support an electrical current. These are supplied in N-type semiconductors by doping a tetravalent semiconductor with a *pentavalent* (five valence electrons) impurity. Four of the five valence electrons of each impurity atom will form covalent bonds with the four valence electrons of nearby semiconductor atoms (**Figure 5-6b**). But this leaves one free electron for each impurity atom. These electrons support electrical current.

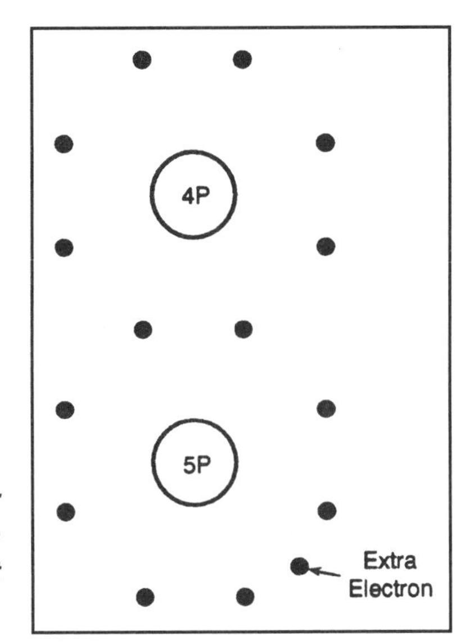

Figure 5-6b. Model for tetravalent impurity atom in covalent bond with pentavalent atom to produce free electrons.

The P-type semiconductor material uses a similar scheme to create a deficiency in the number of atoms available for covalent bonding. This is accomplished by adding some *trivalent* impurities to the tetravalent semiconductor (**Figure 5-6c**). Again, the covalent bonds are formed between the impurity atoms and the semiconductor atoms. Because of the odd number, not every electron will be able to form a pair. As a result, there is a hole in the crystal lattice structure.

A *hole* is a place in the semiconductor crystal lattice network where an electron should be, but is not. The hole does not exist as a physical entity, but it may be treated as though it does. Mathematically, it may be treated as a particle with the same mass as an electron, but with a unit positive electrical charge rather than a negative charge.

The charge carriers in N-type semiconductor materials are electrons, while the charge carriers in P-type materials are holes. Both types of charges exist in both materials, but there is a majority of electrons in N-type materials, and a majority of holes in P-type materials. Current flow via "hole flow" in P-type materials is a misnomer. Actually, hole conduction is an electron flow in reverse. The "hole" only appears_to move; what really moves are electrons. In **Figure 5-6d**, an electron is at point A, while a hole is at point B. When the

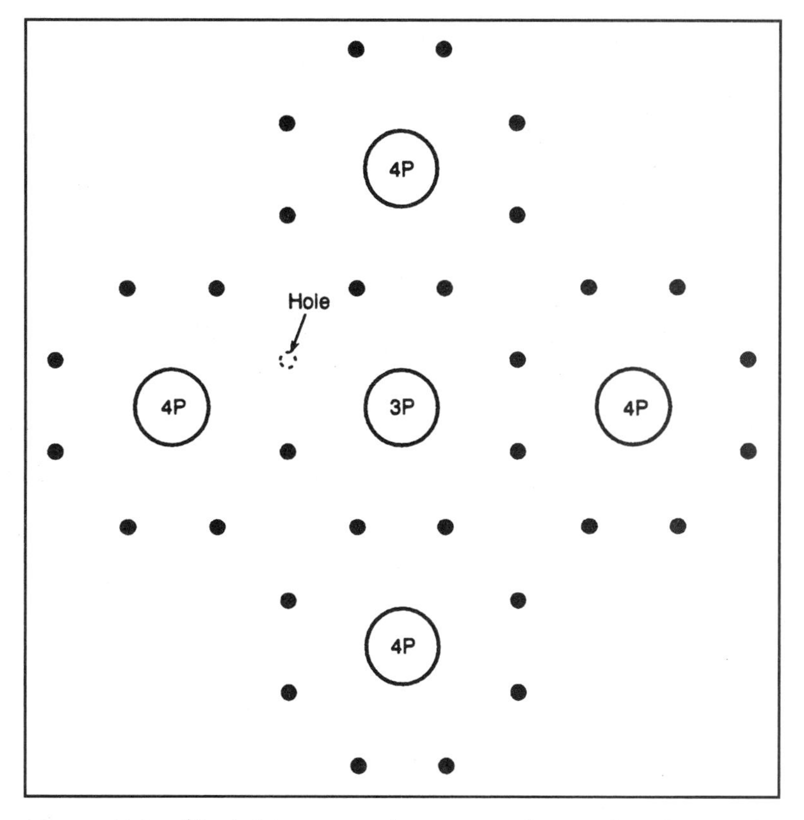

Figure 5-6c. Model for tetravalent atoms in covalent bond with trivalent atoms to create positively charged "holes".

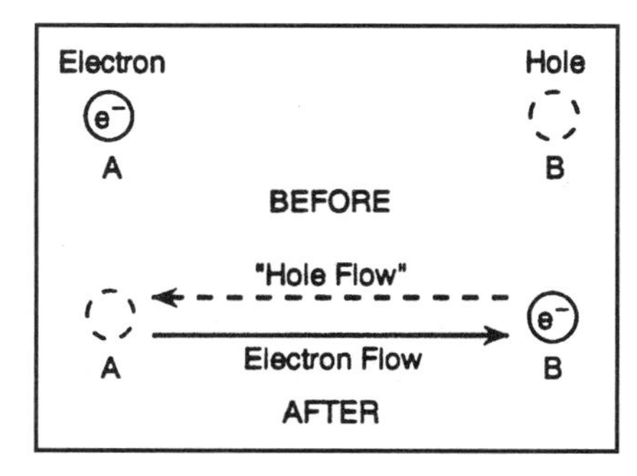

Figure 5-6d. Hole flow is really electron flow in reverse.

electron moves from A to B to fill the existing hole, it leaves a hole in its previous location on the crystal lattice. It appears as if the hole moved from B to A, but it was actually the electron that moved from A to B.

In most PN junctions, the main body of material will be made P-type, and then N-type impurities will be diffused into the region that will become N-type. In **Figure 5-7**, an energy level diagram for the crystal is also shown. When an electrical potential field is impressed across the junction (voltage V), electrons are injected into the N-type material and holes are created on the P-type side. This is called forward biasing the junction. This conduction occurs across the junction when electrons from the N side recombine with holes on the P side in the valence band. In so doing, they collapse back to ground state, and must give up the electrical energy. Thus, a photon is emitted. Through this physical phenomenon, a wide variety of light sources, including LEDs, operate.

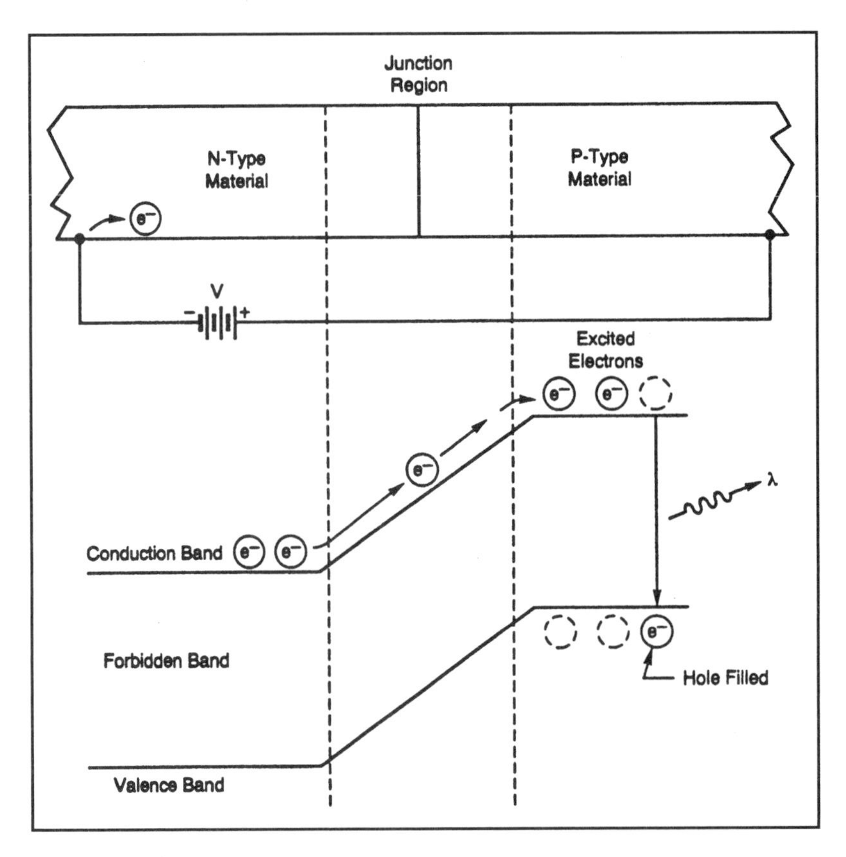

Figure 5-7. Energy diagram for PN junction in solid-state material.

LED Construction

Figure 5-8 shows the planar type construction of an LED. The P-type end of the PN junction is bonded to a metallic mounting tab, which forms one electrode of the PN junction diode. The N-type region is bonded to a thin wire (only several mils in diameter), which forms the other electrode. Light is emitted from the junction and bulk material. In some LEDs, the mounting tab is reflective so the light lost in that direction is reflected upwards.

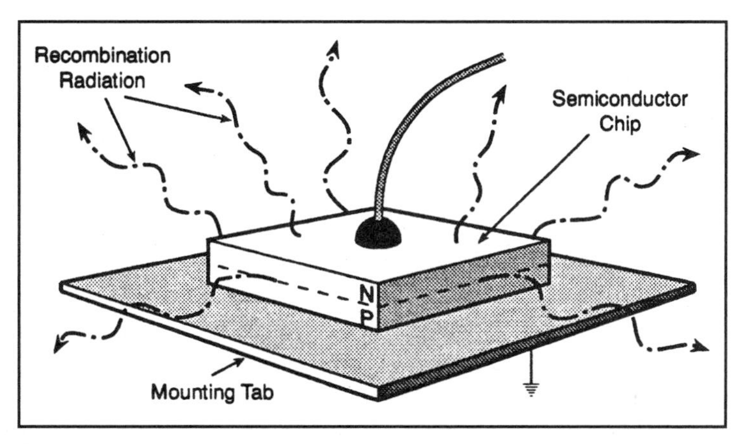

Figure 5-8. Light emitting diode (LED) structure (based on one by Forrest Mims III, used by permission of Mr. Mims).

Figure 5-9 shows several popular packaging styles used for LEDs. The package in **Figure 5-9a** is similar to small transistors in the "TO-18 case" style. The case sides are metal, with a glass lens on top that allows light from the LED chip to exit. The low-cost epoxy package is shown in **Figure 5-9b**. This package style is the most commonly found, and is available in clear, red, yellow, or green colors. A small flat mounting package is seen in **Figure 5-9c**, and a clear plastic flatpack is shown in **Figure 5-9d**. The top of the flatpack is polished to allow light to exit.

Simple, low cost LEDs designed for use as a panel indicator have a diffusing dye in the epoxy material to increase light scattering, so they produce a fuzzy light beam pattern. This pattern is suitable for panel light applications, but may not be adequate for other applications.

Other LEDs focus the light more narrowly, and look more like a pinpoint source. Focused LEDs have a built-in lens section in the region of the package where the light exits from the structure. Unfortunately, these diodes also often exhibit a central beam that is displaced from the boresight optical axis (**Figure 5-9e**), and a halo-effect ring that is caused by internal reflections. Up to 75 percent of the light output of some LEDs can reside in the halo region. An end view of the halo and off-boresight effects is shown in **Figure 5-9f**.

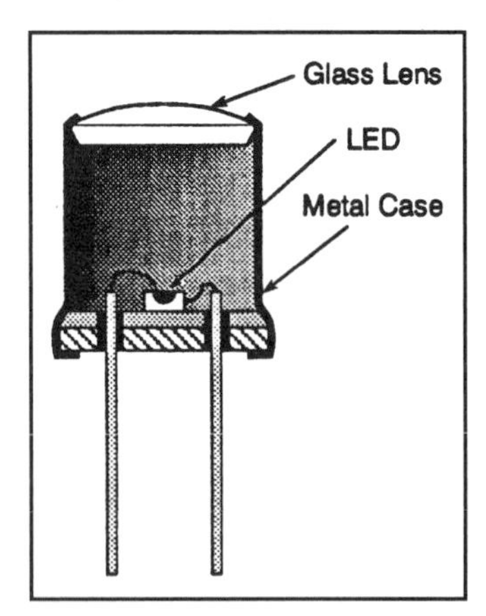

Figure 5-9a. Metal package LED structure.

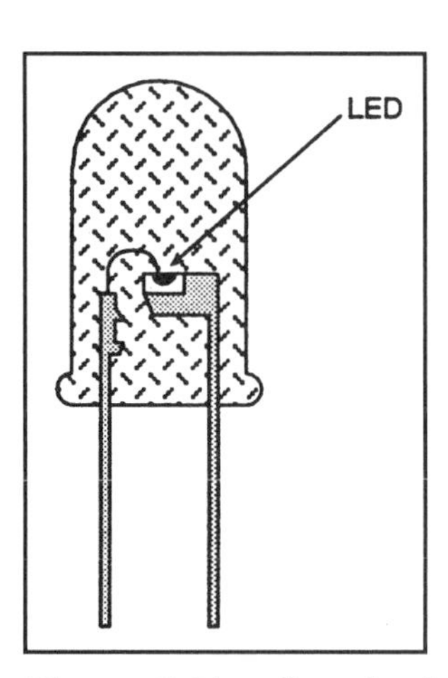

Figure 5-9b. Standard epoxy resin package.

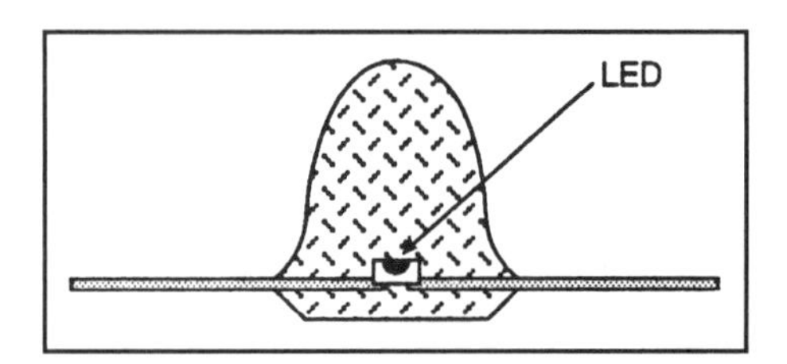

Figure 5-9c. Surface mount LED structure.

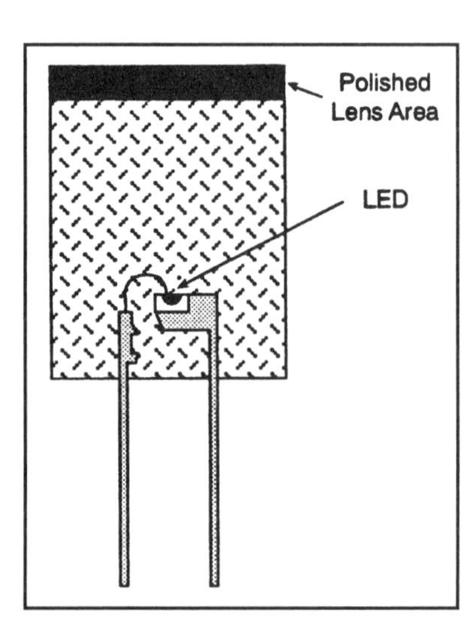

Figure 5-9d. Flatpack LED structure.

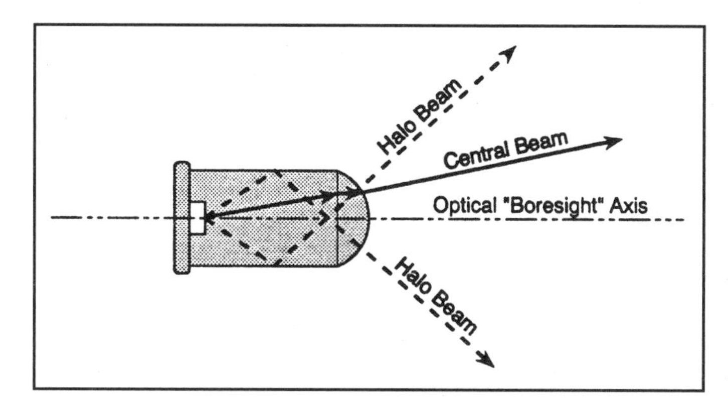

Figure 5-9e. Axis and light zone for LED.

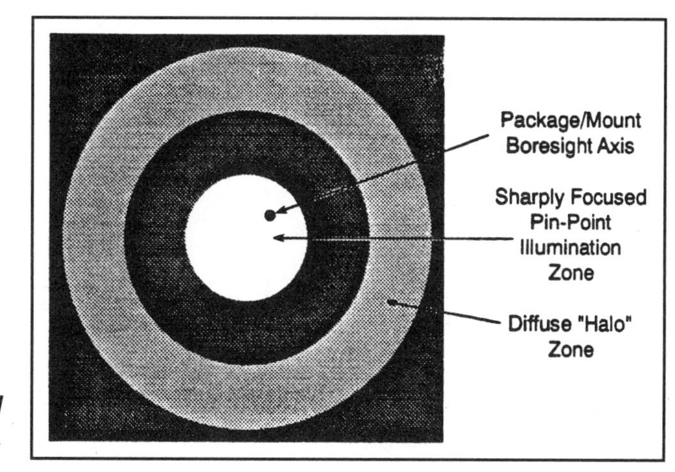

Figure 5-9f. Frontal view of LED light output.

Mounting LEDs

As stated above, LEDs are available in several package styles, but the most common form is the plastic/epoxy package. Mounting can be done using special mounts and retaining rings (**Figure 5-10**). A hole is drilled in a panel, and the mount is pushed through it. The LED is pushed in from the rear until it is seated inside the mount. The retaining ring then slips over the mount until it sets in place. With practice, this is an easy operation.

Simple LED Circuits

Figure 5-11 shows the most basic form of an LED circuit. The diode (D1) is connected across a voltage source, V, such that the positive side of the dc power supply is connected to the LED anode, while the negative side of the

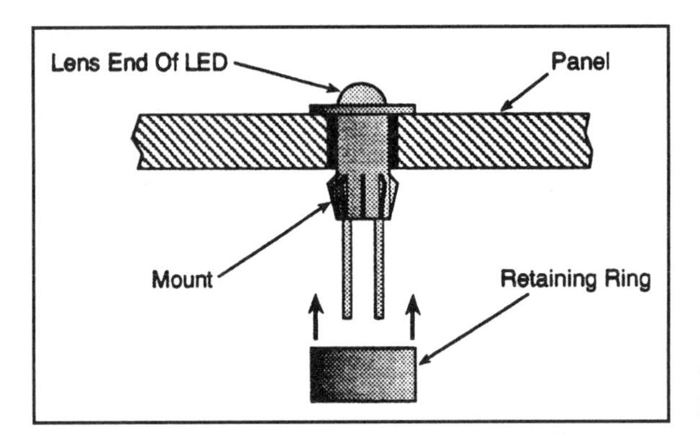

Figure 5-10. Panel mounting an LED.

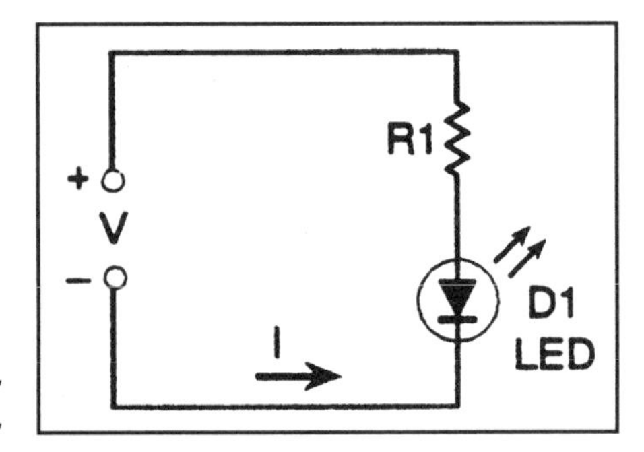

Figure 5-11. Circuit for an LED.

power supply is connected to the cathode. This configuration is called forward biased as we have seen in connection with our discussion of the PN junction. A series resistor (R1) is used to limit the current through the diode to a safe level. The value of the resistor is found from:

$$R1 = \frac{V - V_{LED}}{I} \qquad \text{eq. (5-3)}$$

Where:

R1 is the resistance of the series resistor.

V is the power supply potential in volts.

V_{LED} is the potential across the LED (typically 1.8 volts for a GaAs LED).

I is the LED current in amperes.

Example 5-1

Calculate the resistance needed for R1 in **Figure 5-11** if the power supply voltage is 12 volts dc, D1 is a GaAs type LED, and the desired maximum current is 15 mA.

Solution:

- $$R1 = \frac{V - V_{LED}}{I}$$

- $$R1 = \frac{(12 - 1.8)\ volts}{0.015\ amps}$$

- $$R1 = \frac{10.2\ volts}{0.015\ amps} = 680\ ohms$$

Variable Brightness LED Circuits

Figure 5-12a graphs the LED outputs for various levels of diode current. Because the curve is nearly linear for most of its range, we can make a variable brightness control by controlling the current through the LED. The curves in **Figure 5-12a** were created from data taken from the test set-up shown in **Figure 5-12b**. A silicon solar cell was mounted in a light-tight box that had its interior painted with flat black anti-reflection paint. A common red LED was mounted in the box such that its light fell directly on the solar cell less than 1 cm away.

A solar cell becomes somewhat more linear when it is heavily loaded. As a result, we can linearize the device, as well as provide an output indicator, by shunting a milliammeter across the solar cell output terminals. The result is an output indication in milliamperes that is also a measure of LED light output.

The other scheme shown in **Figure 5-12b** uses a milliammeter as the output indicator. Loading of the solar cell is by a 47 ohm resistor, and it is this resistor that forms the voltage drop for the milliammeter to read. Alternatively, a digital multimeter (with millivolt scales) can be used for the output; it measures the voltage drop across the resistor. **Figure 5-13** shows the use of a variable resistor, R2, for the purpose of controlling LED output brightness.

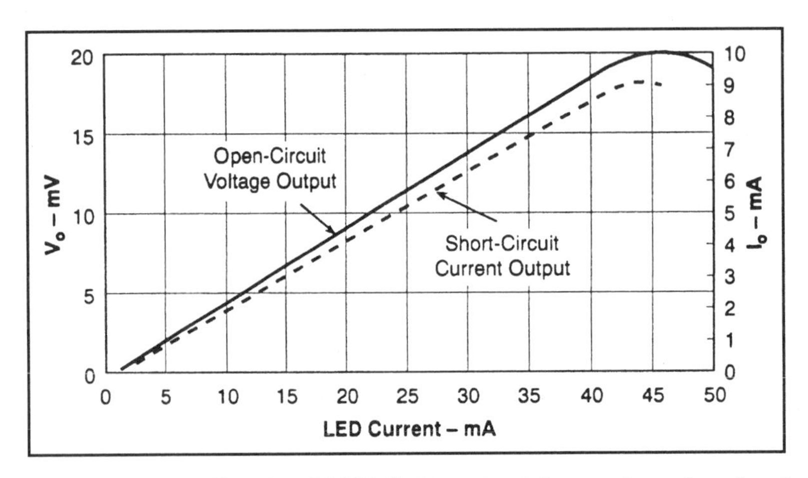

Figure 5-12a. Graph of LED light output for various levels of diode current.

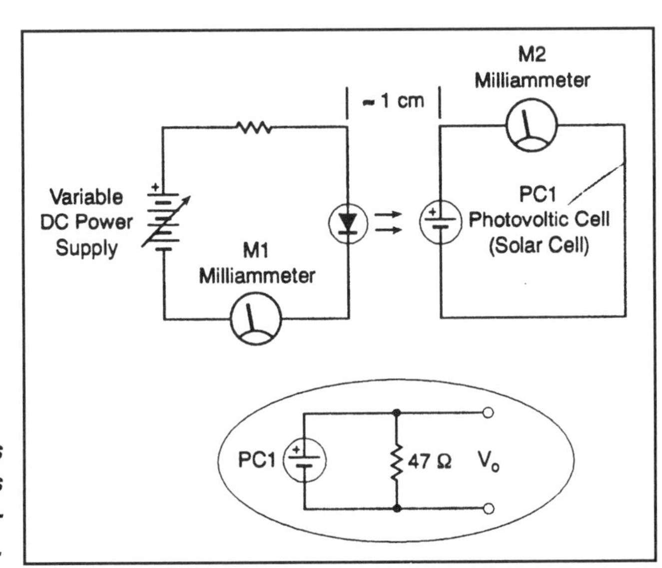

Figure 15-12b. Test circuits for LED brightness tests (inset shows alternate linearized version of sensor circuit).

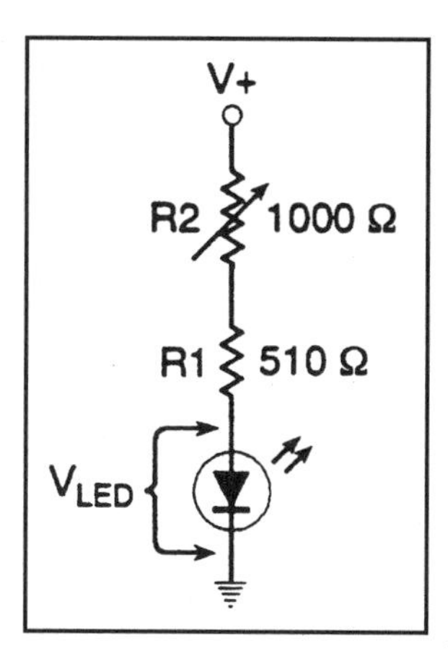

Figure 5-13. Variable brightness LED circuit.

The LED current, I, in **Figure 5-13** is equal to:

$$I = \frac{(V+) - V_{LED}}{R1 + R2} \qquad \textit{eq. (5-4)}$$

The value of fixed resistor R1 should be selected so that current I does not exceed its maximum allowable value, at the highest value of V_{pos}, when the resistance of R2 is zero ohms.

Example 5-2

Assume that $(V_{pos})_{max}$ is 11.8 volts dc, and I_{max} is 20 mA. The value of R1 is found by setting R2 to zero and solving **Equation 5-4** for R1. In the case of Example 5-1 above:

$$R1 = \frac{(V+) - V_{LED}}{I_{\max}}$$

$$R1 = \frac{(11.8) - (1.8)\,volts}{0.02A}$$

$$R1 = \frac{10\,volts}{0.02\,A} = 500\,ohms$$

(As a practical matter, 500 ohms is not a standard commercial value for composition or metal film resistors, so a 510 or 560 ohm resistor may be substituted.)

The minimum power rating of a resistor can be found by using either standard power relationship: I^2R or V^2/R:

- $P = I^2 R1$ *eq. (5-5)*

Example 5-3

Using **Equation 5-5** with **Figure 5-13** and the values in Example 5-2 we can determine:

- $P = (0.02)^2 (500)\ watts$

- $P = 0.2\ watts$

Because only 0.20 watts are dissipated by the resistor, a 1/4-watt rated resistor can be used and still have a 20 percent safety margin. For improved reliability and a wider safety range, a 1/2-watt rated resistor could also be used in this case.

The value of R2 is found by solving **Equation 5-4** for R2 when R1 is set to the selected value of 510 ohms, and I is at the minimum value that corresponds to the minimum desired brightness.

Example 5-4

Assume that I_{min} is 1 mA and the other values in Example 5-2 remain the same:

- $$I_{min} = \frac{(V+) - V_{LED}}{R1 + R2_{max}}$$

- $$0.001\,A = \frac{11.8 - 1.8}{510 + R2_{max}}$$

And, solving for R2:

- $$(0.001)(510) + (0.001)(R2) = (11.8) - (1.8)$$

- $$0.510 + 0.001\,R2 = 10$$

- $$0.001\,R2 = 10 - 0.510$$

- $$0.001\,R2 = 9.49$$

- $$R2_{max} = 9{,}490\ ohms$$

The calculated value for R2 is close enough to 10,000 ohms to make a 10 kΩ potentiometer a good choice.

Note in **Figure 5-13** that R2 has only two terminals. Such a two-terminal variable resistor is called a *rheostat.* Actual rheostats are not often seen today, but they can be made from a three-terminal potentiometer. Use the middle terminal and either end terminal (A-C or B-C), or short the middle terminal (C) and one end terminal (B) together and use this joined terminal, plus the remaining end terminal (A), as the two variable resistor connections.

Driving LEDs from Digital Circuits

Light emitting diodes are often used as indicators in digital circuits. **Figure 5-14** shows four common digital driver situations. **Figure 5-14a** and **Figure 5-14b** show *transistor-transistor-logic* (TTL) drivers, while **Figure 5-14c** and **Figure 5-14d** are *complementary metal oxide semiconductor* (CMOS) drivers. **Figure 5-14e** is a generic LED driver that can be used with almost any form of driver source.

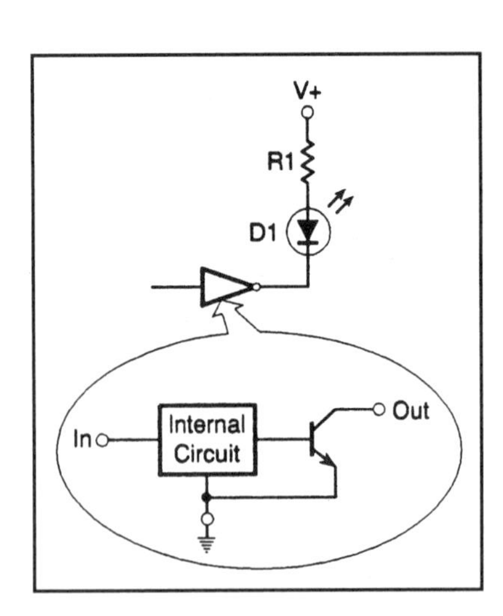

Figure 5-14a. TTL open-collector circuit for driving an LED.

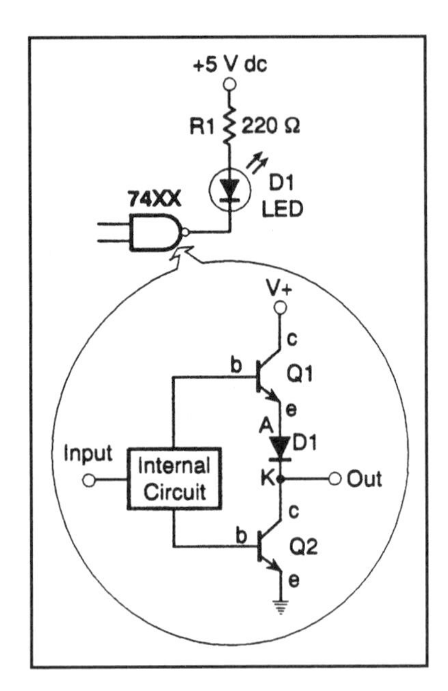

Figure 5-14b. Driving LED from ordinary TTL outputs.

Figure 5-14a shows one form of driver for TTL devices. Certain species of TTL inverters are called *open-collector* devices because the output terminal is the collector of an internal transistor (see inset of Figure 5-14a).

Standard TTL inverter chips can be called hex inverters since they contain six inverters; they also carry a 74XX series part number. Although the power terminal on the inverter chip is +5 volts ±0.15 volts (the standard for all TTL devices), the open-collector output can operate at considerably higher potentials. In some devices (7405, for example) V+ must be +5 volts, while in the 7406 and 7407 devices V+ can be up to +30 volts dc. Output currents up to 30 mA are permitted.

The inverter output state is always opposite the input state. If the input is LOW, then the output is HIGH; if the input is HIGH, then the output is LOW.

Under the output HIGH condition there is no path to ground for the LED, so no current flows. The LED remains turned off. But when the inverter output is LOW, the internal transistor is turned on hard, so its collector terminal is grounded. The cathode end of the LED is grounded through the inverter output transistor, so the LED lights.

The collector-to-emitter voltage of the inverter output transistor is typically very low (less than a few hundred millivolts), so it is usually ignored in calculating the value of the series current limiting resistor. The value of this resistor, referred to as R1, should be:

$$R1 = \frac{(V+) - V_{LED}}{I_{LED}} \qquad \text{eq. (5-7)}$$

Example 5-5

Find a suitable value for R1 using **Equation 5-7** when the LED current is 15 mA, and V+ = +12 volts dc. Assume an LED voltage drop of 1.5 volts.

Solution:

- $$R1 = \frac{(V+) - V_{LED}}{I_{LED}}$$

- $$R1 = \frac{12\ volts - 1.5\ volts}{0.015\ amperes}$$

- $$R1 = \frac{10.5\ volts}{0.015\ A} = 700\ ohms$$

The nearest standard value to 700 ohms is 680 ohms, so normally it would be selected unless the specific LED has a maximum current level of 15 mA.

Digital circuits operate with only two discrete voltage levels, each representing one of the two possible digits of the binary system: 0 and 1. These are also said to be logic 0 and logic 1, or LOW and HIGH, respectively.

When interfacing LEDs and other non-digital components to digital integrated circuits, it is necessary to understand something about how those chips operate, and what the logic levels mean. In the very popular TTL devices, LOW is a voltage between zero and +0.8 volts, while HIGH is a voltage between +2.4 and +5 volts. The output stage of TTL devices is a "quasi-totem pole" circuit. Such a circuit, used in standard TTL output stages (inset to **Figure 5-14b**), consists of three components: two bipolar NPN transistors (Q1 and Q2), and diode D1. All three components are connected in series across the +5 volt dc power supply, and the device output terminal is taken from the junction between the cathode, K, of D1 and the collector, c, of Q2.

An NPN transistor such as Q1 or Q2, is turned on when the base-emitter (b-e) voltage is zero, or actually something less than ~600 mV. In this situation there is no current conduction along the collector to emitter (c-e) path (**Figure 5-14c.1**). But when the b-e voltage is both positive and above the junction potential (~600 mV), then the transistor is turned on, and current flows in the c-e path (**Figure 5-14c.2**). If the value of the base-emitter voltage is sufficiently high, the transistor is turned on very hard, and there is a low resis-

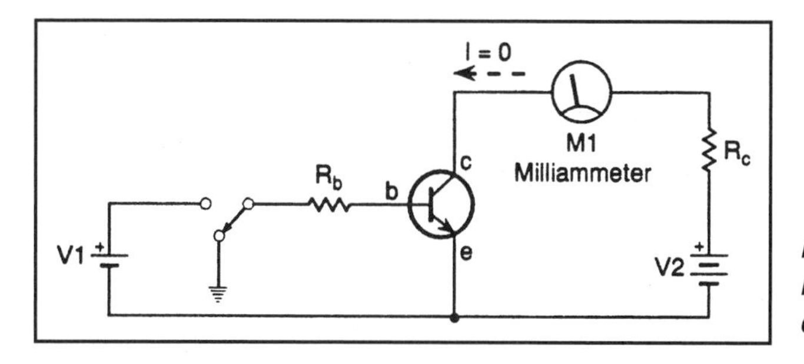

Figure 5-14c.1. NPN transistor in zero-bias condition (no c-e current flow).

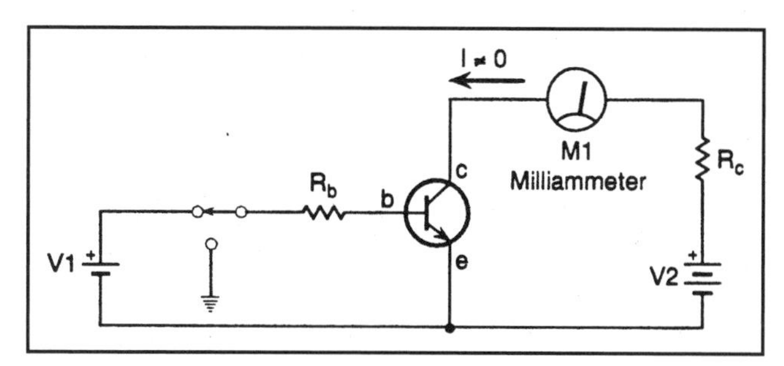

Figure 5-14c.2. NPN transistor in forward bias condition (c-e current flows).

tance between the collector and the emitter. The voltage drop between collector and emitter, under this condition, tends to be very low (e.g. ~100 mV), so the transistor c-e path can be treated as a short circuit.

Now apply this to a TTL circuit. A digital logic family such as TTL is designed to allow the output of one stage to be connected to the input of the following stage, using only a wire or printed circuit track. In other words, there is no external interface circuitry required between them, when only devices of the same logic family (TTL-to-TTL) are used. The connection between the output of Device-A and the input of Device-B (wire OA-IB) in **Figure 5-14c.3** shows this principle.

In order to achieve the goal of easy interconnection, the inputs and outputs of digital devices within a family must be complementary. In TTL devices this goal is accomplished by making the output circuits *current sinks* and the inputs *current sources*. A current sink will accept current, while a current source supplies current. If point IB on Device-B is grounded, a specified current will flow from ground into the emitter of Q3. This current in TTL devices is 1.8 mA, and is taken to be the unit input drive level required to make the device work properly.

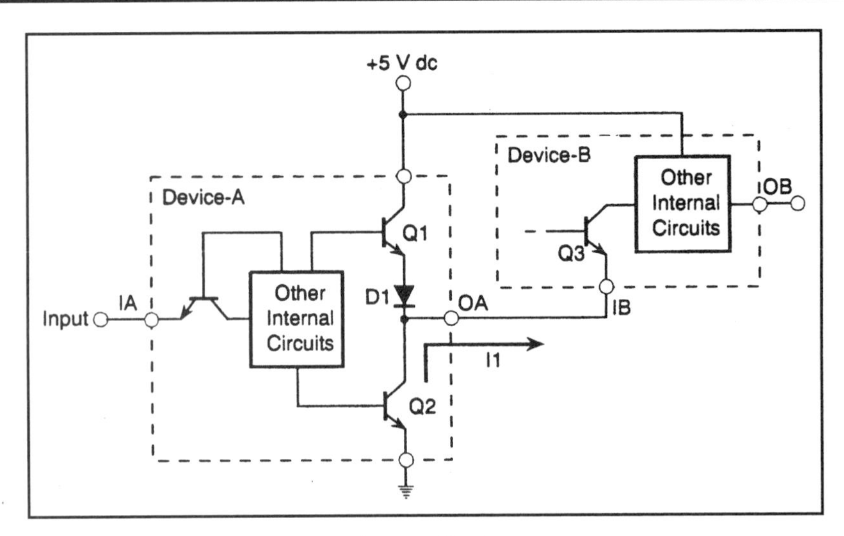

Figure 5-14c.3. Interface of two TTL devices.

Now consider two output states of Device-A. When the output (OA) is HIGH, transistor Q2 is turned off (so there is a high resistance from OA to ground), and Q1/D1 are turned on. This condition places a voltage at the output of ~2.4 to 4 volts (+5 volts dc less any drops across Q1 and D1). Because this voltage matches the voltage at the input of Device-B (IB), no current flows. When the output state of Device-A changes to LOW, Q1 is turned off, and Q2 is turned on hard, placing a short circuit between OA and ground. Since IB is a current sink, its 1.8 mA current is now free to flow to ground.

There is a limit to how much current can be handled by Q2 in Device-A, and that limit sets the output drive capacity of the device. The standard TTL output is rated at 18 mA, or the current supplied by ten standard TTL inputs. The unit of drive is one TTL input (1.8 mA), which is termed a *fan-in* of one. The output capacity, *fan-out*, is the number of standard TTL inputs that the device can drive. The standard 18 mA output stage is said to have a fan-out of ten.

Example 5-6

An LED circuit is naturally a current source, so it can be connected directly to a TTL output. If resistor R1 in **Figure 5-14a** is selected to limit the LED current to 18 mA or less, then it will function well with the TTL device. Assume the LED voltage drop to be 1.8 volts. Using 15 mA to allow for a margin of safety against both resistor value tolerances and voltage variations, the value of R1, in accordance with **Equation 5-7**, should be:

$$R1 = \frac{(+5) - (1.8)\, volts}{0.015\, A} = 213\, ohms$$

This value is very close to the standard value of 220 ohms, so one would ordinarily select 220 ohms as the correct value as shown in **Figure 5-14b**.

The other major class of digital devices is the CMOS family, which carry 4xxx and 45xx part numbers. The output of a CMOS device is a current source when HIGH, and a current sink when LOW. The CMOS output can drive LEDs in either condition. **Figure 5-14d.1** shows a CMOS LED driver that turns on the LED when HIGH. Under this condition, the CMOS output is connected to the V+ power supply through a low resistance. Current flows through the LED, the current limiting resistor, and the CMOS device output terminal to V+.

Figure 5-14d.2 shows the opposite condition where the LED turns on when the CMOS output is LOW. Here, the CMOS output is connected to ground through a LOW resistance. Current flows from ground, through the CMOS output to the LED, R1, and to V+.

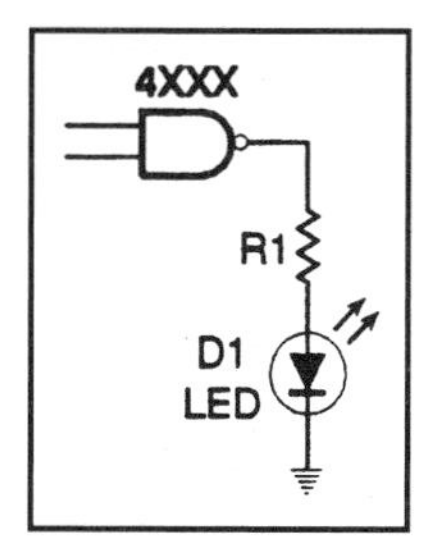

Figure 5-14d.1. HIGH = ON CMOS driver for LEDs.

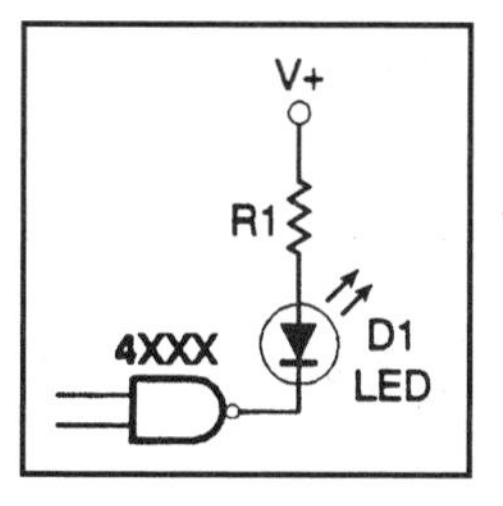

Figure 5-14d.2. LOW = ON CMOS driver for LEDs.

The generic digital LED driver circuit is shown in **Figure 5-14e**. The LED is turned on and off by switching transistor Q1. When point A is HIGH, Q1 is biased hard on and the collector-emitter path is a near-short circuit. The LED cathode is then grounded and the LED turns on.

If the digital source in **Figure 5-14e** is a CMOS device, then the circuit will work as shown. But if the device is any TTL, a pull-up resistor (R2) is needed, because the transistor's base circuit does not provide the current source that a TTL input would normally provide. Resistor R2, connected to V+, would then form the current source required by all TTL outputs.

The transistor selected for Q1 should have sufficient collector current and collector power dissipation ratings to sustain the full LED current indefinitely. It should also have sufficient beta gain (H_{fe}) to turn on hard when A is at a potential of 1 volt lower than V+. The value of R1 can be adjusted to ensure Q1 turn-on without exceeding the base current rating of the device.

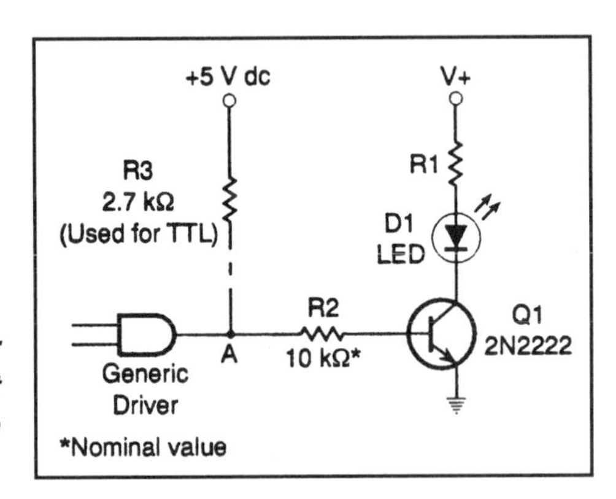

Figure 5-14e. LED driver for generic digital circuit uses transistor to turn on the LED.

Polarity Tester

There are many cases where the polarity of a dc source is critical information. For example, when installing or servicing automobile electronics, it is necessary to know whether the car battery is positive ground or negative ground. Battery powered projects need similar attention.

The circuit of **Figure 5-15** is a simple polarity tester consisting of a pair of back-to-back LEDs and a series current limiting resistor. If point A is positive with respect to point B, then LED D1 will turn on. On the other hand, if point B is positive with respect to point A, then LED D2 turns on. In other words, the polarity of the voltage across the two points determines which LED lights up. (An ac source causes both LEDs to turn on.) This circuit is only useful at low voltages which will not cause an excessive current to flow in the LEDs.

For the resistor value shown in **Figure 5-15** with ordinary LEDs, the maximum safe input voltage is about ±15 volts. If this limit is exceeded, then the LEDs may be destroyed. Higher voltages can be accommodated by increasing the value of R1 proportional to the voltage change, although the peak inverse voltage (maximum ac voltage before damage occurs) of the LEDs should not be exceeded.

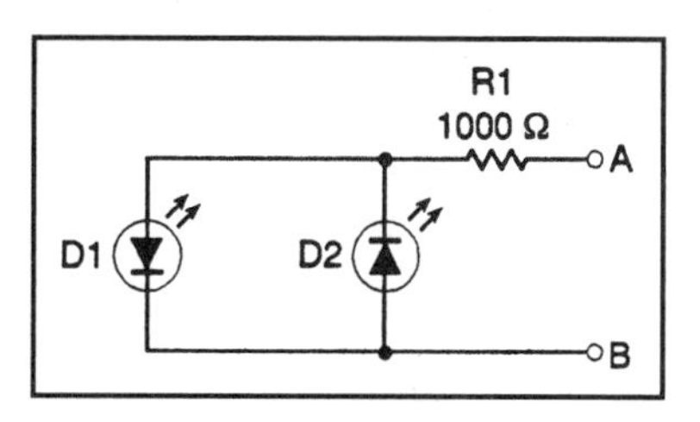

Figure 5-15. LED polarity detector.

Voltage Comparator Output State Indicator

A *voltage comparator* is a circuit that examines two input voltages (V1 and V2), and then issues an output that indicates whether V1 = V2, V1 > V2, or V1 < V2. A simple operational amplifier can be used as a comparator if no negative feedback circuit is used (see Chapter 11).

A comparator was once called "an amplifier with too much gain," and an op-amp fits this requirement nicely in the open-loop configuration because the gain is >200,000. If the maximum available output voltage (V_o) is 2 volts less than the power supply voltage (a typical value), and the dc supplies are ±12 volts dc, then the maximum V_o value is (12) [-] (2) volts, or ±10 volts. At a gain of 200,000 (A_v) then, the maximum input voltage, the input voltage level that causes output saturation, is:

$$V_{in(\max)} = \frac{V_{o(\max)}}{A_v} \qquad \text{eq. (5-8)}$$

$$V_{in(\max)} = \frac{\pm 10\ volts}{200{,}000}$$

$$V_{in(\max)} = 5\ x\ 10^{-5}\ volts = 50\ \mu V$$

Thus, when the difference between V1 and V2 is greater than 50 mV, the output of the op-amp will be saturated. This is the basis for voltage comparator circuits.

Figure 5-16 shows an op-amp voltage comparator circuit that uses the polarity tester circuit as the output state indicator. The output of op-amp A1 (point A) will be close to V+ when $V1 < V2$, close to V- when $V1 > V2$, and zero when $V1 = V2$. By connecting LEDs D1 and D2 back-to-back we obtain an indicator of which state exists.

There is one ambiguity state in this circuit. The LEDs are both turned off when $V1 = V2 \neq 0$, but also when $V1 = V2 = 0$. If V1 and V2 are never zero except when a failure exists (or some other circuit is not turned on), then one may erroneously believe that $V1 = V2 \neq 0$.

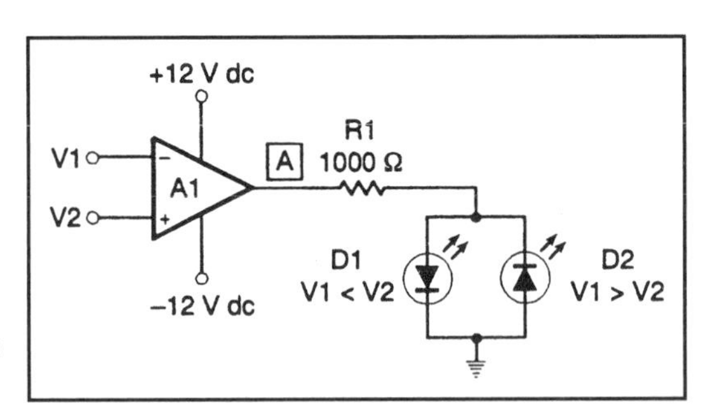

Figure 5-16. LED/op-amp polarity detector.

Voltage Level Indicators

One application for the voltage comparator circuit of **Figure 5-16** is as a voltage level indicator. If the noninverting input (+), or V2, is set to a predetermined value, then D1 will turn on as long as $V1 < V2$, and D2 will turn on when $V1 > V2$. The reference voltage for V2 can be provided by a potentiometer, resistive voltage divider, zener diode or an IC voltage regulator device.

A simpler voltage level indicator is shown in **Figure 5-17**. This circuit consists of a zener diode (D1) and an LED (D2). As long as applied voltage V is less than the sum of the zener diode potential of D1 (V_z) and forward bias potential of D2 ($\approx$1.8 volts), then the LED is off. Once $V > (V_z + V_{LED})$, the zener begins to conduct and the LED turns on.

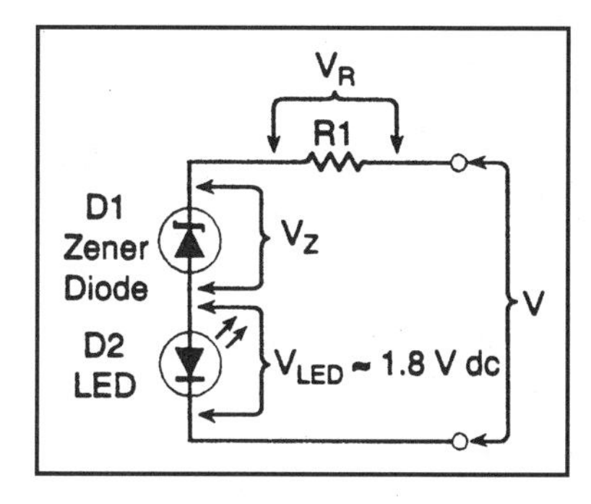

Figure 5-17. Voltage level threshold detector.

Homemade LED Optocoupler

An *optocoupler* is a device that provides ac and dc isolation between two circuits, while still coupling a signal between them. **Figure 5-18** shows both the circuit and the mechanical configuration method for making one at home. An LED is positioned so that it touches the optical window of a phototransistor (Q1). Black heat shrink tubing is placed over both of these components, heated until it shrinks, then sealed at the ends with black silicone seal adhesive.

A principal use for the optocoupler is to isolate one circuit from another. For example, a circuit might operate at a high dc voltage that is incompatible with the low-voltage operation of the other circuits. The optocoupler will permit signal to pass from one side to the other without having an electrical connection. Additional optocoupler circuits are described in Chapter 9.

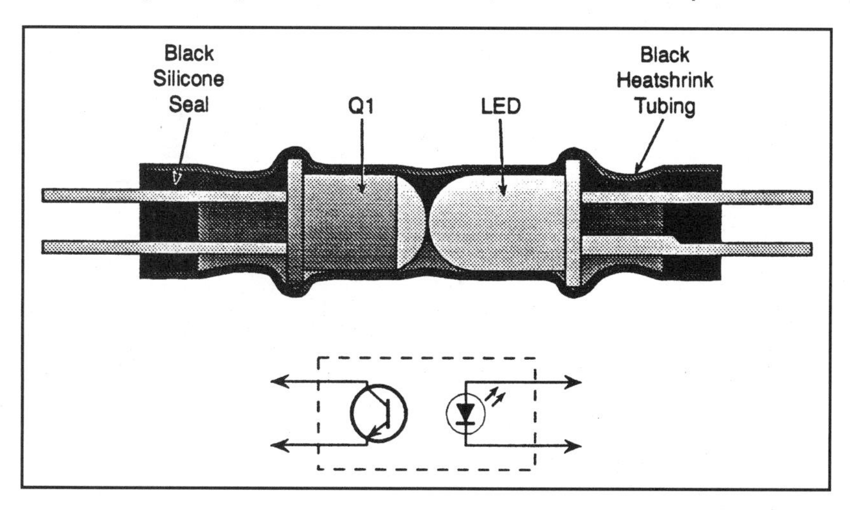

Figure 5-18. Crude LED/phototransistor optoisolator.

Simple Digital Logic Probe

A *digital logic probe* is a simple device that allows a circuit troubleshooter to determine whether a logic level is 1 or 0 (HIGH or LOW). As noted earlier, in TTL circuits, a logic 1, or HIGH, is represented by a voltage of +2.4 to +5.2 volts; logic 0, or LOW, is represented by zero to +0.8 volts. CMOS circuits may or may not follow the same pattern as TTL circuits. To troubleshoot these circuits, it is necessary to know which level is present at any given point.

Figure 5-19 shows the circuit for a simple logic probe based on an LED indicator and a pair of 2N2222 NPN transistors. The two transistors are connected in the Darlington configuration. The emitter of the input transistor directly drives the base of the output transistor; the collectors of the two transistors are tied together. The result of the Darlington configuration is an extremely high beta gain. The overall gain of a Darlington pair is the product of the gains of the individual transistors. If the two devices are identical, then the overall gain is the square of the individual gain. For example, if two transistors with beta gains of 100 are connected into the Darlington configuration, the overall gain is 100^2 or 10,000. The Darlington pair can be treated as if it were a single transistor with a beta equal to the composite product gain.

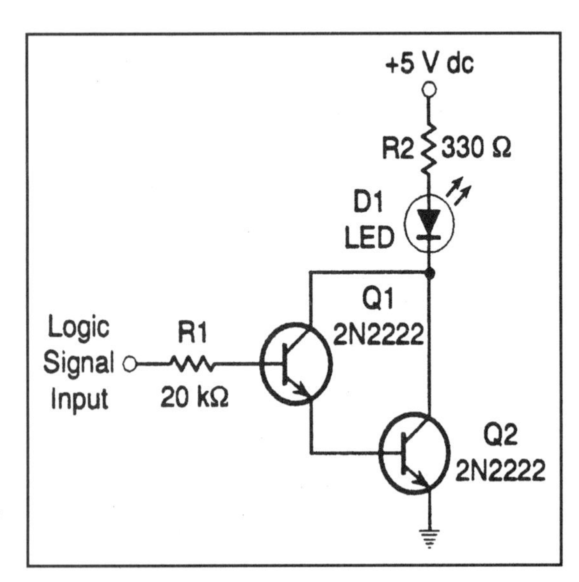

Figure 5-19. High gain "Darlington" driver for LED.

The circuit of **Figure 5-19** shows the Darlington pair being used as a transistor switch. When the transistors are turned on hard, the LED is grounded so current will flow through the LED. The transistors are turned on by applying a HIGH logic level to the input. A resistor of approximately 18 to 24 kW is connected in series with the base of transistor Q1. When a HIGH is applied to the input of this circuit, the LED is lighted. In order for this device to measure properly, it must be connected to the ground of the logic circuit that it is measuring.

Photoconductive Cells

A *photoconductive cell* (photoresistor) is a device that changes electrical resistance when light is applied. **Figure 6-1** shows the usual circuit symbol for photoresistors. It is a normal resistor symbol enclosed within a circle, and given the Greek letter lambda (λ) to denote that it responds to light.

The active region of a photoconductive cell (**Figure 6-2**) is a thin film of silicon, germanium, selenium, a metallic halide or a metallic sulfide (e.g., cadmium sulfide, CdS). When these materials are illuminated, free electrons are produced as photon energy drives them from a valence band into a conduction band. As in any conductor, free electrons mean that current can flow if an electrical potential is applied. Because increased illumination creates additional free electrons, the resistance of the photoconductive cell decreases with increases in illumination level.

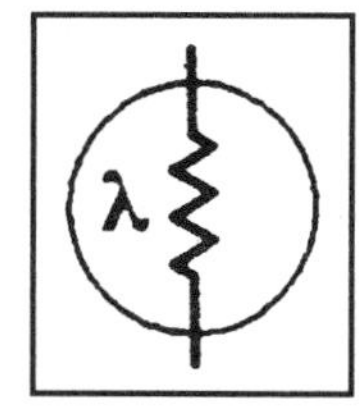

Figure 6-1. Schematic symbol of the photoresistor.

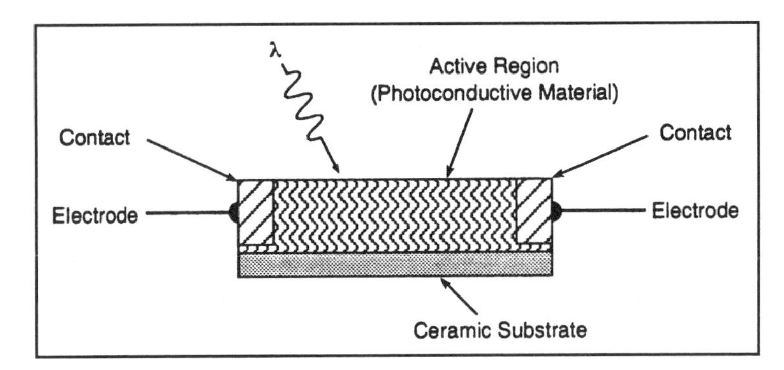

Figure 6-2. Photoresistor structure.

When photoresistors are specified, it is typical that a *dark resistance* is given, as is a *light/dark ratio*. Changes of millions of ohms when dark, to hundreds of ohms (**Figure 6-3**) under maximum illumination, are common. In most common varieties, the resistance is very high when dark, and drops to very low under intense light. Since intensity affects the resistance, they can be used in photographic lightmeters, densitometers, colorimeters and other measurement apparatus.

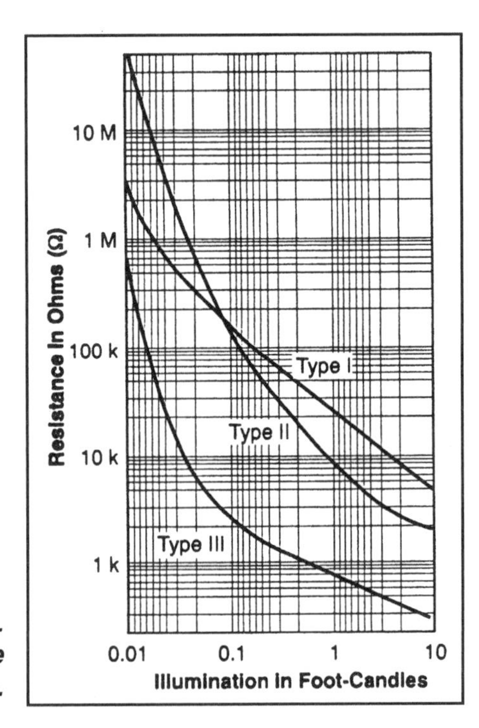

Figure 6-3. Resistance vs. illumination curve for three photoresistors.

Photoconductive cells are used in a wide variety of instrumentation, control and measurement applications. In selecting this type of device it is helpful to keep several factors in mind:

Material used, which affects spectral response.

Physical configuration.

Electrical parameters.

Spectral Response

The *spectral response* of any photosensor relates its sensitivity to light to the wavelength of electromagnetic radiation (IR, visible, UV) impinging on it. In most cases, the response peaks at a certain wavelength. This point is taken as unity, and all other wavelengths are compared to it. When the peak response wavelength is designated either 1 or 100 percent, then the response curve is said to be *normalized.*

Figure 6-4 shows the normalized spectral response curves for three different photoconductive cells. The material represented in **Figure 6-4a** shows a sharply peaked response from at about 5,100 angstroms (A), although from the 60 percent point the response broadens a bit. A smoother response centered around 7,300 A is shown in **Figure 6-4b**. An even broader response peaked near 5,500 A is shown in **Figure 6-4c**. This response is largely in the visible spectrum and is particularly well suited for certain photometric instrumentation applications.

With a monochromatic light source, the spectral response of the photoresistor is not important in many applications. In those cases, a good response at the wavelength of interest is needed. In other cases, however, a wider spectral response is required because multiple wavelengths are used in the light source, or the light source is white light, which is inherently a multiple wavelength source. An ideal response would be flat, having the same (or nearly the same) response over a wide range of wavelengths.

In other cases, the photoresistor operates in an environment where there is a high ambient light level that differs in color temperature from the light source. In that case, the desirable spectral response is one that has a peak at the

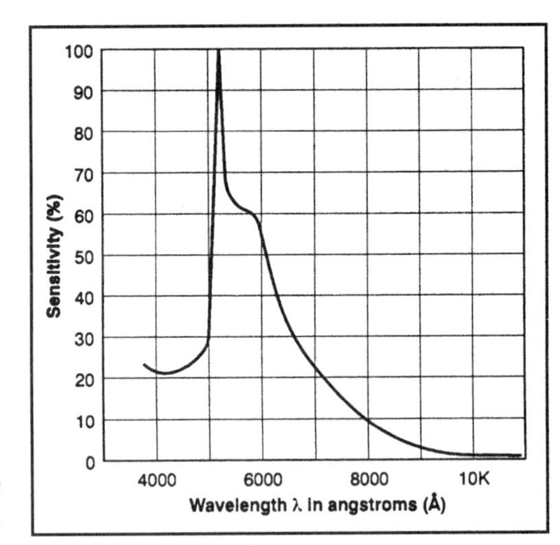

Figure 6-4a. Spectral response of a photoresistor peaked at 5,100 A.

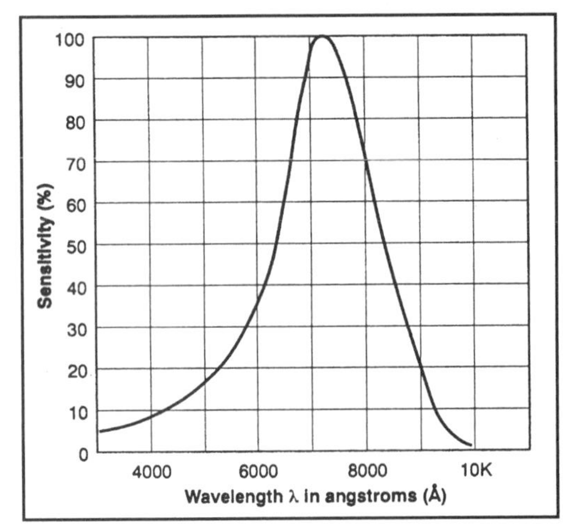

Figure 6-4b. Spectral response of a photoresistor peaked at 7,300 A.

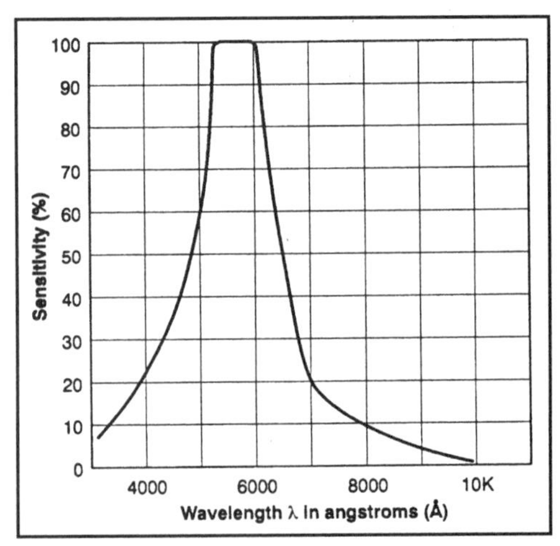

Figure 6-4a. Broad response peaked at 5,500 A.

"Light History" Errors

Photoconductive cells exhibit a kind of *hysteresis effect* in which the resistance change that will be caused by a given level of illumination depends on the light levels applied in the cell's past. According to one source: "The present or instantaneous conductance of a cell at a specific light level is a function of the cell's previous exposure and duration of that exposure."

The specific nature of this *light history effect* depends on whether the previous light level was greater or less than the present level. If the cell is tested for conductance at a specific light level and stored in relative darkness, then the new conductance seen when it is re-tested at the original level will be greater than before. The conductance will then decay exponentially until it is equal to the original test conductance (**Figure 6-6a**).

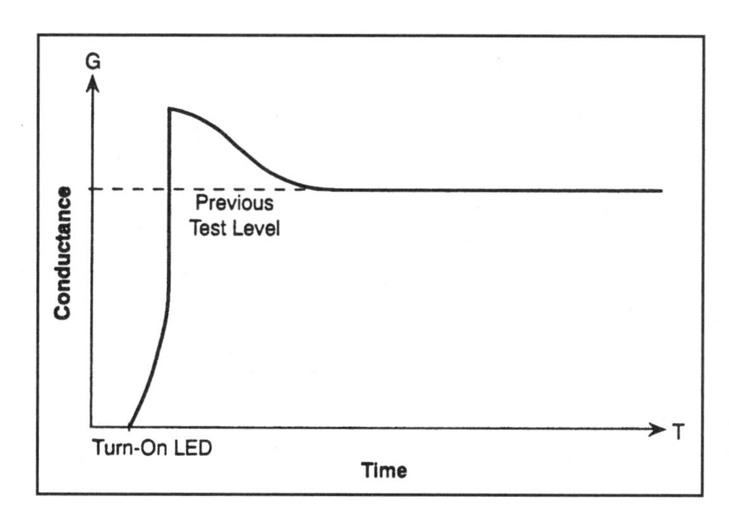

Figure 6-6a. Light response graph of the "Light history" effect.

Temperature Coefficient

The *temperature coefficient* (τ) of any resistor is the change in resistance per unit change in temperature, usually specified in degrees Celsius.

$$\tau = \frac{\Delta R}{\Delta °C} \qquad \text{eq. (6-3)}$$

In photoresistor cells, the temperature coefficient is a function of both light level and the material used. As the light level changes, the temperature coefficient tends to vary inversely. Using a photoresistor in as high a light level as possible is one method of dealing with this characteristic.

Photoresistor cells are designed with both positive temperature coefficient (PTC) and negative temperature coefficient (NTC). In the PTC case, the resistance increases with increases in temperature, while in the NTC case the resistance decreases with temperature increases.

Response Time

The response time of a photocell refers to the time required for a specified current change following a light change. The *rise time* of the cell is the time required for the current to increase after the cell changes from a low ambient light level to full illumination. The general rule is to measure the time required for the cell to reach 63 percent of its final value following turn-on of the light (curve A in **Figure 6-6b**). The figure 63 percent is selected because the rise time phenomenon is exponential, and 63 percent represents one time constant unit.

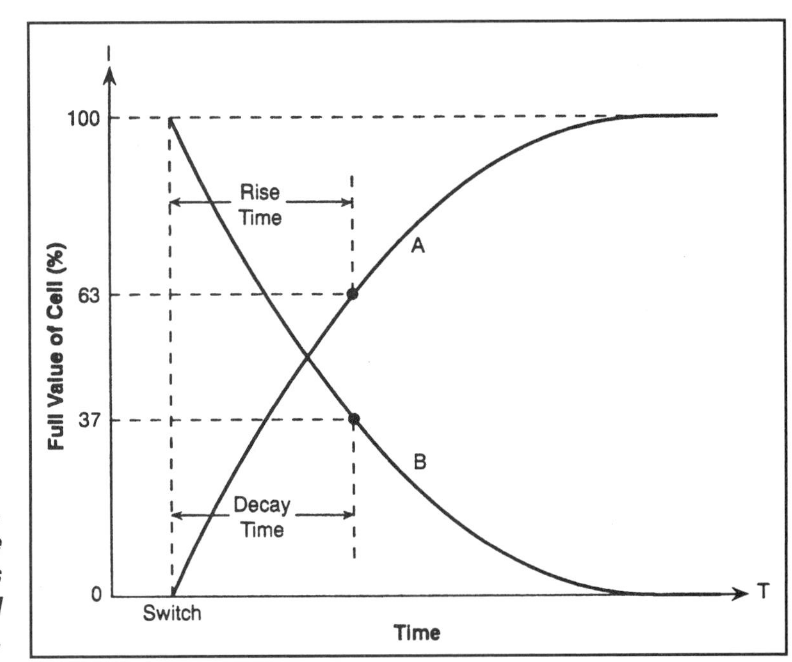

Figure 6-6b. Exponential rise and decay curves showing 63% and 37% points.

The *decay time* (curve B in **Figure 6-6b**) of the cell is the time required for the cell current to change a specified amount when the light illumination is dropped to zero. Because this phenomenon is also exponential, it is common practice to measure the decay time as the time required for the current flowing in the cell to drop from the maximum light level to 37 percent of that value.

Photoresistor Circuits

Figure 6-7 shows three circuits in which photoresistors can be used. The half-bridge circuit in **Figure 6-7a** has the photoresistor connected across the output of a voltage divider made up of R1 and PC1. The output voltage is based on Kirchoff's voltage divider rule, expressed as:

$$V_o = \frac{V \times PC1}{R1 + PC1} \qquad \text{eq. (6-4)}$$

Where:

V_o is the output potential in volts.

V is the applied excitation potential in volts.

R1 and PC1 are resistors, measured in ohms.

In this circuit, the output potential does not drop to zero, but always has an offset value.

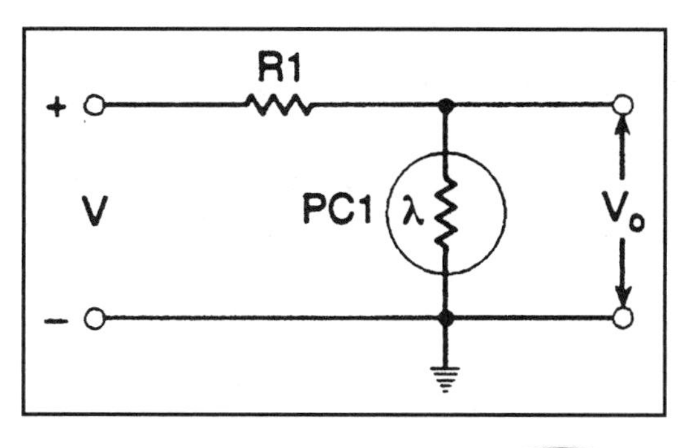

Figure 6-7a. Half-bridge photoresistor circuit.

In **Figure 6-7b**, the photoresistor is the feedback resistor in an operational amplifier inverting follower circuit. The output voltage, V_o, is found from:

$$V_o = (V_{ref})\left(-\frac{PC1}{R1}\right) \qquad \text{eq. (6-5)}$$

If $V_{ref} > 0$ volts, then $0 > V_o > V-$.

If $V_{ref} < 0$ volts, then $0 < V_o < V+$.

This circuit provides a low-impedance output, but like the half-bridge circuit (**Figure 6-7a**), the output voltage does not drop to zero. In addition, with some photoresistors the dynamic range of the operational amplifier may not match the dark-light ratio of the photoresistor at practical values of $-V_{ref}$ potential.

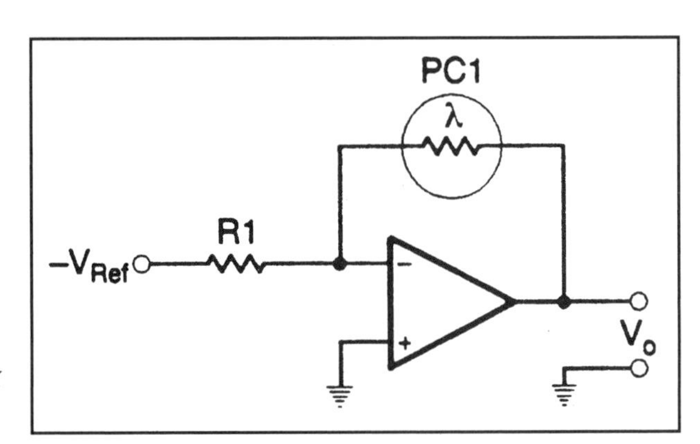

Figure 6-7b. Feedback loop of inverting amplifier.

The Wheatstone bridge is shown in **Figure 6-7c**. This circuit allows the output voltage to be zero under certain circumstances, and is the circuit favored by most designers. If a low impedance output is required, or additional amplification is needed, then a differential dc amplifier can be connected across the output potential, V_o.

The Wheatstone bridge can be considered as two half-bridges in parallel. The output voltage is equal to the difference between the respective half-bridge output voltages (at points A and B in **Figure 6-7c**). The voltages at these points are derived from the same equation as for the half-bridge in **Figure 6-7a**, so the output voltage from the bridge is equal to:

- $$V_o = V \times \left(R \frac{3}{R1 + R3} \right)$$ *eq. (6-6)*

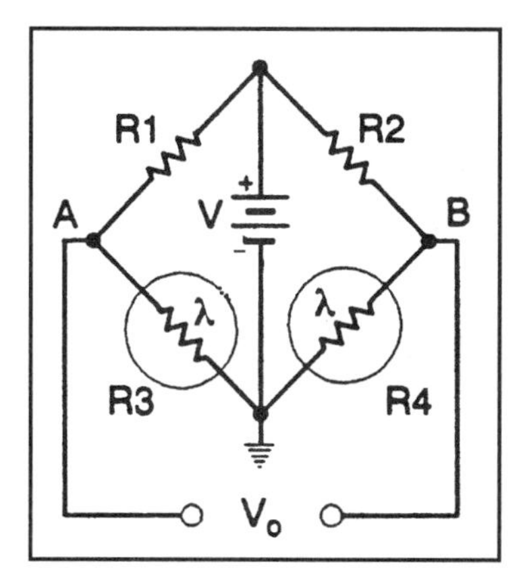

Figure 6-7c. Wheatstone bridge.

Example 6-1

Consider a Wheatstone bridge that is excited by a potential of +7.5 volts dc. Find the output voltage, V_o, when R1 = 10,000 ohms, R2 = 10,010 ohms, R3 = 9,990 ohms and R4 = 10,020 ohms.

Solution:

- $$V_o = V \ x \left(\frac{R3}{R1 + R3} \right)$$

- $$V_o = (7.5V)\ x\left(\frac{9{,}990\,\Omega}{(10{,}000\ +\ 9{,}990)\,\Omega}\right)$$

- $$V_o = (7.5V)x(0.4998\ -\ 0.5003)$$

- $$V_o = (7.5V)(-5x10^{-4}) = -3.75\ x\ 10^{-3}\ volts = -3.75mV$$

These circuits will be seen later in this book, as they are popular in instrument designs.

Light Meter Circuits

A photoconductive cell can be used as a light meter for photographic purposes; many low-cost photographic meters are based on CdS photoresistors. **Figure 6-8** shows the simplest circuit for these meters. The sensing element is a CdS photocell (PC1) in series with a calibrating variable resistor and a dc current meter (M1). Depending on the sensitivity required, the meter will be rated at a full-scale of 50 mA to 1 mA. A battery from 1.5 to 9 volts dc can be used as the power source.

In the circuit of **Figure 6-8** the resistance of the photocell is inversely proportional to the light intensity. By Ohm's law, the current is found from I = V/R. As resistance drops in response to increasing light levels the current will go up. Thus, the deflection of the meter pointer is proportional to the applied light level. In order to get the best reading from the meter, care must be taken to ensure that the resistive vs. light level curve is reasonably linear over the range of interest.

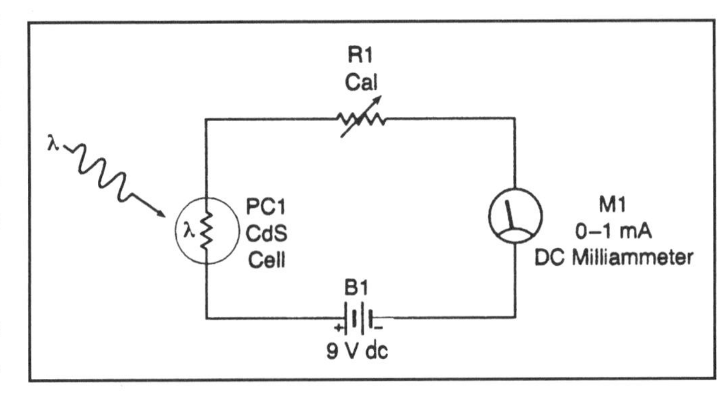

Figure 6-8. Simple light meter circuit.

An alternate form of light meter circuit is shown in **Figure 6-9**. This circuit is based on the Wheatstone bridge circuit in **Figure 6-7c**. Resistors R1 and R2 are fixed, while R3 is a variable resistor; PC1 is a variable resistance photoconductive cell. When the ratios R1/R3 and R2/PC1 are equal, then the output voltage across the meter (M1) is zero. In this light meter, the meter sensor (PC1) is exposed to light and then R3 is varied to null the meter. A calibrated dial reads either the light level or the parameters needed for that application. In some meters, R1 and R2 are also variable resistors calibrated to outside factors, such as the film speed or shutter speed in photography applications.

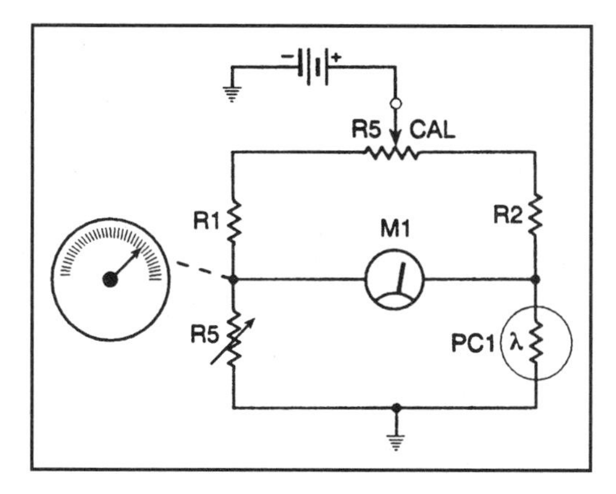

Figure 6-9. Wheatstone bridge light meter.

Light-Operated Tone Generator

Figure 6-10 shows the circuit for a tone generator that will increase in pitch (frequency) as light intensity increases. The basis of this circuit is the popular 555 IC timer. This chip can be used in either monostable (one-shot) or astable (continuous output pulse) circuits. For the tone generator example circuit in **Figure 6-10**, the astable multivibrator configuration is selected.

The 555 IC timer uses a resistor-capacitor network consisting of PC1, R1, and C1 to set the operating frequency. The frequency is found from:

$$f = \frac{1.44}{(R1 + 2\,PC1)\,C1} \qquad \text{eq. (6-8)}$$

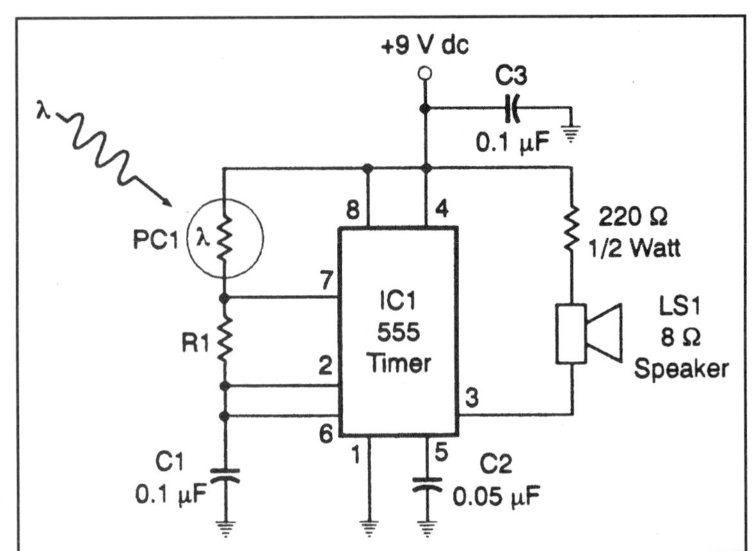

Figure 6-10. Light-operated tone generator.

From **Equation 6-8** it is shown that decreasing the resistance of PC1, which occurs when light intensity increases, will cause the frequency of the 555 IC timer to increase.

The output signal is coupled directly to an 8-ohm loudspeaker. A 220-ohm resistor in series with the loudspeaker acts in a pull-up function. In most applications, it will be sufficient to use a small commercial speaker (such as for a transistor radio or cassette recorder) for LS1.

Light-Controlled Relays

A light-controlled relay will turn a circuit on or off in response to changes in light level; two varieties are shown in **Figure 6-11**. A relay is an electromechanical switch that uses the electromagnetism of an inductor coil (K1 in **Figure 6-11a**) to actuate the switch elements. A certain minimum current is required to actuate the relay (pull-in current), although the coil will hold in this energized state at a lower current level.

Figure 6-11a shows a *dark-activated* relay circuit; that is, one that will activate the relay when the light level drops below a preset level. The switching element that actuates the relay is an NPN transistor (Q1). This transistor will turn on when the voltage applied to the base is more positive than the voltage applied to the emitter—assuming that the collector voltage (V+) is more positive than either. In the circuit of **Figure 6-11a**, the emitter is grounded through the coil resistance of K1. In most cases, this resistance (for 5- to 12-volt dc relays) is on the order of 400 to 800 ohms. The emitter voltage, V_e, found at Q1 is the product of the emitter current and the K1 coil resistance. When the base voltage, V_b, is 0.6 volt higher than the emitter voltage, the transistor will turn on, causing a higher current to flow in K1, thus forcing it to energize as soon as the pull-in current is exceeded.

The base voltage at Q1 is set by a bias network voltage divider consisting of PC1 and potentiometer R1:

$$V_b = \frac{(V+)PC1}{PC1 + R1} \qquad \text{eq. (6-9)}$$

Equation 6-9 shows that decreasing the resistance of PC1 will decrease V_b, while increasing the resistance of PC1 will increase V_b. At high light levels, the resistance of PC1 is low, so the voltage V_b is below the threshold (V_e [+] 0.6). Once this threshold is exceeded, the transistor will turn on and apply current to the relay coil, K1. This event takes place when the resistance of PC1 rises in response to decreasing light levels. Potentiometer R1 can be used to set the specific light level that will actuate the relay.

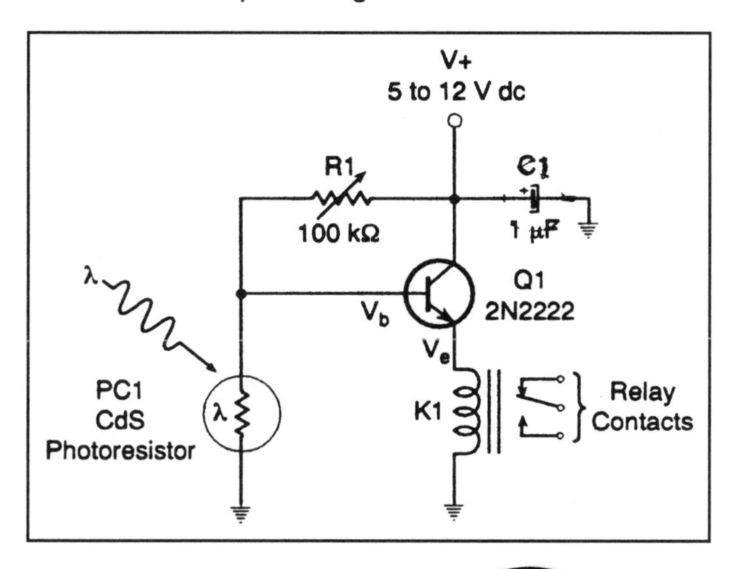

Figure 6-11a. Light-operated relay circuits: Dark = active.

A *light-activated* relay circuit is shown in **Figure 6-11b**. The principal functions in this circuit are the same as the one shown in **Figure 6-11a**, but the positions of the photocell and resistors are reversed. A low light level (which increases the resistance of PC1) will reduce the voltage at the base of Q1, and prevent the transistor from turning on. At higher light levels, the resistance of PC1 drops, forcing V_b higher. If V_b exceeds the threshold, then the transistor turns on and energizes the relay.

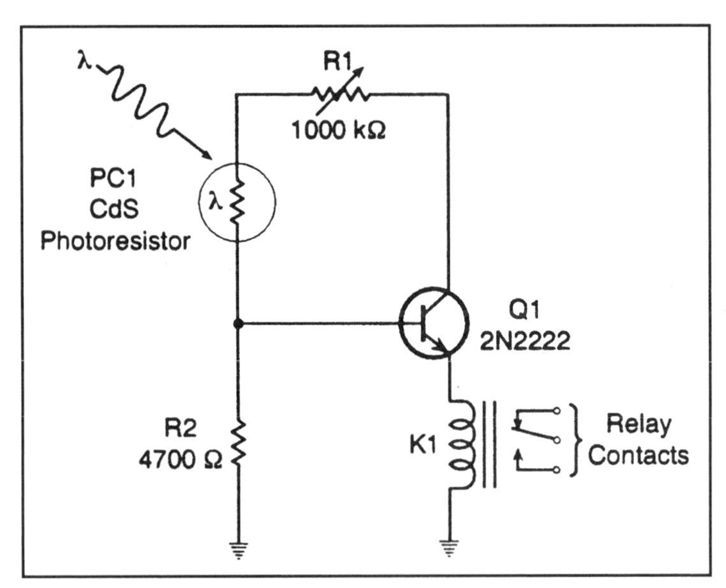

Figure 6-11b. Light-operated relay circuits: Light = active.

Light-Operated Audio Switch

Some medical, audiological and psychological tests use tones to test a patient's hearing or other responses. In one test used by neurologists and researchers, for example, an audio tone is repetitively turned on and off while the electroencephalogram (EEG) signals from the brain are time-averaged. After a sufficient number of tone bursts, the portion of the brain wave that results from the tone is enhanced, while all other components are reduced. In this manner, the scientist or physician can study the effects of the tone on the brain wave.

One problem seen in these tests is that a switching transient will also create a response, corrupting the test. **Figure 6-12** shows a switching system that will turn the audio signal on and off without creating a switching transient.

Figure 6-12 is a basic high-ratio attenuator based on the voltage divider principle. When the resistances of PC1 and PC2 are high, then the output signal is very low—nearly zero. Conversely, when PC1 and PC2 are in the low resistance state, the attenuation ratio is very low, so the maximum audio signal passes from input to output.

The photoswitch in this circuit consists of a pair of photoconductive cells in a light-tight enclosure, mounted in such a way that light from an incandescent bulb I1 falls on them. When switch S1 is closed, the lamp is turned on and the photocells are illuminated. Under this condition, the audio signal passes. When S1 is open, the lamp is turned off and the photocell resistance is very high; the audio signal is not passed.

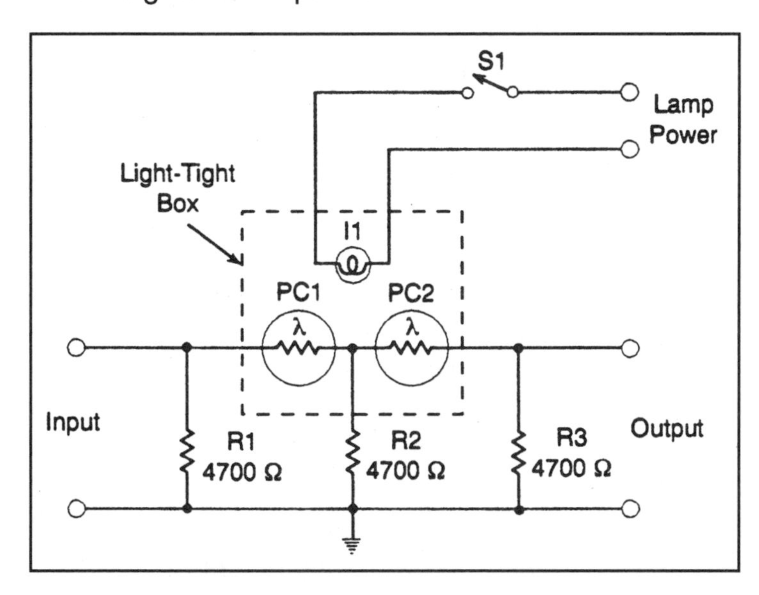

Figure 6-12. Light-operated attenuator.

Optical Chopper for Low-Drift Amplifiers

One of the keys to designing a low-drift amplifier is to use a *chopper* at the input to interrupt the signal, and then process the chopped signal in a feedback-controlled ac amplifier. Many different choppers exist, including electromechanical vibrators. One of the earliest all-electronic choppers was the photocell circuit. One design approach is to use the photoattenuator in **Figure 6-12**, replacing the incandescent lamp with a neon glow lamp. Another approach is shown in **Figure 6-13**. In this circuit, a simple voltage divider attenuator network consisting of PC1 and PC2 is used to turn the signal path on and off. These cells are actuated by a pair of neon glow lamps, I1 and I2.

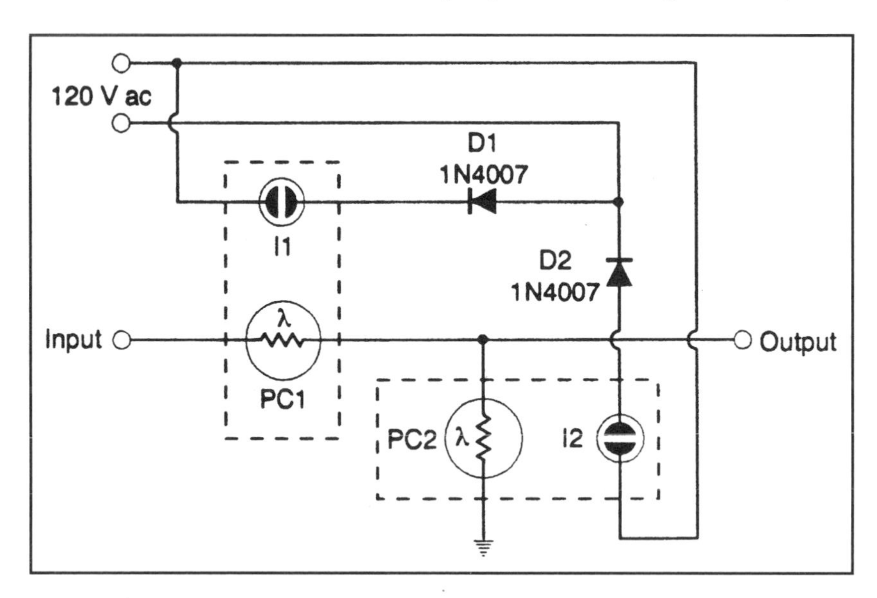

6-13. Light-operated chopper.

A pair of 1N4007 diodes is used to direct the current in a manner that turns I1 and I2 on and off opposite one another. In other words, when I1 is on, I2 is off, and vice versa. If I1 is on and I2 is off, PC1 has a low resistance and PC2 has a high resistance, so the output signal has nearly the same amplitude as the input signal. On the other hand, when I1 is off and I2 is on, then PC1 has a high resistance and PC2 has a low resistance. In this case, the signal is largely attenuated and is effectively turned off. The signal chopping occurs as PC1 and PC2 switch between high and low resistance states.

References

Optoelectronic Designers Handbook, Clairex Electronics, Mount Vernon, NY.

Chapter 7

Photovoltaic Cells

A *photovoltaic cell* (also called *photogalvanic cell* or *self-generating cell*) is a light sensor that generates an electrical potential when illuminated. A current can be made to flow in an external circuit by shining light onto the surface of the cell. The solar cell, which is used to generate electrical power, is the most common (or at least most widely known) example of the photovoltaic cell. There are also instrumentation photovoltaic cells available that are used to measure light levels rather than to generate power.

Figure 7-1 shows two common symbols for the photovoltaic cell. The symbol in **Figure 7-1a** is a standard diode symbol with two arrows aiming into it. This symbol is similar to that used for light emitting diodes, except for the direction of the two parallel arrows, in the LED symbol the arrows are aimed outward to indicate an emitting device.

The symbol shown in **Figure 7-1b** is a battery symbol inside a circle, with the Greek letter lambda (λ) superimposed. Both symbols are used interchangeably today, but for clarity's sake **Figure 7-1a** might better apply to only the PN junction form of photovoltaic cell.

Figure 7-1a.
"Diode" version.

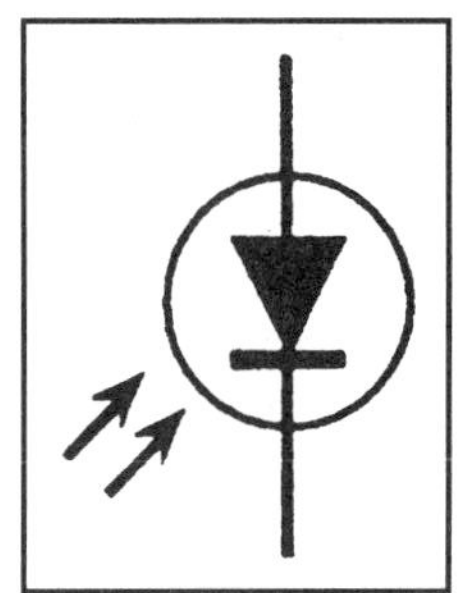

Figure 7-1b.
"Battery" version.

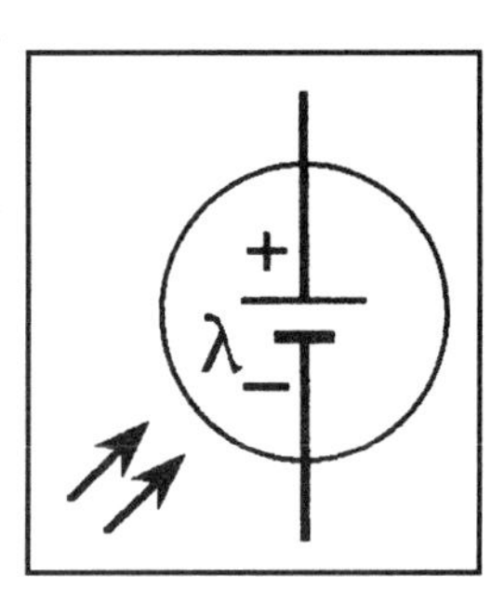

Photovoltaic Phenomena

There are two broad classes of photovoltaic phenomena: metal-oxide-semiconductor and PN junction. The earliest known method is the metal/semiconductor bond. In these devices, a metal such as copper is interfaced with a semiconductor such as selenium or germanium across a thin oxide layer.

Metal-Oxide/Semiconductor Devices

In **Figure 7-2** a metal disk - copper, gold or platinum - is coated with a layer of copper oxide, which is in turn covered with a layer of a semiconductor material that passes light and collects emitted electrons. The copper-oxide cell was invented before World War I by Bruno Lange, and eventually marketed by Westinghouse under the trade-name "Photox cell."

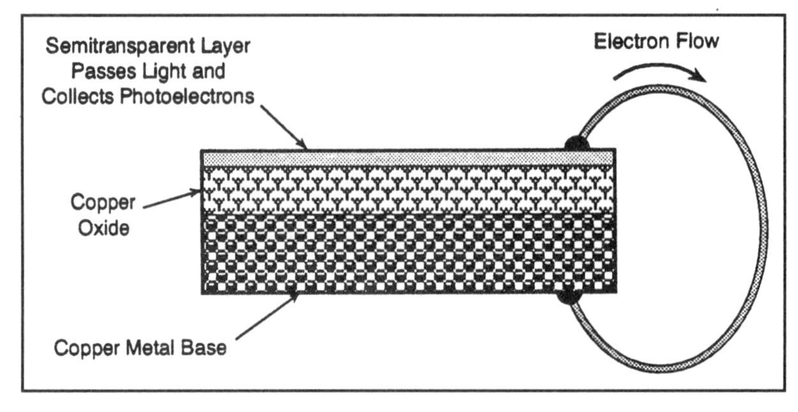

Figure 7-2. Metal-oxide/semiconductor (MOS) photovoltaic cell.

A similar photovoltaic cell made of selenium is shown in **Figure 7-3**. This cell was invented in the 1930s, and was marketed by Weston Instruments under the trade name "Photronic cell." In selenium cells, an iron, steel or aluminum plate is coated with a thin layer of photosensitive selenium. In both forms of metal photovoltaic cell, a thin insulator forms a barrier layer. When light illuminates the barrier layer, the impinging photons are absorbed, and free electrons are emitted. The presence of free electrons causes a difference of electrical potential to appear across the barrier layer. The selenium layer is negative, while the transparent thin metal film side is positive.

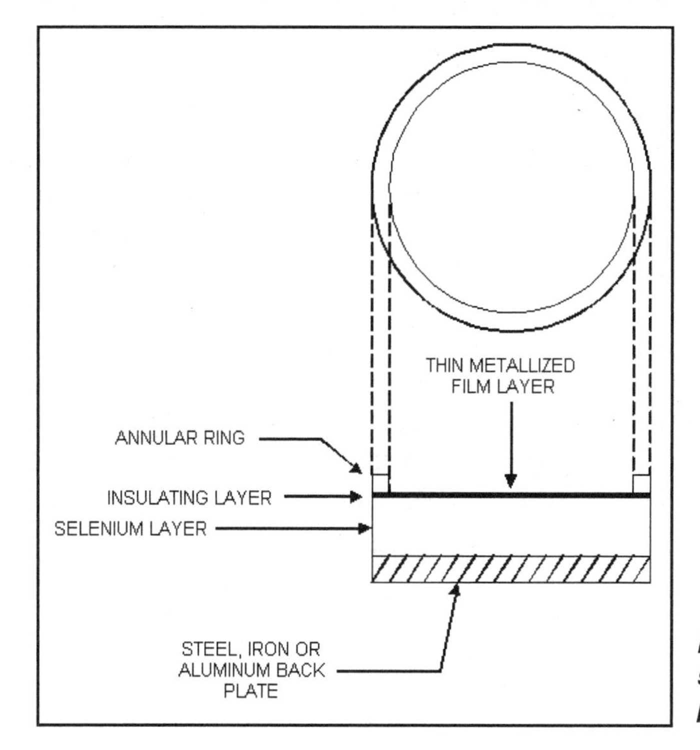

Figure 7-3. Mechanical structure of the MOS photovoltaic cell.

Selenium cells produce an output potential in the range of 0.2 to 0.6 volt dc (0.45 volt dc under 2,000 footcandles of illumination is a common standard). Photovoltaic cells designed for power applications produce between 20 and 90 milliwatts of dc power per square inch of photoactive surface exposed to light. The selenium cell covers the electromagnetic spectrum from 300 to 700 nm, with its peak spectral response near 560 nm. It is common practice to alter the response of selenium cells and other sensors by placing filters over the transparent window. For most instrumentation purposes, the selenium cell must be loaded with a resistor, or the characteristic curve is highly non-linear.

Silicon PN Junctions

Perhaps the most successful of the modern forms of photovoltaic sensor is the Y device. There are two recognized forms of junction photovoltaic sensor: *heterojunction* and *homojunction*. The heterojunction device uses two different materials in a PN junction; an example is the isotype germanium/silicon (nGe-pSi) device. Other types include the anisotype thin-film pCu_2-nCdS, pSi-nCdS, and pGe-nGaP devices. The latter two can be formulated to offer a response spectrum of 0.5 to 24 mm. Most heterojunction devices were developed for solar power generation, but some are available as light sensors. Homojunction devices use but one type of material, of which the popular nSi-pSi device is an example.

The PN junction photovoltaic cell works because there is an inherent electric field across the PN junction (**Figure 7-4**). When photons strike the cell, electron-hole pairs are created by the photon energy. The electric field forces the holes (with a positive charge) to migrate towards the P-type material, and electrons to migrate to the N-type region. This migration results in an imbalance that creates a potential difference, ΔV, of several hundred millivolts. This potential is the open-circuit potential of the photovoltaic cell.

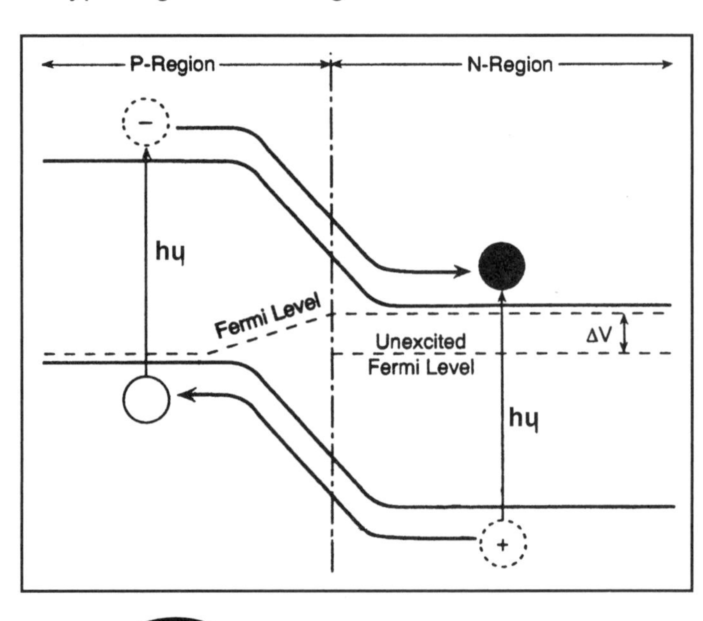

Figure 7-4. Energy diagram for PN junction photovoltaic cell.

Figure 7-5a shows the structure of a silicon photovoltaic cell, which was invented in 1958 by scientists working at Bell Telephone Laboratories. An equivalent circuit is shown in **Figure 7-5b**. This circuit contains two current sources, the reverse saturation (or leakage) current, I_s, and the junction current, I_j, both of which contribute to the output current, I_L. A shunt resistance, R_j, represents the resistance of the junction when forward biased. A series loss resistance, R_s, tends to drop the available output voltage to the load, V_L, because output current, I_L, causes a voltage drop equal to $I_L R_s$. The output voltage, V_L, is the product of the output current and the load resistance, R_L.

The silicon cell consists of a PN junction of P- and N-type silicon. In the P-on-N form shown in **Figure 7-5a**, a thin layer (~0.5 mm) of arsenic-doped N-type silicon is deposited onto a metallic substrate, which also forms the negative terminal of the cell. The P-type boron-doped layer is diffused into the silicon, and forms the surface exposed to light. The positive electrode is an annular ring deposited onto the exposed surface of the P-type silicon region. These cells output a potential of 0.27 to 0.6 volt under illumination of 2,000 foot-candles. Response tends to peak at wavelengths around 900 nm.

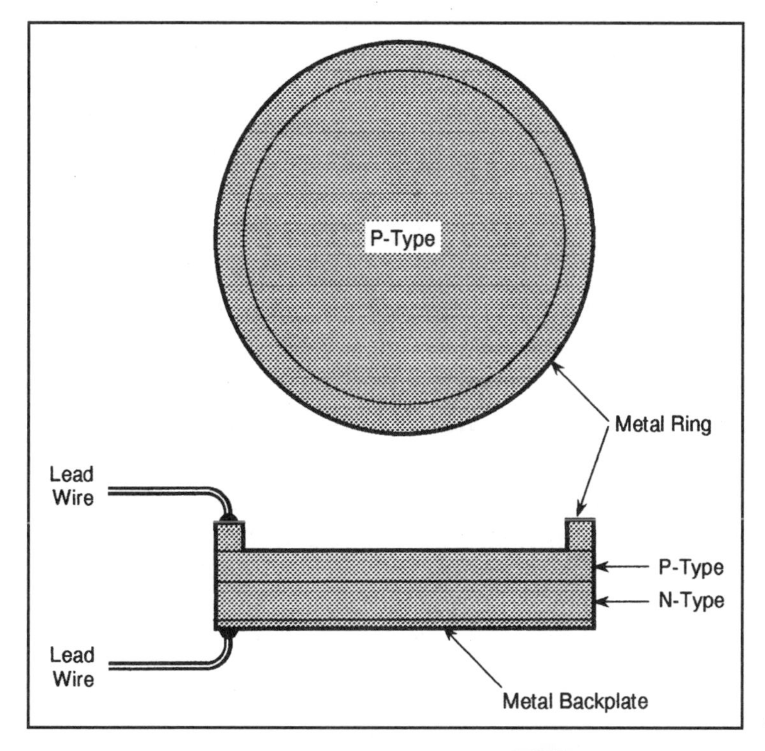

Figure 7-5a. Mechanical structure of PN junction photovoltaic cell.

An N-on-P silicon cell exists in which a thin layer of phosphorus impurities are diffused into boron-doped P-type silicon. These cells show peak response near 800 nm.

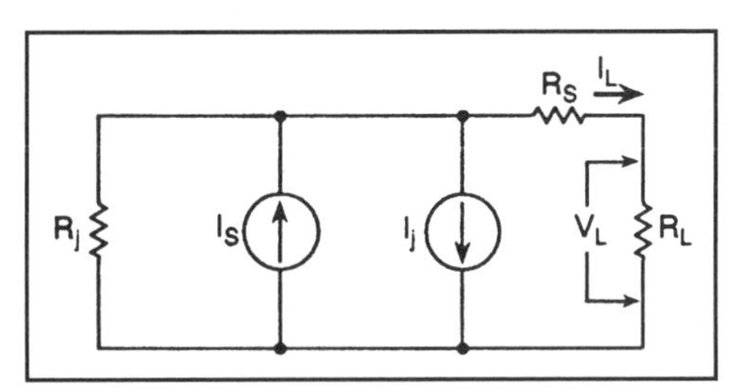

Figure 7-5b. Equivalent circuit model.

Infrared Photovoltaic Sensors

Infrared (IR) sensors are designed to be sensitive to electromagnetic radiation at wavelengths longer than those of visible light. Because IR wavelengths are also created by heat sources, it is often necessary to cool IR sensors. Some sensors are operated at room temperature, which is about 27°C or 300 K (Note: 0 ºC = 273.16 K), but most of them require a cooled environment. A typical arrangement embeds the sensor in a Dewar flask (a double-walled glass vacuum bottle, loosely called a "thermos bottle") that is filled with liquid nitrogen to cool the sensor to 77 K. In other cases, dry ice packs are used to cool the device to 196 K.

A more recent development is the use of thermoelectric Peltier-effect cooling devices. These thermocouples are made from dissimilar metallic or semiconductor sections that are excited by a dc current, causing one junction to absorb heat, and the other to give up heat.

The silicon photovoltaic cell is used in the visible and very near-infrared regions, and has no sensitivity at all in the mid-infrared (to 3 mm) and far infrared. Thus, when infrared wavelengths must be detected, materials other than silicon are needed. Several different materials are used in the IR region:

Germanium (Ge). This material is used in metal/semiconductor cells at room temperature (300 K). This sensor material exhibits a narrow response centered around 1500 nm.

Indium Arsenide (InAs). The InAs cell is used at 300 K, and at colder temperatures down to 77 K (see **Figure 7-6**). The impedance these cells varies with temperature, around 100 W at 300 K and up to 10 MΩ at 77 K. The response time of the InAs cell is on the order of 1 µs.

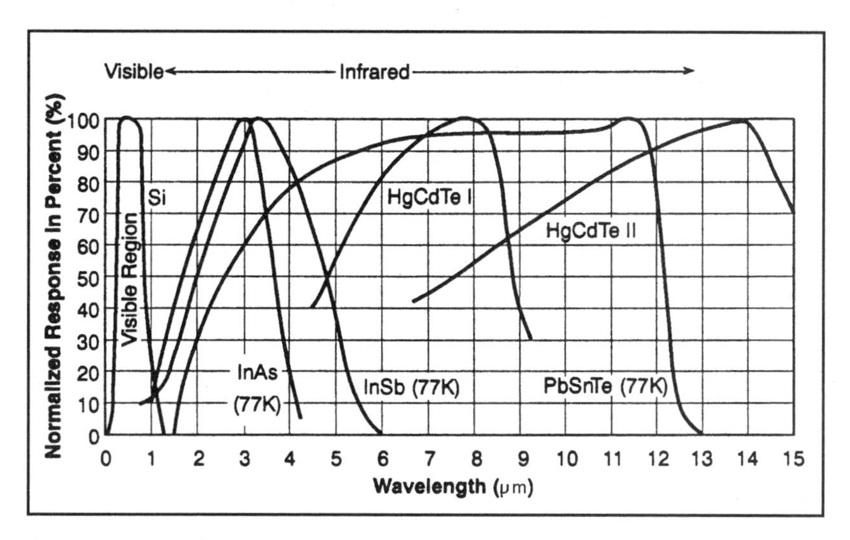

Figure 7-6. Spectral responses of several different types of photovoltaic cells.

Indium Antimonide (InSb). The InSb IR sensor offers a peak response at wavelengths around 3.5 µm when operated at 77 K. It has a response time similar to InAs cells (1 µs). The impedance of this cell varies with temperature, as with the InAs cell, but to a considerably lesser extent (20 kΩ to 50 kΩ).

Lead-Tin-Telluride (PbSnTe). The PbSnTe cell is capable of far-infrared response from 6 to 15 µm, and offers some response in the near IR region around 2 µm. This material operates at 77 K, with response times in the 50 to 100 ns range.

Mercury Cadmium Telluride (HgCdTe). This material can operate at temperatures between 77 K and 120 K, in a spectrum over a range of 2 to 14 µm. This covers the entire IR spectrum from near-IR to far-IR. The actual response of any given cell can be custom tailored by changing the proportions of the three materials used. This phenomenon is shown in **Figure 7-6** by the two different HgCdTe curves labelled HgCdTe-I and HgCdTe-II.

The spectral response of a typical silicon photovoltaic cell is shown in **Figure 7-7**. The cell is usable at ultraviolet wavelengths below in the 400 nm range, but the response is quite poor, and exhibits a significant wavelength sensitivity. Response peaks above 500 nm in the blue-green visible range, and falls off gently to a wavelength around 900 nm, where it drops off rapidly.

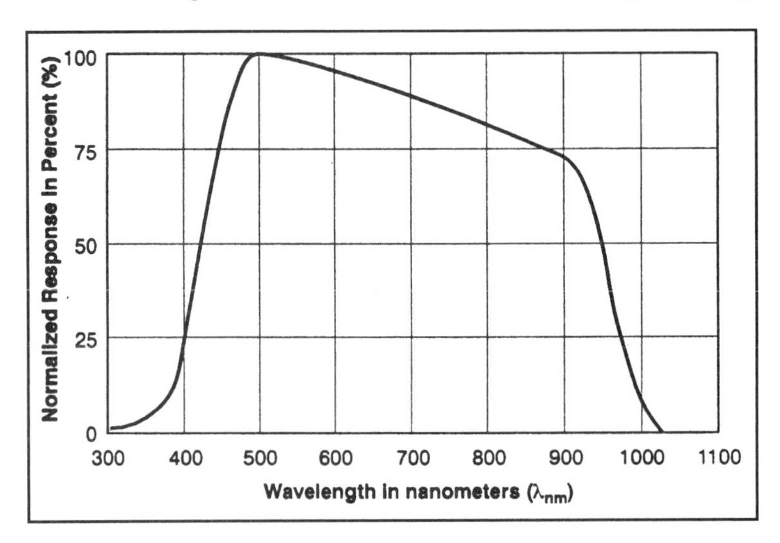

Figure 7-7. Spectral response of the "blue" silicon cell.

Photovoltaic Cell Characteristics

The photovoltaic cell is an electrical power source that converts light energy into electrical energy. It should not be surprising, therefore, to find that some of the characteristics of the photovoltaic cell resemble the characteristics of other forms of dc power supply.

Output Voltage, Voc. This parameter is usually measured in an open circuit (with the load disconnected), and is defined as the output potential when the cell is illuminated by a standard 100 footcandle light source. The V_{oc} values for common photovoltaic cells are 0.2 to 0.45 volt for selenium cells, and 0.3

to 1.5 volts for silicon cells. Higher voltages are produced by connecting two or more cells in series; cell arrays of 12 volts are common, while arrays of 24, 28, 32, and 120 volts dc are also available. **Figure 7-8** shows such an array being used to charge nickel-cadmium (NiCd) cells. The diode D1 in series with the photovoltaic cells is used to prevent the battery cells from discharging backwards through the photovoltaic cells (which would otherwise happen, especially when the light level is reduced). The diode shown here is satisfactory for very low current levels, but for larger arrays a diode with a larger current rating is required. One disadvantage to using the diode is that it will cause a voltage drop of 0.6 to 0.7 volt because of its junction potential. This loss is easily overcome by using one or two more solar cells.

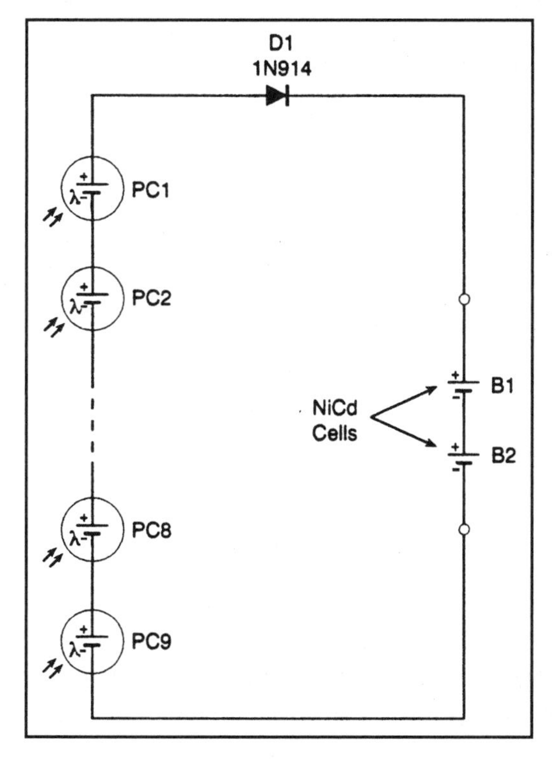

Figure 7-8. Stacking photovoltaic cells in series to increase voltage.

Loaded Output Voltage, V_{ol}. Because of internal resistance losses in the cell, the output voltage drops from V_{os} to a lower value (V_{ol}) when current is drawn by an external load.

Output Current, I_o. The output current rating is the short-circuit value under standard illumination conditions (100 fc). Selenium cells vary from around 15 µA to almost 800 µA per cell, while silicon cells vary from about 5 mA to almost 40 mA per cell. Current rating is increased in an array of solar cells by connecting two or more cells in parallel. A higher current array is shown in **Figure 7-9**. In this case two cell stacks are used, and an isolation diode is used for each stack. Again, there will be a 0.6 to 0.7 volt drop across each diode, so the number of solar cells in each stack needs to be increased in order to overcome the loss.

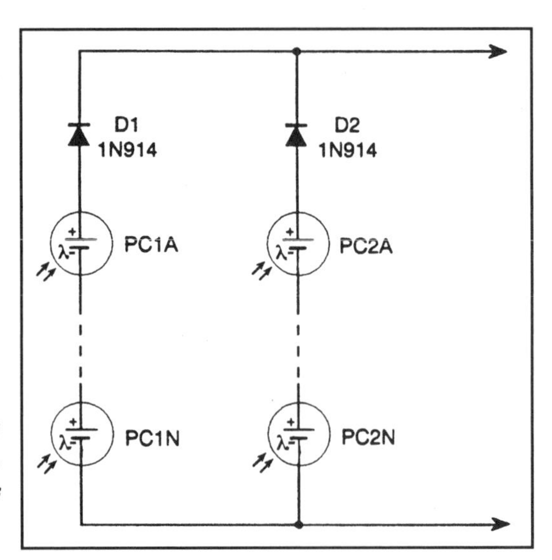

Figure 7-9. Parallel connection for increasing current capacity of photovoltaic cell array.

Internal Resistance, R_s. All electrical current sources possess a certain inherent internal resistance, or *source resistance*, R_s. This resistance represents a loss because a voltage is dropped across R_s when current is drawn (**Figure 7-10**). This voltage drop must be deducted from the open-circuit output voltage in order to find the output voltage under load. The internal resistance can be calculated from:

$$R_s = \frac{V_{oc}}{I_{sc}} \qquad \text{eq. (7-1)}$$

Where:

R_s is the internal resistance in ohms (Ω).

V_{oc} is the open-circuit potential in volts (V).

I_{sc} is the short-circuit current in amperes (A).

Internal resistance, R_s, is the source of voltage regulation problems.

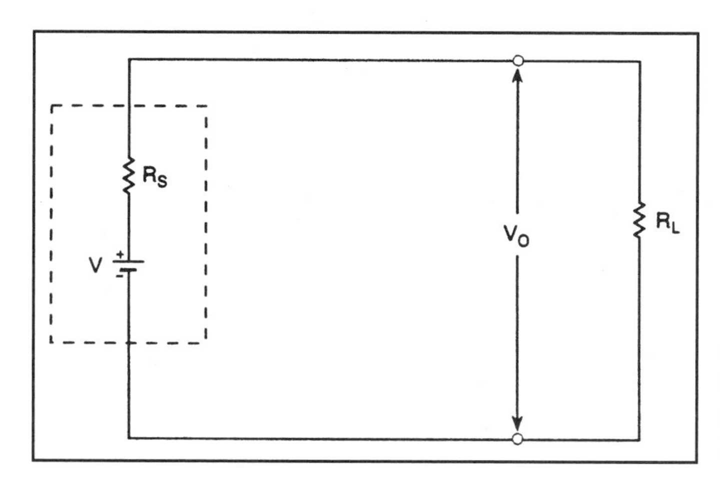

Figure 7-10. Equivalent circuit including cell internal resistance.

Output Power, P_o. Three factors affect the output power available from a specific photovoltaic cell: light illumination level, active area actually exposed to the light and the load resistance, R_L. The output power can be found from any of the following:

- $$P_o = V_{ol} I_o \qquad \text{eq. (7-2)}$$

- $$P_o = \frac{(V_{ol})^2}{R_L} \qquad \text{eq. (7-3)}$$

or,

- $$P_o = I^2 R_L \qquad \text{eq. (7-4)}$$

The maximum power transfer between the photovoltaic cell and the load is found, as on any electrical power source, when the load resistance is matched to the internal resistance of the cell, that is, when R_L equals R_S.

Photovoltaic Cell Applications Circuit

In this section we will examine some applications circuits for photovoltaic devices, including both sensors and solar cells. Although the circuits use only one style of symbol, the same circuits work as well with PN junction devices as with metal-semiconductor devices.

Instrumentation/Communications Applications

Figure 7-11 shows a basic light level meter circuit for use in photographic and other light measurement applications. The photovoltaic cell is used to supply a photocurrent, I_o, to an external circuit consisting of a calibration resistor and a dc microammeter. The calibration resistor is adjusted to produce the desired meter deflection at standard light level.

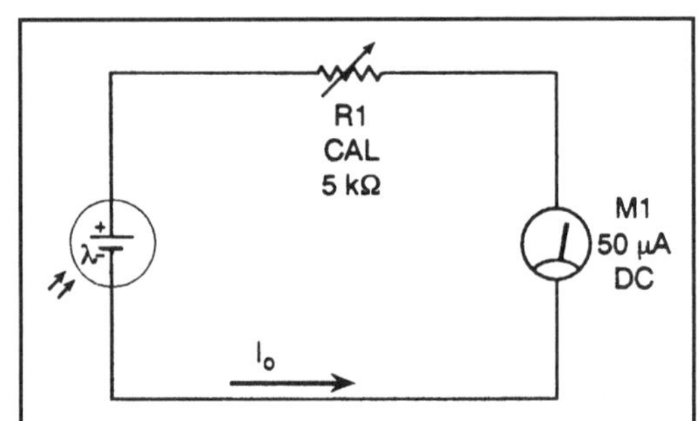

Figure 7-11. Photovoltaic cell light meter circuit.

Figure 7-12 shows two circuits that will work with small, low-voltage piezoelectric ceramic buzzers. These devices are widely used as alarm elements in various applications, and are readily available at electronics supply outlets. When a photovoltaic cell is used as the power source, as in **Figure 7-12a**, the buzzer will sound when the light level reaches a certain point, set by the characteristics of the photocell.

An alternative circuit, shown in **Figure 7-12b**, is used when the light level varies naturally, is modulated, or is intentionally chopped. This last technique is often used in instrumentation applications, while modulation is used in communications applications. The transformer is an output transformer intended for transistor audio amplifiers. The 8-ohm winding is normally used as the secondary in radio circuits, but here it is the primary; the 1000-ohm winding is used as the secondary. The use of the transformer in this manner produces a step-up action that will increase the voltage fluctuations caused by a varying light source. The principal application of this circuit is in communications, where a light beam is modulated with an audio signal.

One application of ac-coupled circuits is in burglar alarms. If a pulsed light or IR source is used for the source, then only signals from the true source will be passed to the secondary. If an intruder attempts to defeat the alarm by shining a flashlight into it, then the alarm would sound anyway. One would not use a buzzer in this case, but rather a detector circuit that will turn on when the pulsed source disappears.

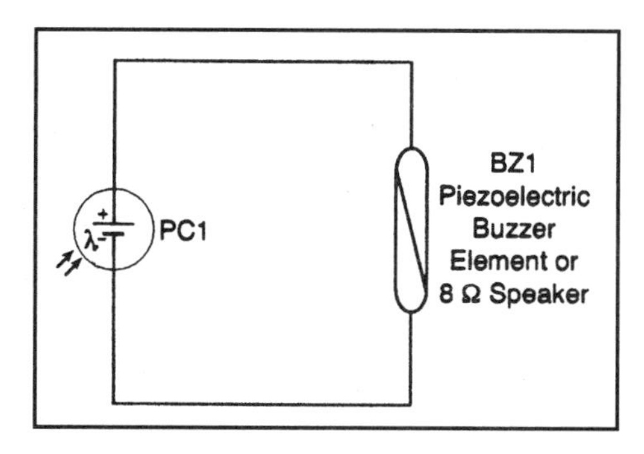

Figure 7-12a. Buzzer/alarm circuit.

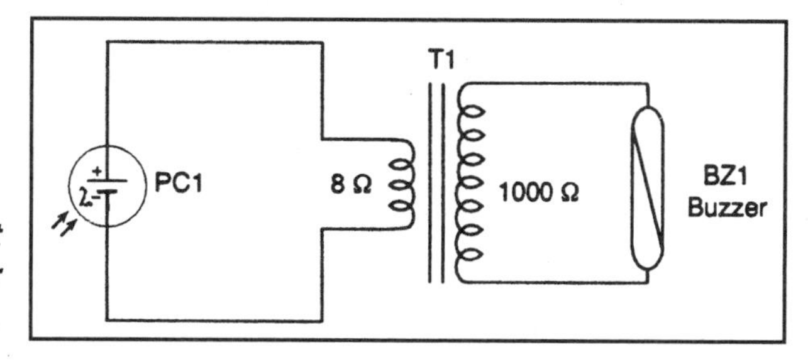

Figure 7-12b. Alarm circuit to detect modulated or pulsed light beam.

Figure 7-13 shows a typical circuit for instrumentation applications of the photovoltaic cell. The cell is connected across the input of a high impedance amplifier, such as the noninverting operational amplifier shown. The amplifier output voltage is found from:

- $$V_o = V_{C1}\left(\frac{R2}{R1} + 1\right) \qquad \textit{eq. (7-5)}$$

Where:

V_o is the output voltage.

V_{c1} is the voltage across the capacitor C1.

R1 and R2 are resistors in the feedback network.

Example 7-1

Consider a circuit such as **Figure 7-13a** in which R1 = 1 kΩ and R2 = 15 kΩ, which are typical values. Find the amplifier output voltage, V_o, when the photocell output voltage, V_{C1}, is 0.38 volt.

Solution:

- $$V_o = V_{C1}\left(\frac{R2}{R1} + 1\right)$$

- $$V_o = (0.38\ volt)\left(\frac{15\ k\Omega}{1\ k\Omega} + 1\right)$$

- $$V_o = (0.38\ volt)(15 + 1) = (0.38\ volt)(16) = 6.08\ volts$$

The operational amplifier provides a high input impedance buffer between the cell and the outside world, as well as providing voltage gain. The operational amplifier can be almost any common type, depending on the application. For most applications, this includes common forms such as the 741, LM-301, or CA-3140 (or their dual derivatives).

A variant of the basic circuit is shown in **Figure 7-13b**. Because this circuit is ac-coupled, it will not pass the static dc level created by ambient light levels, but will pass changing voltages caused by modulation of the light beam, chopping, or natural variations. When the light beam is truly static, the alternate circuit will not pass a signal.

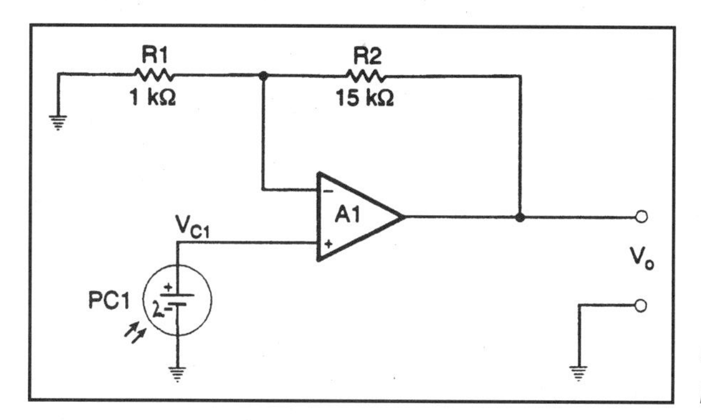

Figure 7-13a. DC coupled photocell amplifier.

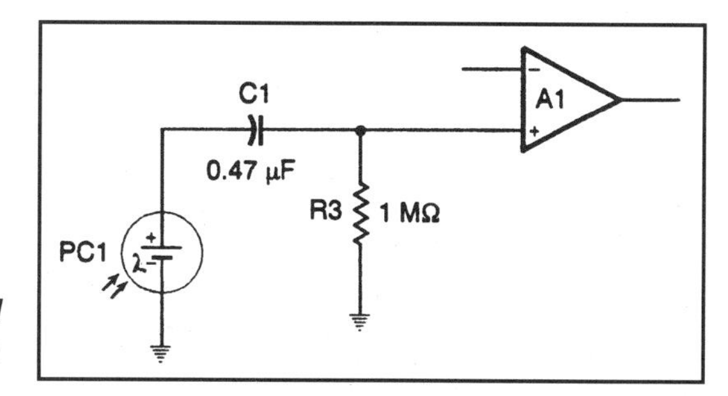

Figure 7-13b. Ac-coupled photocell amplifier.

A practical light-beam communicator receiver is shown in **Figure 7-14**. This circuit uses the ac-coupled variant of **Figure 7-13b** with a 741 operational amplifier. The circuit will operate from +9 volt dc power supplies. The audio output stage is a single-chip audio amplifier that contains both preamplifier and power amplifier stages in one 8-pin miniDIP package. With the configuration shown, the LM-386 will provide a gain of 20; if a higher gain (´200) is required, then connect a 10 μF capacitor between pins 1 and 8 on the LM-386 ("+" terminal of the capacitor to pin 1). Two controls are provided. The GAIN control adjusts the circuit sensitivity by varying the gain of the operational amplifier. The VOLUME control is adjusted to produce a comfortable listening level.

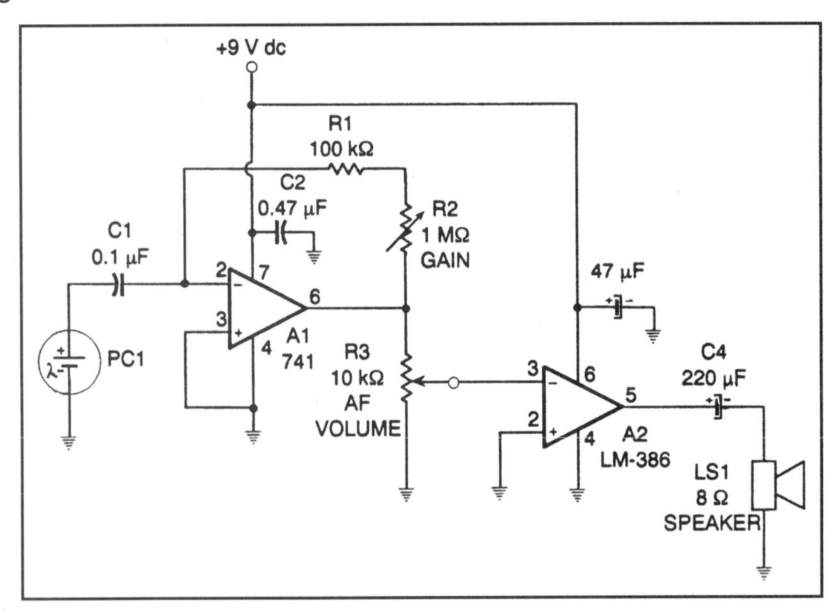

Figure 7-14. Light communications receiver circuit.

Solar Cell Applications

One application of photovoltaic cells is the generation of dc electrical power from sunlight. These cells are used in a variety of power situations, but most involve charging batteries from sunlight. Batteries can power electronic equipment at night or when the sunlight fails. Examples include radio equipment in the tropics or deserts, on boats, or on remote sites where servicing fuel-run generators or running electrical power lines is not practical or cost effective.

One fellow I met served in a medical unit in the Sudanese desert. He told me that they charged automobile batteries from solar cells for their communications equipment. Another person I know owns a traditional (no electrical system) sailboat. He generates battery power for his radio and navigation equipment using a solar cell array.

Still another person I know created a small three-home TV cable system in the mountains of southwest Virginia. He needed 90 mA to run a wideband antenna amplifier at the headend of an 800 foot long coaxial cable. The power problem was solved by a 24 volt dc, 15 ampere/hour, gel-cell battery that is charged with a solar array that can produce nearly 1 ampere of dc when needed. This battery served for several years before needing service!

Figure 7-15a shows the basic solar cell device. These are thin silicon units on a rectangular, metal-backed plate. These cells are extremely brittle, so great care is needed when handling them. Actual solar cells are shown in **Figure 7-15b**. The two smaller units (a penny is provided for size comparison) are both 0.45 volt, 40 mA units, while the larger is a 70 mA, 1.5 volt flexible solar cell.

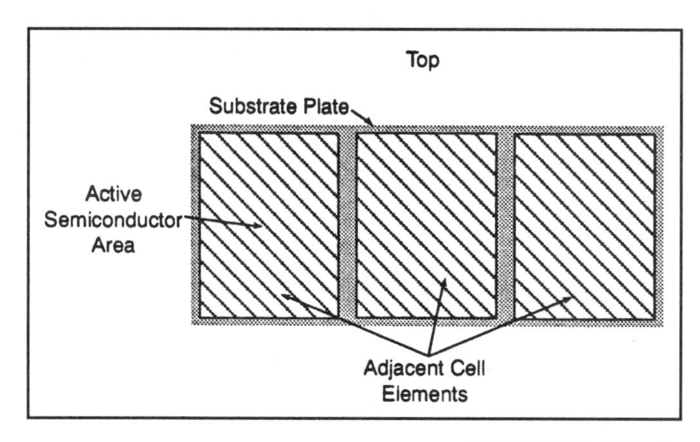

Figure 7-15a. Schematic of typical structure of the solar cell.

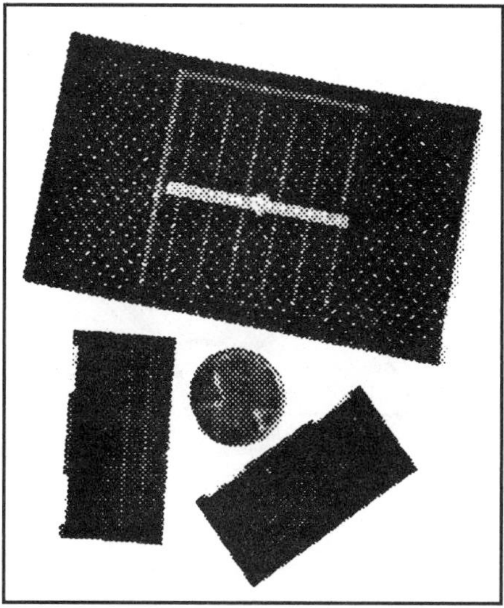

Figure 7-15b. Actual commercial solar cells.

The basic photovoltaic charger is shown in **Figure 7-16a.** The solar cell array is aimed at the sun (in North America a southern exposure is best). The negative terminal of the solar array is connected to the battery's negative terminal, while the positive side is connected to the battery's positive side through an isolation diode. (Note: some arrays contain built-in diodes, and do not require an additional isolation diode.) The load is connected in parallel with the battery.

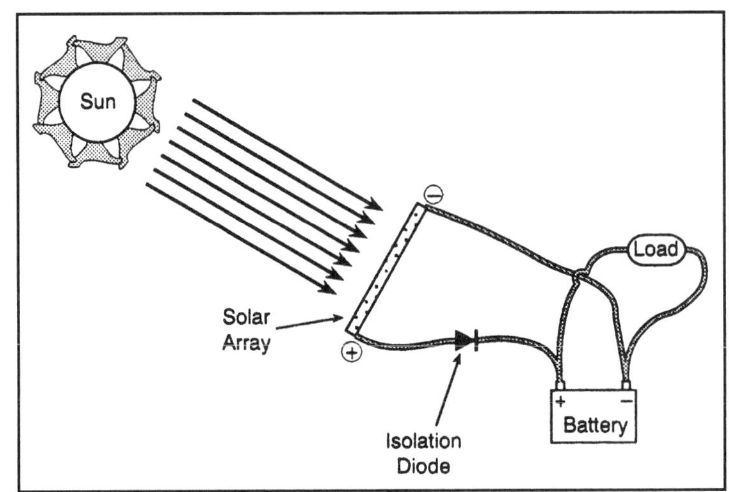

Figure 7-16a. Simple version of circuit may experience overcharging of battery.

One danger of the circuit is the possibility of overcharging the battery during periods of prolonged sunlight. If that is a problem, then a voltage regulator can be inserted into the circuit in place of the isolation diode. A suitable example is shown in **Figure 7-16b**. With the battery disconnected, the output voltage is adjusted to the exact voltage that will be expected when the battery is fully charged.

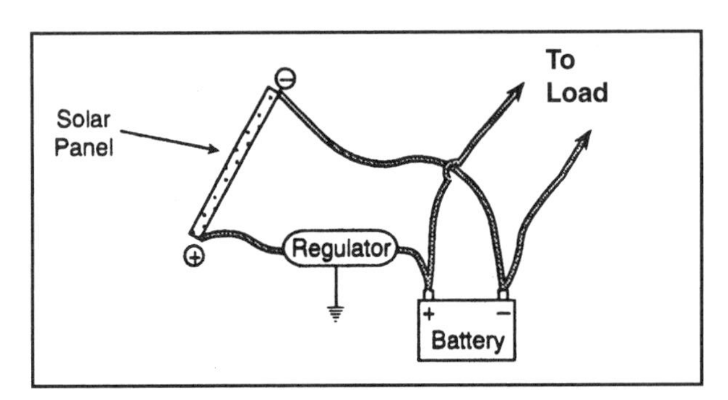

Figure 7-16b. Version of circuit with a voltage regulator to prevent overcharging.

Another problem with solar powered devices is that the sun does not stand still over the course of a day, or as the seasons change. It is sometimes necessary to mount the solar cell array on a movable platform that is connected to a motor/tracker circuit. **Figure 7-17** shows a sun position sensor and motor/tracker designed by science writer Forrest Mims III for the premier edition of *Science Probe!* magazine (November 1990, p.75). Two matched solar cells are mounted on a metal backplate that can be made from printed circuit copper clad stock.

A light shade painted flat black is mounted vertically on the backplate so that the sun will fall unevenly on the two sensors, except when the panel is pointed directly at the sun. The two solar cells are connected in opposite polarity across a low-voltage dc motor, which is mechanically linked to a bearing sleeve on the underside of the backplate. Construction details are found in Mims' article.

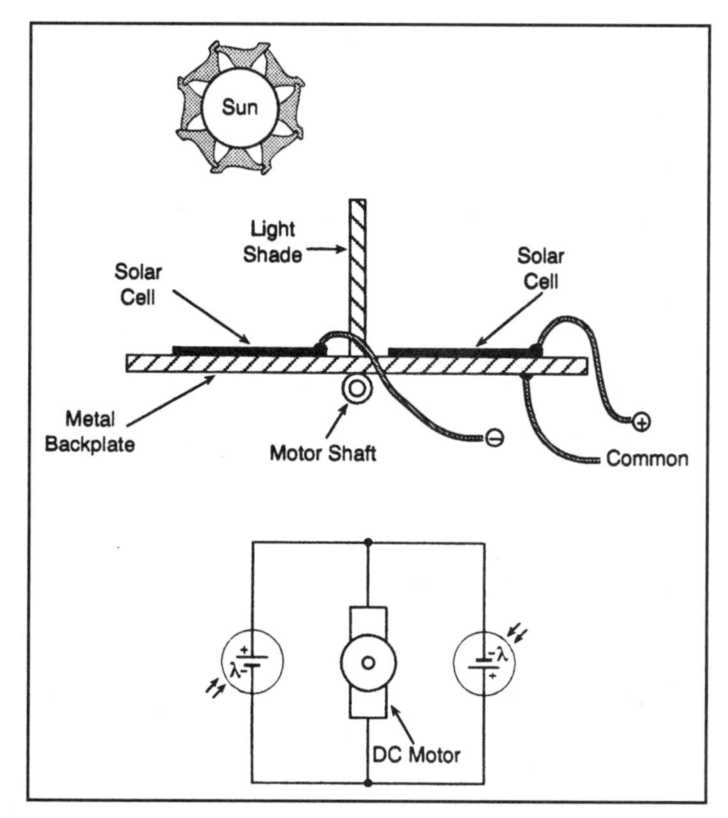

Figure 7-17. Sun tracker designed by Forrest M. Mims III for **Science Probe!** ***magazine's premier issue (Courtesy*** **Science Probe!** ***magazine).***

A potential problem with this simple circuit is that it will not drive larger arrays. The motor is simply too low-powered to move an array more than a foot square. This problem can be overcome by using a more powerful motor, and an outboard power supply (which may require its own solar array and battery set). The operational amplifier shown in **Figure 7-18** is a power operational amplifier, which may be found in retail catalogs. The V_{in} signal is derived from a differential sun sensor circuit.

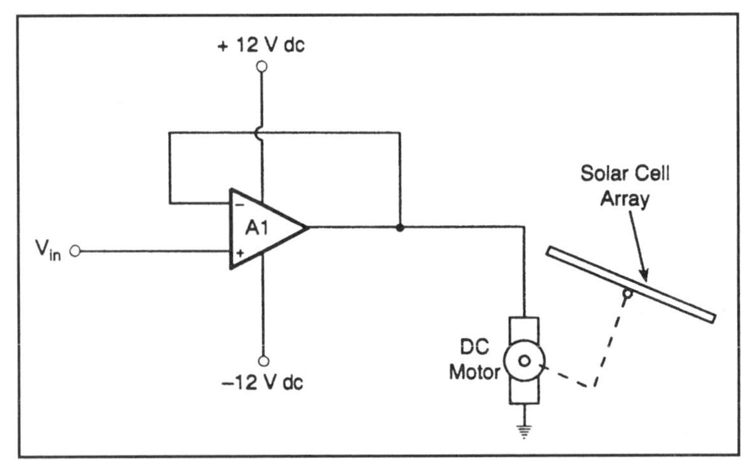

Figure 7-18. Alternative sun tracker circuit.

Two differential sun sensor circuits are shown in **Figure 7-19**. These circuits can use simple solar cells such as the Radio Shack 276-124, or equivalent. In both cases the electronic arrangement shown is used for mounting the sensors. The passive version shown in **Figure 7-19a** uses a pair of solar cells whose output voltages are opposing each other, and combined in a resistive summer circuit. The output voltage, V_o, will be zero when the two cells are equally illuminated because the two cells' contributions cancel each other. When the light level is uneven, V_o will take on the voltage polarity of the more highly illuminated cell.

Figure 7-19b shows a pair of cells applied to the -IN and +IN inputs of a dc differential amplifier made from an op-amp. The voltage gain, A_v, of the op-amp can be adjusted as needed by varying the ratios of the feedback and input resistors. When R1 = R2, and R3 = R4, the gain is:

$$A_v = \frac{R3}{R1} \qquad \text{eq. (7-6)}$$

The output voltage from the operational differential amplifier is:

$$V_o = (V2 - V1)\left(\frac{R3}{R1}\right) \qquad eq.\ (7\text{-}7)$$

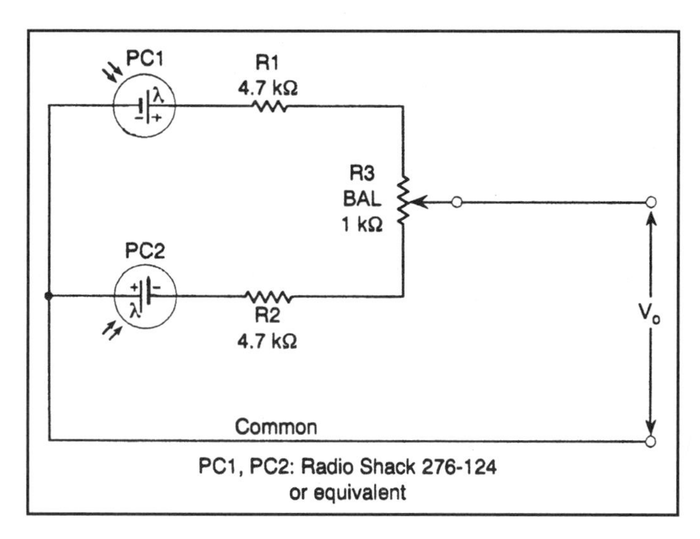

Figure 7-19a. Sensor array for a sun tracker sensor.

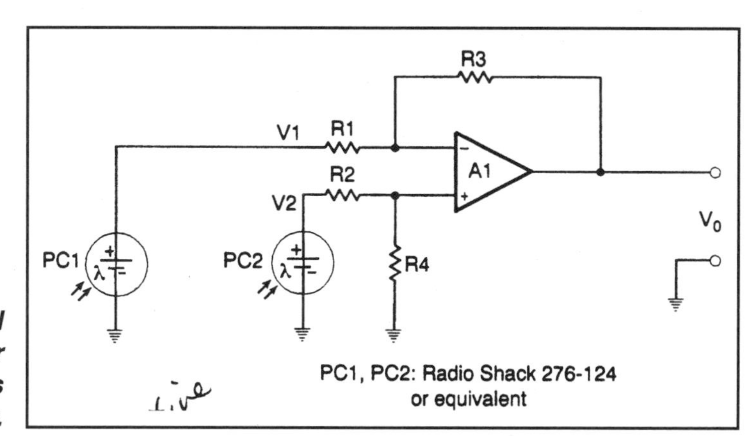

Figure 7-19b. Differential amplifier sun tracker sensor. Resistor values are left to the designer.

The output of the differential amplifier is connected to the input of the circuit in **Figure 7-18**. In **Figure 7-19b**, the two cells are connected in the same polarity, but the inputs of the differential amplifier have opposite polarities. When the two cells are equally illuminated, then the two voltages, V1 and V2, are equal, and by **Equation 7-7** the output voltage is zero. Thus, the motor will not turn in either direction. But when the illumination is unequal, the output voltage is proportional to the difference between V1 and V2.

A crude four-quadrant light sensor is shown in **Figure 7-20**. This sensor will detect two dimensional movement. Similar circuits are used in toy cars and robots that will follow a light source. This sensor is used with two circuits such as **Figure 7-19b**, one for the UP-DOWN motor and one for the LEFT-RIGHT motor.

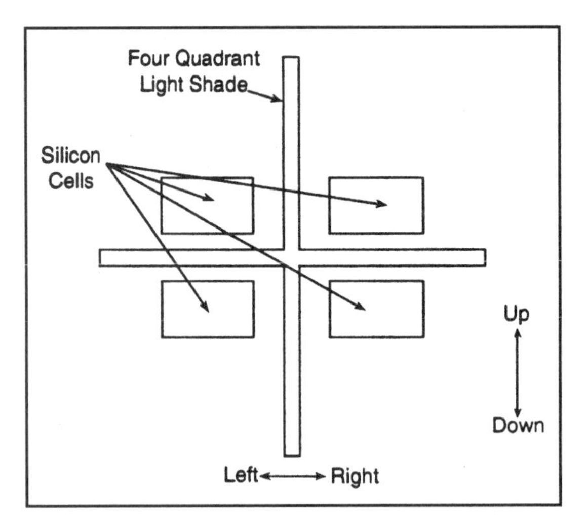

Figure 7-20. Four quadrant sun tracker array.

Chapter 8

Photodiodes and Phototransistors

Semiconductor technology revolutionized the electronics industry in the 1950s, and now dominates all areas except the highest power radio frequency segments of the field, where tubes are still used. Electro-optics especially benefited from the solid-state revolution. There are now scores of devices on the market that sense light - visible, infrared, or ultraviolet - or X-radiation, and these devices often possess attributes that are superior to previous optical sensors. We have already discussed the silicon photoconductive and photovoltaic cells, and now we will examine more complex devices such as photodiodes and phototransistors.

In Chapter 5 the conduction mechanism in semiconductor materials was discussed, as was the basic PN junction diode, the basis for the LEDs. Now the PN junction diode will be viewed as a light sensor.

A PN junction diode (**Figure 8-1a**) consists of N-type and P-type material joined together at the junction. As we have seen, if a positive electrical potential is applied to the P-type side, and a negative potential to the N-type side, the respective positive and negative charge carriers are repelled from the

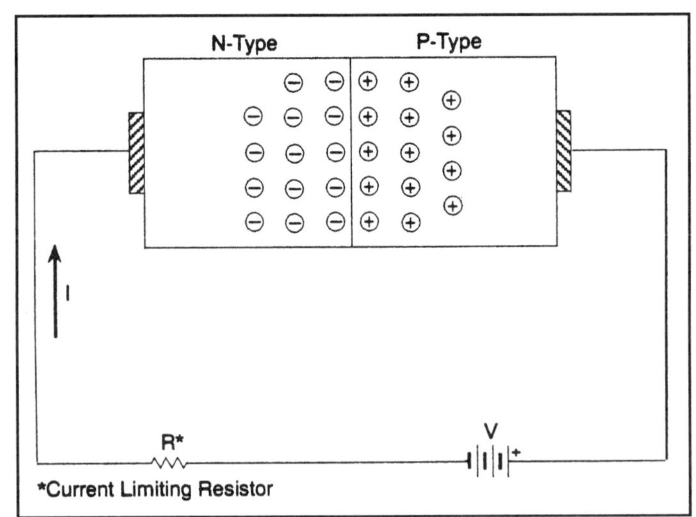

Figure 8-1a. Forward biased PN junction diode.

ends of the structure towards the PN junction region. Under this condition a large amount of electron/hole recombinations can occur, and a current flows across the junction. But when the potentials are reversed (**Figure 8-1b**), the charge carriers are attracted away from the junction toward the end electrodes. This action forms a *depletion zone* between the N-type and P-type materials in which few charge carriers, ideally none, exist.

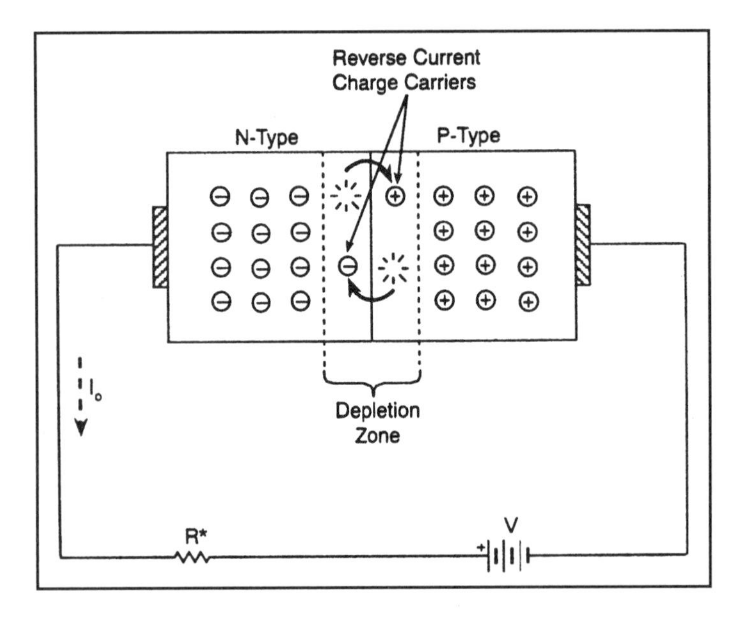

Figure 8-1b. Reverse biased PN junction diode.

In the idealized cases shown in **Figure 8-1** only the majority charge carriers are shown. In real devices, however, there will be minority charge carriers in each material. That is, there are positively charged holes in the N-type semiconductor and negatively charged free electrons in the P-type semiconductor. Recombination of those few minority charge carriers that do exist in the depletion zone forms a minute reverse bias *leakage current*, I_o, across the PN junction; this current is frequently labelled the *reverse bias saturation current*. Ideally, the reverse saturation current is zero, but in working diodes it is about 10^{-7} amperes.

Figure 8-1c shows the I-vs-V curve for a PN junction diode. In the reverse bias region (V), the reverse saturation current is shown as a very tiny current until the reverse bias voltage reaches a critical point, V_z. At this point *avalanching* occurs, and the diode breaks down, causing a very large reverse current flow. This point on the V range is called the *zener potential*; we will discuss both of these phenomena later in this chapter.

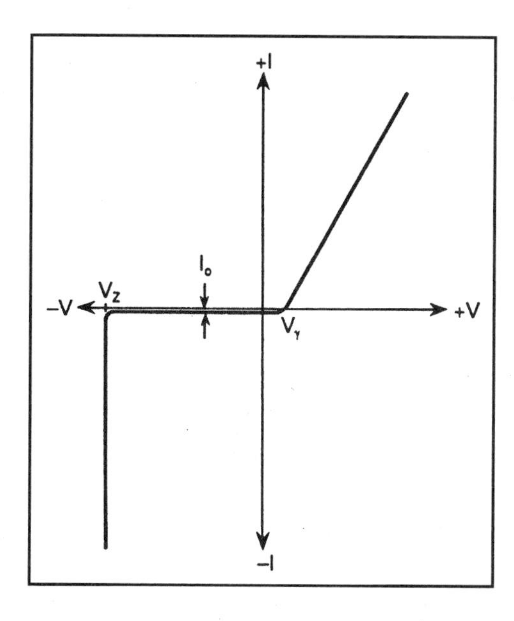

Figure 8-1c. I-vs-V curve shows graphically the forward and reverse bias conditions.

In the forward bias region (V), a certain *junction potential* ($V\gamma$) also exists across the PN junction. In germanium based semiconductors, $V\gamma$ is normally on the order of 200 to 300 millivolts, while in silicon based semiconductors it is normally 600 to 700 mV. The leakage current at low values of V is essentially I_o, but at the critical potential the current begins to increase until it is essentially ohmic. The general expression for the PN junction current is:

$$I = I_o\left(e^{\frac{V}{\eta V_T}} - 1\right) \qquad \text{eq. (8-1)}$$

Where:

I is the PN junction current in amperes.

I_o is the reverse saturation current (typically 10^{-7} amperes).

V is the applied potential in volts.

η is a constant that is not temperature dependent (typically 1 for Ge devices and near 2 for Si devices).

V_T is the quantity kT/q, or about 26 mV at room temperature; T/11,600 at other temperatures.

T is the junction temperature in Kelvins (K), q is the electron charge (1.602×10^{-19} coulombs), and k is Boltzmann's constant (1.381×10^{-23} joules per Kelvin).

Photodiodes

In certain PN junction diodes, illumination of the junction region by visible light (or other EM waves such as IR or UV radiation or X-rays), will increase the level of reverse leakage current available. These diodes are built with an extremely thin N-type region, configured so that light can illuminate it. The photon energy causes electrons to strip away from their respective atoms to become negative charge carriers. Their former locations in the semiconductor crystal lattice become holes, which are seen as positive charge carriers in the P-type material.

If a reverse bias voltage is applied (**Figure 8-2a**), then a very large electrical field exists in the region of the junction. This field prevents electrons and holes from recombining with each other, and drives them towards the oppo-

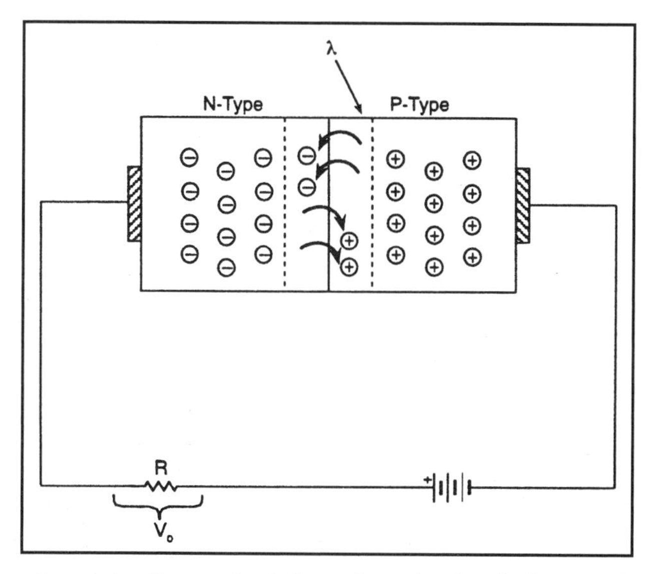

Figure 8-2a. The mechanism for leakage current across any PN junction diode is stray electrons and holes in the depletion zone recombining across the junction.

site polarity end of the electric field. As explained earlier, a leakage current flows across the reverse biased region. The leakage current can be increased by light illuminating the junction, as shown by the three different curves in **Figure 8-2b**. The difference between the illuminated and nonilluminated currents is a measure of the impinging radiation. Note that there are two levels of photocurrent, I_{ph1} and I_{ph2}, corresponding to the different light levels.

Figure 8-2c shows the basic circuit used for PN junction diode light sensors. The diode is normally reverse biased, with a current limiting resistance (R) in series. In practical instrumentation applications, either the leakage current I_o or the voltage drop (V_o) across load resistance R can be measured to find the light level. Voltage V_o is the product of the current and the resistance (I_oR).

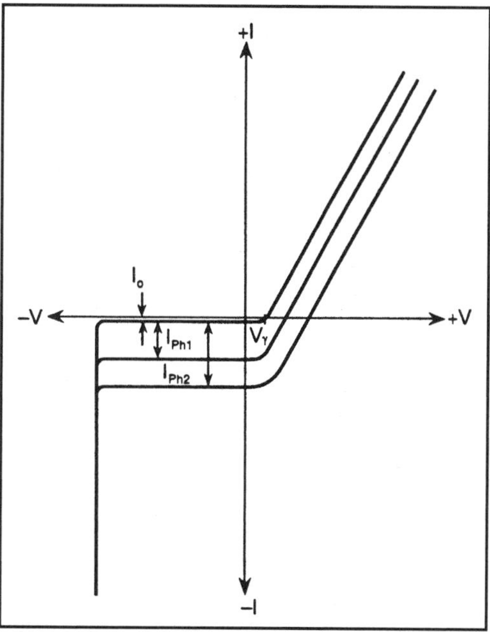

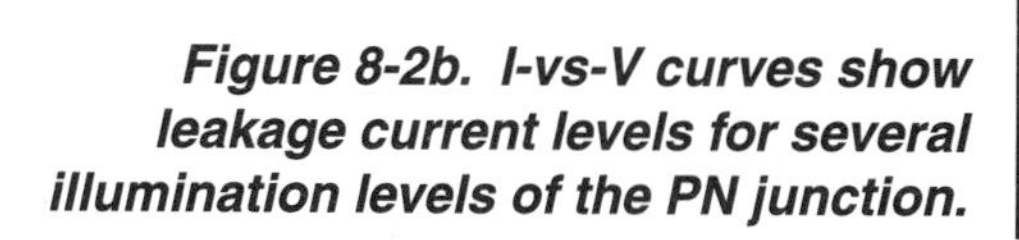

Figure 8-2b. I-vs-V curves show leakage current levels for several illumination levels of the PN junction.

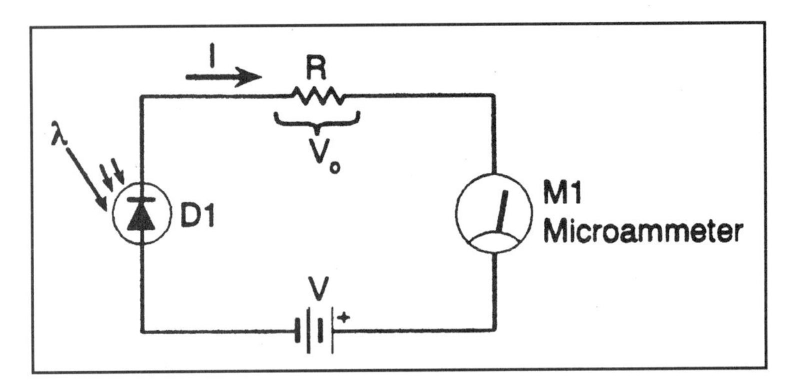

Figure 8-2c. PN photodiode light meter based on the photosensitivity of PN diode leakage current.

The dark current of the photodiode is found from the same equation as the leakage current in a PN junction:

$$I_{dk} = I_o \left(e^{\frac{V}{V_T}} - 1 \right) \qquad \text{eq. (8-2)}$$

or,

$$I_{dk} = I_o \left(e^{\frac{qV}{\eta KT}} - 1 \right) \qquad \text{eq. (8-3)}$$

The photocurrent, I_{ph}, is the reverse current that flows when the PN junction is illuminated, and is found from:

$$I_{ph} = \frac{qNP}{hf} \qquad \text{eq. (8-4)}$$

Where:

I_{ph} is the photocurrent in amperes (A).

q is the electron charge (1.6×10^{-19} coulombs).

h is Planck's constant (6.63×10^{-27} ers-s).

N is the quantum efficiency.

P is the power of the radiation applied to the junction (watts).

f is the frequency of the applied radiation (1/l).

PIN Photodiodes

The PN junction photodiode is basically an electro-optical version of the standard PN junction diode concept, which is widely used in electronics. Another class of diode used extensively in E-O applications is the PIN diode (**Figure 8-3**). This type of diode consists of P-type and N-type regions superimposed across an *intrinsic* (I) region of bulk resistive semiconductor material of width, W, hence the designation PIN.

The depletion zone of the PIN diode is contained within the P-type and N-type materials, so there is a wide effective width consisting of I-region width, W, and the widths of the depletion sectors. Thus, the capacitance of the PIN diode is considerably less than the capacitance of ordinary PN junction diodes. The lower capacitance translates into faster operating times in response to light level changes.

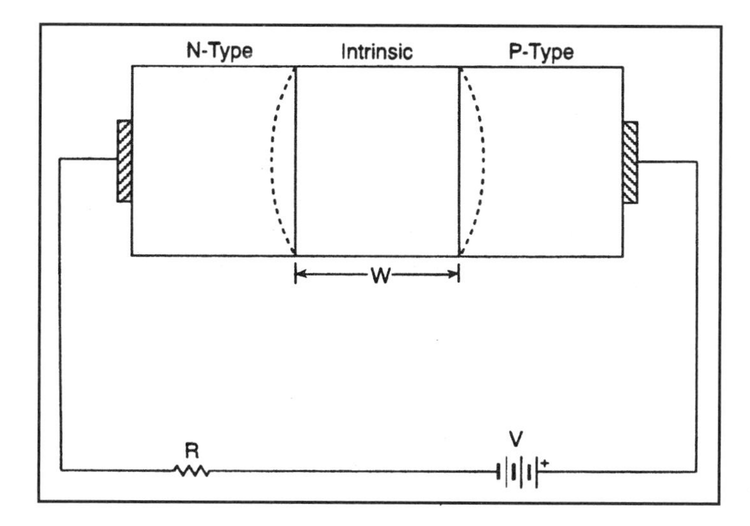

Figure 8-3. PIN diode structure.

Avalanche Photodiodes

The *avalanche effect* in PN junctions is a current multiplication effect in which charge carriers dislodge others by giving up energy. When an energetic electron strikes an atom it may dislodge one or more orbital electrons, producing a cascade of new electrons, as well as holes where those electrons were located. These new electrons, in turn, dislodge other electrons, and the process multiplies until a very large current flows. Avalanching tends to occur when the reverse bias potential is high enough to impart high velocity, and hence high kinetic energy, to the carriers; the avalanche potential tends to be very abrupt.

Uncontrolled avalanching leads to diode breakdown and destruction of the device. Controlled avalanching, on the other hand, is quite useful. For example, the zener potential of zener voltage regulator diodes occurs at the avalanche breakdown potential. Avalanching is also used in switching diodes, microwave diodes and other applications where speed enhancement is desired. In *avalanche photodiodes* (APD), the operating speeds are sufficient to allow 100 GHz gain-bandwidth products.

The gain of an APD tends to be higher than in ordinary PN and PIN diodes because the dc bias is set just below the avalanche point. The small added energy of dim light is sufficient to push the system into avalanche condition. Hence, a large photocurrent is produced by a relatively low light level.

The current multiplication factor, M, of the avalanche PIN diode is a function of the width, W, of the intrinsic region and the absorption coefficient of the semiconductor materials. Values of M are usually 1 to 100. Assuming the simple case in which the absorption coefficient, α, of the P-type and N-type materials are equal ($\alpha_n = \alpha_p$), we can find the multiplication factor by:

- $$M = \frac{1}{1 - \alpha_n W}$$ *eq. (8-5)*

- $$M = \frac{I_{phm} - I_{dkm}}{I_{phu} - I_{dku}}$$ *eq. (8-6)*

$$M = \frac{1}{1 - \left(\frac{V_r}{V_b}\right)^n} \qquad \text{eq. (8-7)}$$

Where:

α_n is the absorption coefficient of the material.

W is the width of the intrinsic region.

I_{phm} is the multiplied photocurrent.

I_{dkm} is the multiplied dark current.

I_{phu} is the unmultiplied photocurrent.

I_{dku} is the unmultiplied dark current.

V_r is the reverse bias potential.

V_b is the diode breakdown voltage.

n is a constant of the material.

Phototransistors

The same electro-optical principle for photodiodes also governs a class of bipolar NPN or PNP transistors called *phototransistors*. These transistors are called "bipolar" because they use both N-type and P-type semiconductor material in one unit. Phototransistors combine the theoretical foundations of NPN/PNP bipolar transistors and photodiodes. We have already discussed the PN junction and its sensitivity to light in Chapter 5. Here, we will review some basic bipolar transistor principles.

Figure 8-4 shows the schematic representation of two forms of bipolar junction transistor: NPN and PNP. Both consist of two PN junctions. In **Figure 8-4a**, we see the NPN transistor in which a section of P-type material is sandwiched between two sections of N-type material. In the PNP transistor (**Figure 8-4b**) just the opposite situation is found—a single N-type section sandwiched between two P-type sections. These transistors are the same except for their polarities.

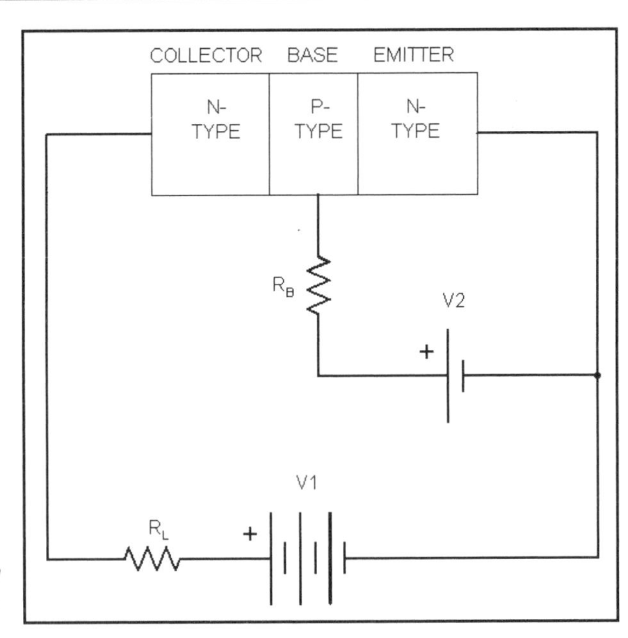

Figure 8-4a. Block diagram for an NPN transistor.

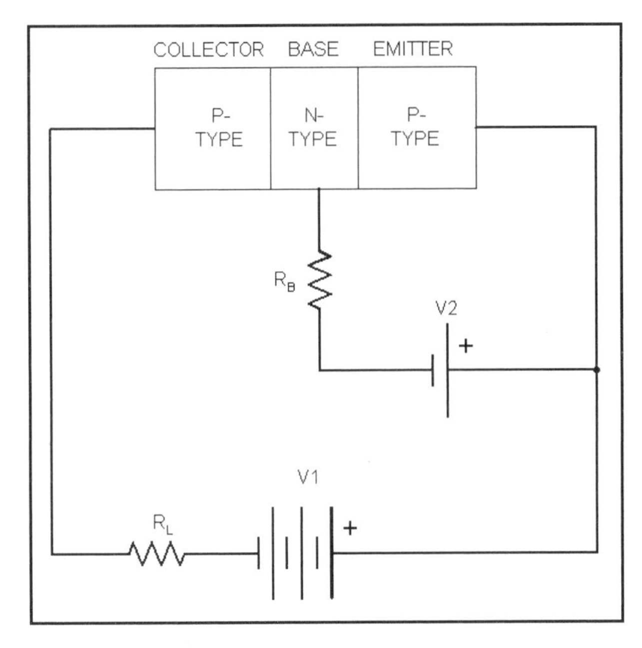

Figure 8-4b. Block diagram for a PNP transistor.

Note the polarities on batteries V1 and V2 in **Figure 8-4** are opposite for the two transistors. This is due to the opposite natures of the two transistors. In a NPN transistor (**Figure 8-4a**) the collector is positive with respect to both the emitter and the base, with the base slightly more positive then the emitter.

In the PNP transistor (**Figure 8-4b**) the collector is more negative than the base and emitter, while the base is slightly more negative than the emitter.

The voltage polarity relationships shown are in their normal mode of operation as amplifiers. You will, however, see other polarities when the transistor is being used as a switch, or for some special purpose. *Amplification* can be defined as the control of a larger current or voltage by a smaller current or voltage. Ideally, the waveshape of the smaller input signal will be reproduced in larger amplitude at the output of the transistor.

The base-emitter junction controls the current flowing in the collector-emitter path. Since the current flowing in the base circuit is only to 2 to 5 percent of the current flowing in the collector-emitter path, the transistor amplifies the base current.

Gain is the amplification factor in any type of amplifier device. In the simplest case we could define gain as the ratio of output over input:

$$A = \frac{output}{input} \qquad \text{eq. (8-8)}$$

or, for voltage gain:

$$A_v = \frac{V_{out}}{V_{in}} \qquad \text{eq. (8-9)}$$

and, for current gain:

$$A_i = \frac{I_{out}}{I_{in}} \qquad \text{eq. (8-10)}$$

When talking in less generalized terms, we can further define gain for transistors. Two gain definitions are usually given as alpha (α) or beta (β).

In **Figure 8-5**, a simple "common emitter" transistor amplifier shows the respective base, emitter, and collector currents. Keep in mind that I_b is approximately $0.05I_e$, while I_c is approximately $0.95I_e$. By Kirchoff's current law it is also true that:

$$I_e = I_b + I_c \qquad \text{eq. (8-11)}$$

Alpha gain is defined as the ratio of collector current to emitter current. Mathematically,

$$\alpha = \frac{I_c}{I_e} \qquad \text{eq. (8-12)}$$

This number will always be less than one, and where $I_c = 0.95I_e$, it will be exactly 0.95:

$$\alpha = \frac{I_c}{I_e} = \frac{0.95\,I_e}{I_e} = 0.95 \qquad \text{eq. (8-13)}$$

Beta gain is defined as the ratio of collector current to base current:

$$\beta = \frac{I_c}{I_b} \qquad \text{eq. (8-14)}$$

or, in the dynamic case:

$$\beta = \frac{\Delta I_c}{\Delta I_b} \qquad \text{eq. (8-15)}$$

Beta gain will always have a value greater than one.

The relationship between beta and alpha gains can be summarized as:

- $$\alpha = \frac{\beta}{1+\beta}$$ eq. (8-16)

and,

- $$\beta = \frac{\alpha}{1-\alpha}$$ eq. (8-17)

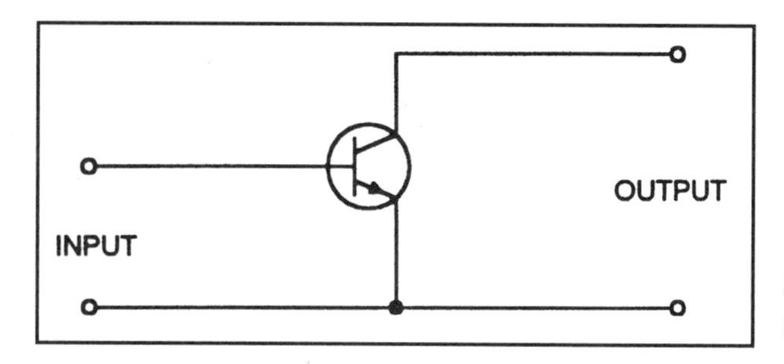

8-5. Common-emitter transistor amplifier.

Most texts label transistors as current amplifiers, which is true, but unfortunately; it supports the erroneous idea that current amplification is the only possible mode for a bipolar transistor. Transistors may also offer substantial amounts of voltage gain, as well as current gain.

Voltage amplification is possible in transistors, even though the device is basically a current amplifier. The transistor can function as a voltage amplifier if a resistance is connected in series with the collector (**Figure 8-6**). Output voltage, V_o, is the difference between the power supply potential, V_{cc}, and the voltage drop at V2 across collector resistor R_c. Assume that V_o is set to a point that is equal to ½V_{cc} when the input signal, V1, is zero. When V1 goes positive, the transistor collector current increases. The voltage across R_c increases. But V_{cc} is a constant, and according to Kirchoff's voltage law $V_{cc} = V2 + V_o$. We must, therefore, expect an increase in V2 to cause a decrease in V_o. The minimum V_o occurs when V1 is maximum.

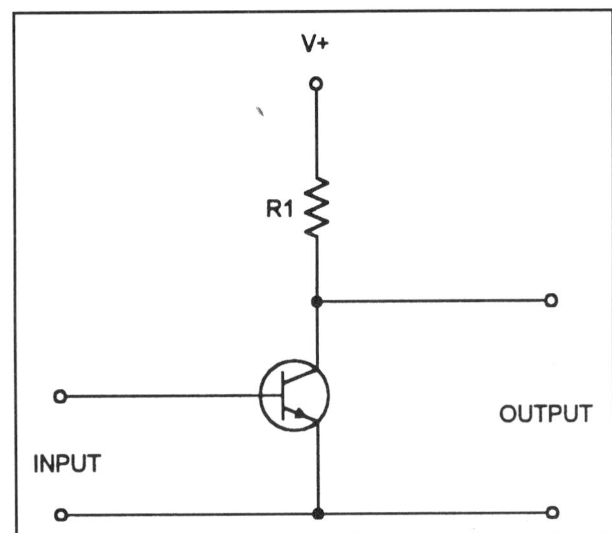

Figure 8-6. Common-emitter amplifier connected as a voltage amplifier.

The inverse occurs when V1 goes negative. The transistor collector current drops, which reduces the voltage drop across R_c, and increases V_o. By varying the collector current flowing in R_c, the output voltage is proportional to either the input current or the input voltage.

The phototransistor is very similar to other NPN or PNP bipolar transistors, except that the device is physically constructed so that light can fall on the base region (**Figure 8-7a**). When photons strike the exposed base region, energy is given up and the region frees electrons, producing a minute current in the region. This current is analogous to the base current injected into ordinary bipolar transistors by an external signal. The current flowing in the c-e path of the phototransistor is controlled by the photon-produced current in the base region.

Figure 8-7b shows three electronic circuit symbols that are commonly used in schematics for phototransistors. They are basically the NPN or PNP transistor symbol with a pair of "light lines" or the wavelength symbol, λ (lambda).

In phototransistors, collector-to-emitter current flows when the base region is illuminated, so it can be used as a light sensor in a variety of applications. These devices are the heart of optoisolator and optocoupler integrated circuits, as well as various linear instrumentation sensor applications.

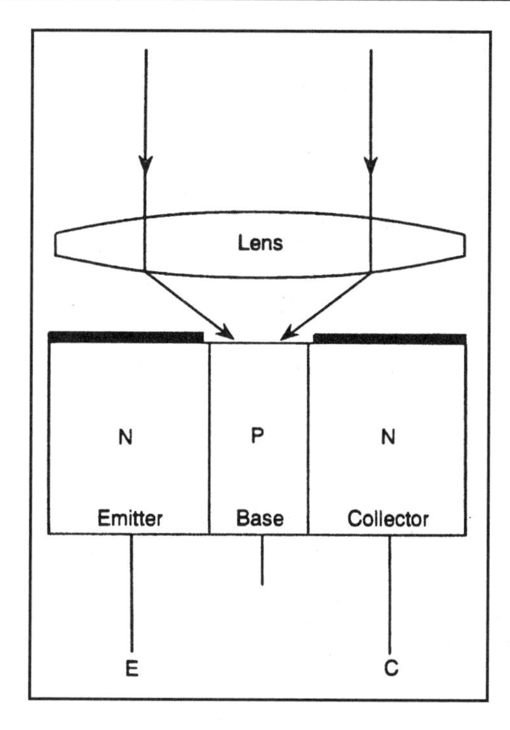

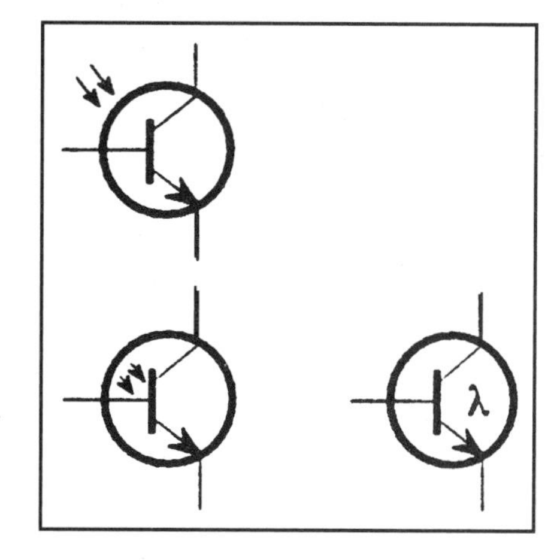

Figure 8-7b. Phototransistor symbols.

Figure 8-7a. Lens used to focus light on active region of phototransistor.

Application Circuits

Figure 8-8a through **Figure 8-8d** shows different ways in which photodiodes and phototransistors are used. The current flow through the device under light conditions is the transducible event, so we must connect them into a circuit that makes use of that property. The inverting follower operational amplifier circuit (**Figures 8-8a** and **8-8b**) serves that function. In all such cases, the output voltage, V_o, is equal to the product of the input current, I_L, and the feedback resistance, R1:

$$V_o = -I_L R1 \quad \text{eq. (8-18)}$$

In the case of **Figure 8-8b**, a zero control is added, and can also be added to the other circuit as well.

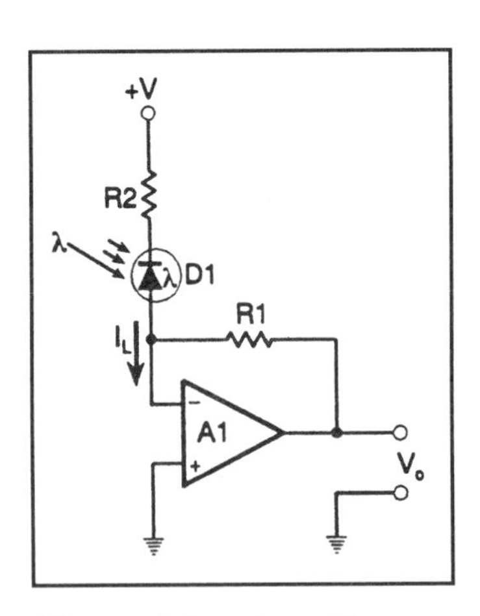

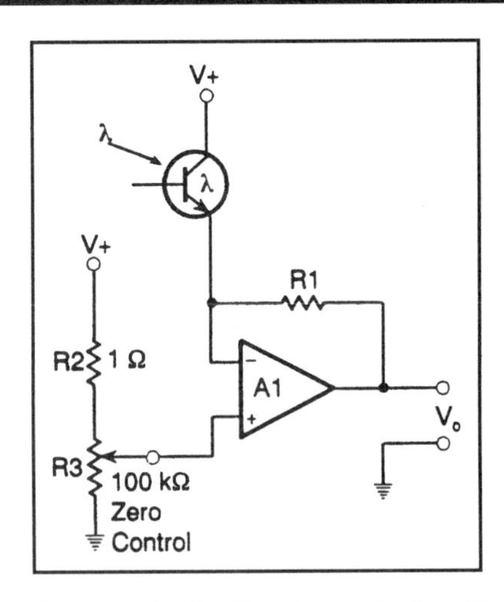

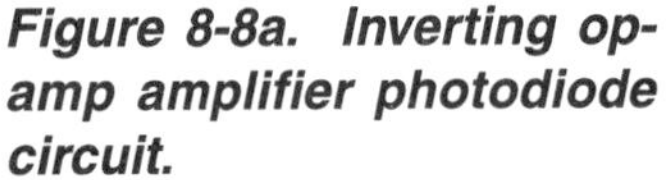

Figure 8-8a. Inverting op-amp amplifier photodiode circuit.

Figure 8-8b. Version of circuit with dc offset control.

The noninverting follower version of this circuit is shown in **Figure 8-8c**. In this circuit, a phototransistor or diode is connected to the noninverting input of the operational amplifier. A voltage drop, V1, is produced by transistor output current, I_L, flowing in resistor R3. This voltage becomes the input signal of the op-amp, and is proportional to the light level applied to the transistor base region. The operational amplifier input voltage for **Figure 8-8c** is:

$$V1 = I_L R3 \qquad \text{eq. (8-19)}$$

and output voltage, V_o:

$$V_o = V1\left(\frac{R2}{R1} + 1\right) \qquad \text{eq. (8-20)}$$

or,

$$V_o = I_L R1\left(\frac{R2}{R1} + 1\right) \qquad \text{eq. (8-21)}$$

Figure 8-8d shows a circuit for use when the light signal is varying naturally or by modulation. An example of the former will be seen in medical electronics in the form of pulse or heart rate meters. These devices, called *photoplethysmographs*, use a red light source transmitting through tissue of a thumb, finger, or ear lobe; a phototransistor senses the light on the other side of the path. When blood pulses through the tissue, it changes its optical density to red light a small amount, and this change can be used for taking the pulse waveform. Modulated light sources are used in alarms and other applications in order to help the circuit distinguish between light from a pulsed emitter or ambient light.

In burglar alarm circuits, if the intruder attempts to defeat the alarm by shining a light into a sensor with a pulsed source emitter, then a constant light, or a light pulsed at the wrong frequency, will sound the alarm.

Figure 8-8d consists of a phototransistor (Q1) driving an ac-coupled noninverting follower operational amplifier circuit. The sensitivity of Q1 can be determined by the resistor R3 in series with the collector and V supply. A high value, 100 kΩ to 1 MΩ, will increase sensitivity, while a low value, 10 kΩ to 100 kΩ, will increase operating speed.

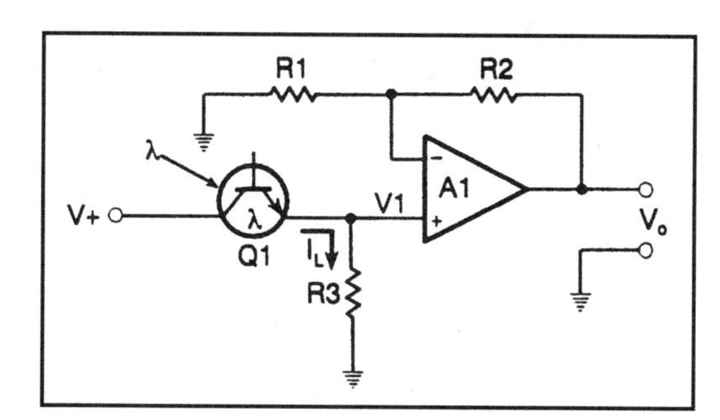

Figure 8-8c. Noninverting amplifier circuit.

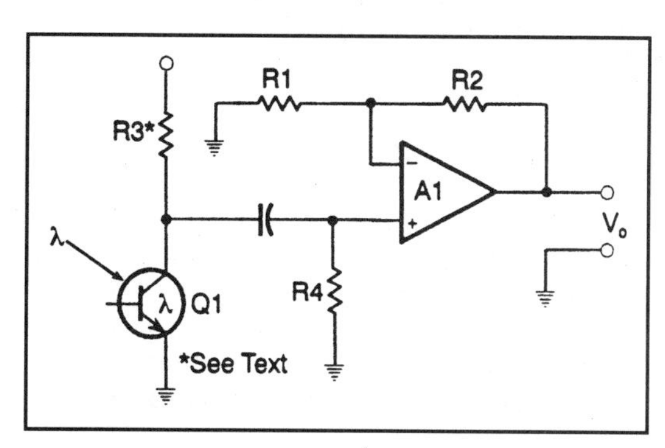

Figure 8-8d. Ac-coupled phototransistor to noninverting amplifier.

The collector of Q1 has a large dc offset, with variation due to stimulus light levels being superimposed on the dc component. The capacitor C1 strips off the dc component, allowing only the variations to pass on to the amplifier. The low frequency response of this circuit is dependent on the capacitor value and the resistance of R4:

$$F_{-3dB} = \frac{1}{2\pi R4\, C1} \qquad \text{eq. (8-22)}$$

Selecting a Sensor

Photodiodes and phototransistors, like other electro-optical sensors, exhibit a marked response curve that is a function of wavelength. **Figure 8-9** shows a typical silicon device response curve in which the maximum response (100

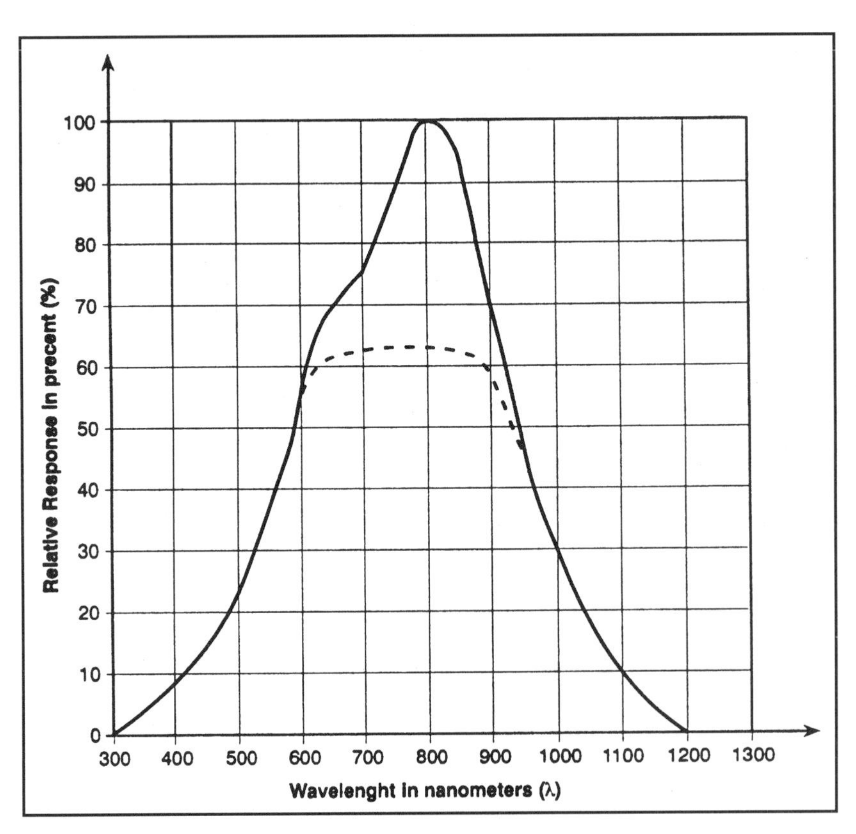

Figure 8-9. Spectral response of phototransistor.

percent) occurs around 880 nanometers. The device achieves maximum sensitivity at wavelengths near this point, but the sensitivity falls off rapidly at points removed. When choosing sensors in these regions, you should not overlook whether the sensitivity is adequate for the application at hand.

The response can be flattened somewhat by placing transmissive filters in front of the sensor that attenuate narrow bands in the vicinity of the peak response (dotted line in **Figure 8-9**). This filtering will broaden the response, but at the expense of sensitivity.

Figure 8-10 shows the spatial response of the sensor. In this case, the peak response occurs when the light source is boresighted to the sensor optical axis. Although the actual region is quite broad, the response is centered to ±20 degrees of the axis. This is another important factor in choosing a sensor.

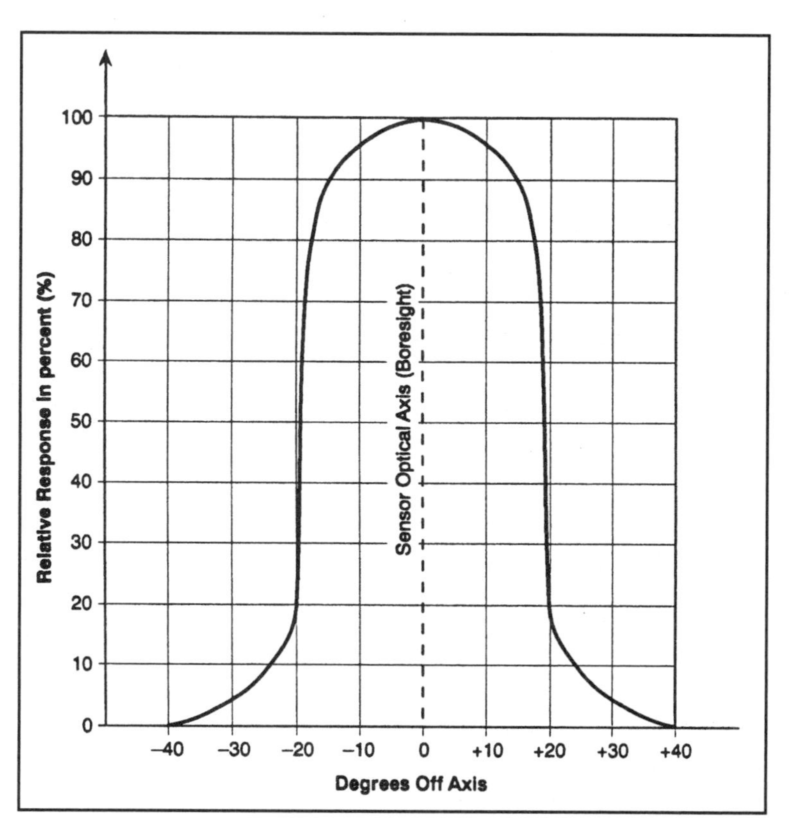

Figure 8-10. Illumination spacial sensitivity of phototransistor.

Chapter 9

Optoisolators, Optocouplers and Optoswitches

Optoisolators, optocouplers, and optoswitches form overlapping categories of electro-optical devices that have a wide range of applications. They are especially useful in situations where the circuit is at high voltage (and therefore damaging to low-voltage electronics), or there is some inherent reason why two circuits should be kept apart from one another.

An example of a high voltage application is one in which a 115 volt ac load is controlled either by low-voltage electronics or a digital computer, which are low-voltage. The high voltage ac line is inherently damaging to the low-voltage electronics, so one can connect the two together for control purposes using an optoisolator.

RS-232C serial communications circuits for computers should be kept apart from either TTL circuits or 20-mA current-loop communications circuits. In the latter circuit, high-voltage transients found in some older teletypewriter circuits will damage digital computers, so optocouplers are used for the interface circuit.

Medical electronics offers yet another example of the need to keep circuits apart from each other. For patient safety, the amplifiers used for electrocardiograph (ECG), electroencephalograph (EEG), and other instruments are isolated from the ac-powered dc power supply by a very high impedance. Industrial and medical amplifiers called *isolation amplifiers* use optoisolators as the isolation element.

Optoisolators and Optocouplers

In common usage the terms optoisolator and optocoupler are interchangeable, although different usages are found in some retail catalogs. All of these devices include a light source, such as an incandescent lamp, neon glow lamp or light emitting diode, juxtaposed with an electro-optical sensor, such as a photoconductive resistor, photodiode, phototransistor or other device. These are usually packaged so that only the light from the internal source finds its way onto the active surface of the sensor.

Figure 9-1 shows four different optoisolators. In **Figure 9-1a**, a small incandescent low-voltage lamp (IL1) is positioned to shine its light on a photoconductive CdS cell. When the lamp is off (no current), the photoconductive cell is dark, so its resistance (measured across terminals Y1—Y2) is very high. But when the lamp is on, the photocell is illuminated, and its resistance drops very low.

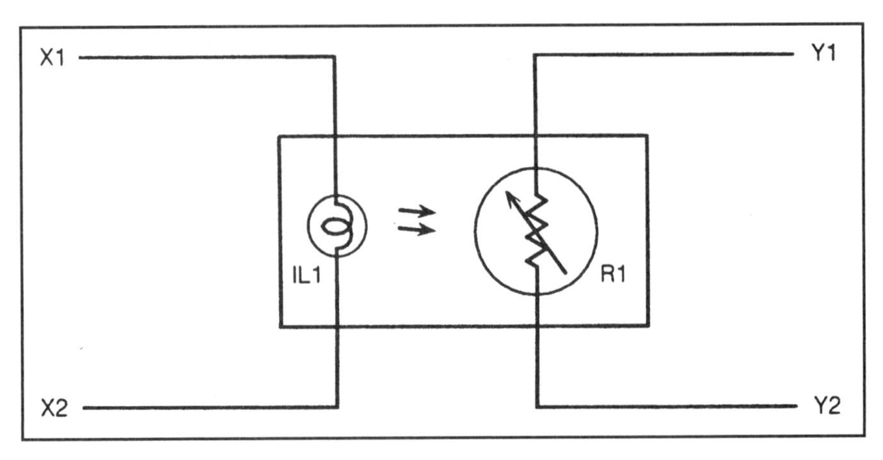

Figure 9-1a. Incandescent/photoresistor optoisolator.

The optoisolator of **Figure 9-1a** is its oldest form. Commercial examples are nearly as old as photoconductive cells. They are also inefficient and unreliable. The inefficiency is due to the amount of current (40 to 500 mA) the lamp needs to glow to incandescence. Therefore, a large amount of heat is generated. The lack of reliability is due largely to the fact that the lamp in it has a high burn-out rate.

The problems of **Figure 9-1a** are solved nicely in the device of **Figure 9-1b**. This device uses D1, an LED, to shine on the photoconductive cell, R1. As current through the LED increases, its brightness also increases, and the photoconductive cell's resistance will drop.

Both **Figure 9-1a** and **Figure 9-1b** are sometimes used as voltage variable resistors. As the voltage across the lamp or LED increases, the current and light output also increases, and the resistance of R1 drops. Therefore, the resistance seen across Y1-Y2 in both cases is inversely proportional to the voltage applied across X1-X2.

In **Figure 9-1b**, the positive terminal of the dc power supply must be connected to X1, and the negative to X2. A current limiting resistor should be connected in series with the LED to keep current below the maximum permissible level (typically 15 to 20 mA).

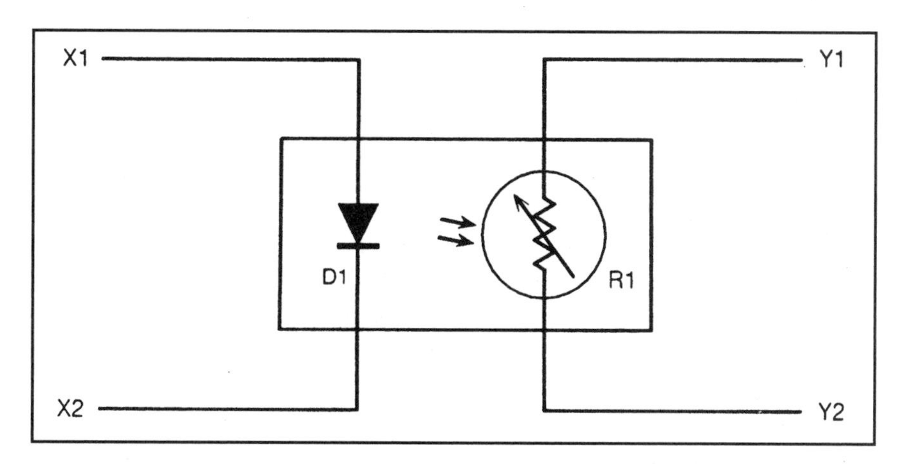

Figure 9-1b. LED/photoresistor optoisolator.

An optocoupler based on an LED, D1, and a phototransistor, Q1, is shown in **Figure 9-1c**. The LED shines its light onto the active base region of the phototransistor. When the LED is bright, then a base current flows in the transistor, causing current in the emitter-collector path.

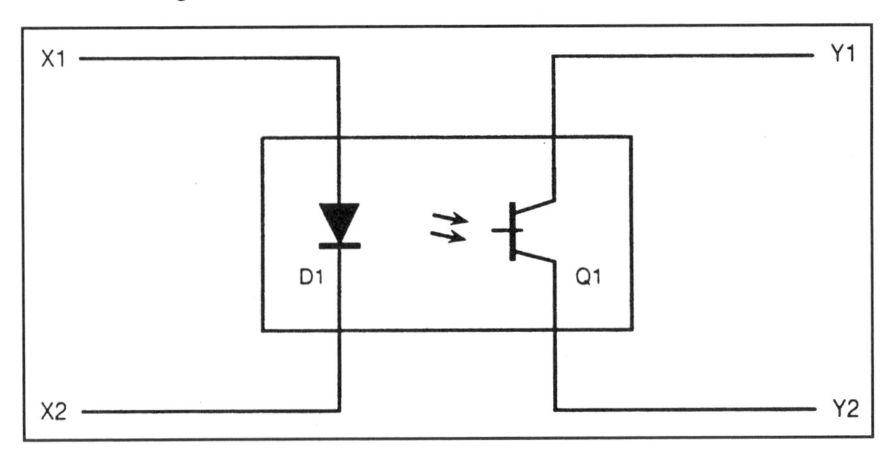

Figure 9-1c. LED/transistor optoisolator.

The next device (**Figure 9-1d**) is an optocoupler in which the active output device, Q1, is a *junction field effect transistor* (JFET). This device will act as an amplifier and a voltage variable resistor. The JFET channel resistance appearing between Y1 and Y2 is a function of the current flowing in the LED.

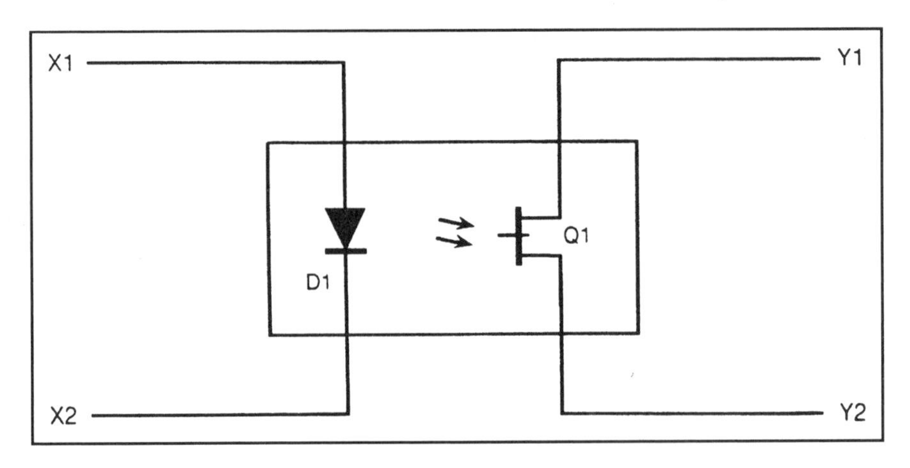

Figure 9-1d. LED/FET optoisolator.

Figure 9-2 shows other devices that are generally classified as optoswitches. In all three cases, the light source is an LED. In **Figure 9-2a**, the sensor is a light sensitive *silicon controlled rectifier* (SCR). These devices are gated PN diodes that don't pass current in either direction until a trigger current flows in the gate terminal. After triggering, the device operates like any PN junction diode, passing current in only one direction. In **Figure 9-2a**, the gate current is created by the light source. Many of these devices are photodiodes connected to an internal SCR gate.

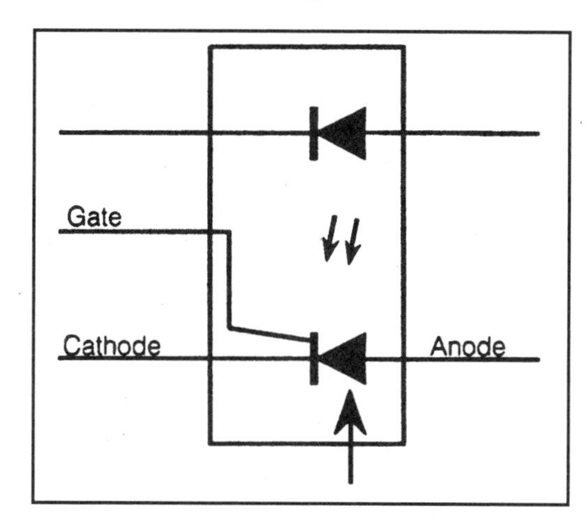

Figure 9-2a. LED/SCR switch circuit.switch.

A *triac* is a special form of SCR-like device. Where the SCR is a PN junction diode, passing current in only one direction, the triac uses a pair of back-to-back SCRs with a common gate. **Figure 9-2b** shows an optical switch that uses a photosensitive triac as the sensor element. This device is used when the load needs ac power.

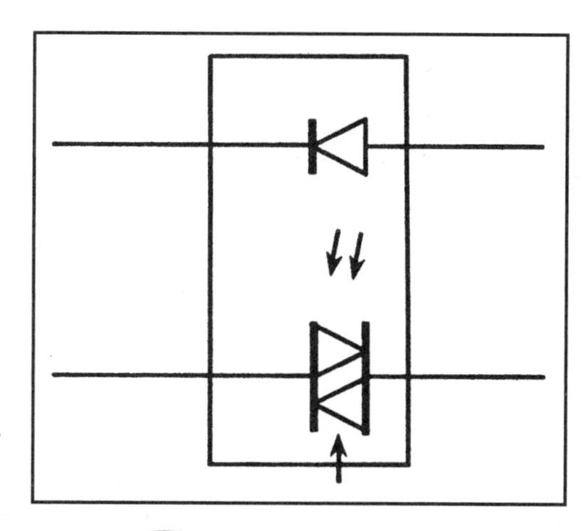

Figure 9-2b. LED/ triac switch circuit.

Finally, in **Figure 9-2c** we see an optoswitch that uses a digital logic element called a *Schmitt trigger*. These devices will not produce a HIGH output until the input (in this case, a light level) rises about a certain ON threshold. The output remains HIGH until the light drops below an OFF threshold (which may be different from the ON level). The Schmitt trigger is used to clean up signals in noisy environments, where spurious emissions and random signals are prevalent. If the trigger levels are too low, as they might be under some circumstances in TTL devices, then some noise pulses might trigger a false output.

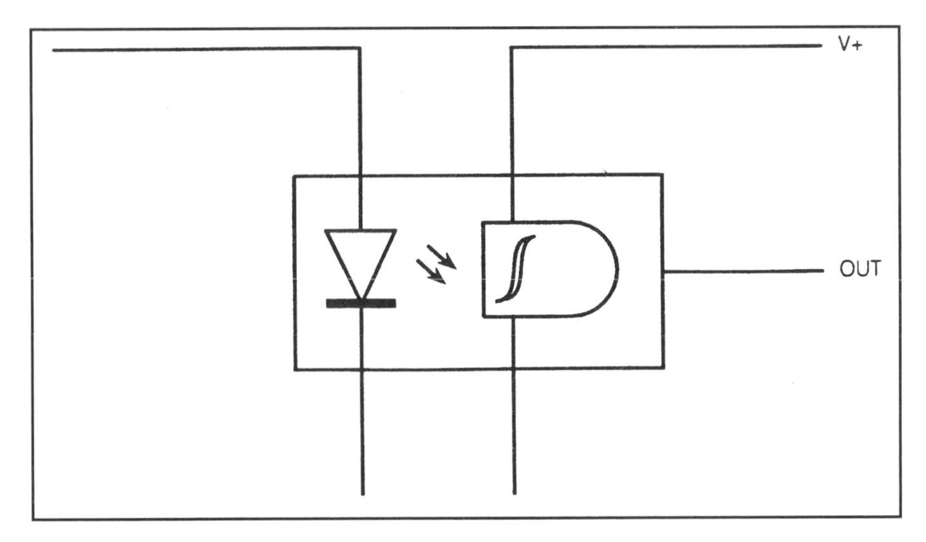

Figure 9-2c. LED/Schmitt trigger circuit.

Styles of Optoisolator

Figure 9-3 shows three different styles of optoisolator or optocoupler. The simplest type is the *closed-pair* in **Figure 9-3a**. A light source (an LED) is placed so that its principal axis is shining into the lens window of a phototransistor device. When the LED is ON, the phototransistor is illuminated. These devices are typically used in circuits where isolation between two circuits is required.

A *transmission-slot pair* is shown in **Figure 9-3b**. In this type of device, a slot is cut in the light transmission path between the LED light source and the phototransistor. When an opaque object is placed in the slot, light transmis-

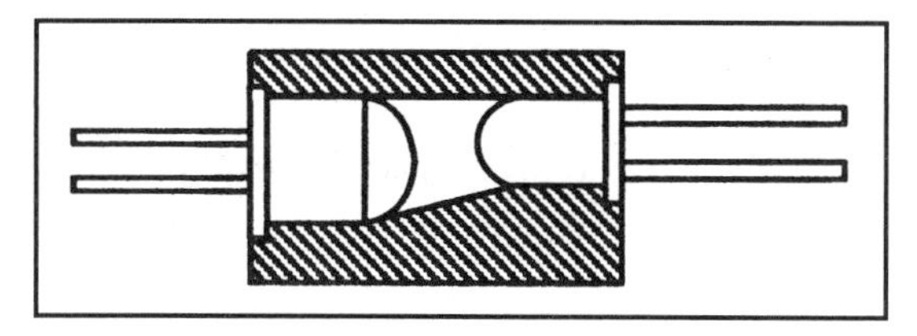

Figure 9-3a. Direct optocoupler construction.

sion is interrupted so the phototransistor is not illuminated. When the object is removed, the transistor is illuminated. Such devices are used in "paper out" alarms in printers, in alarm circuits indicating when a window or door is opened and other applications.

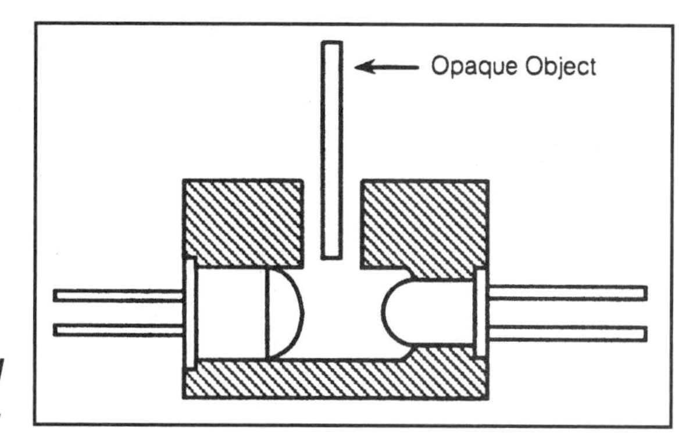

Figure 9-3b. Interrupted optocoupler construction.

A *reflective pair* is shown in **Figure 9-3c**. This device is used in proximity detector circuits where the sensor/pair is not supposed to touch the object. The reflective pair is also used in a type of sensor for blood pressure waveform called a photoplethysmograph (PPG), described in detail in Chapter 11.

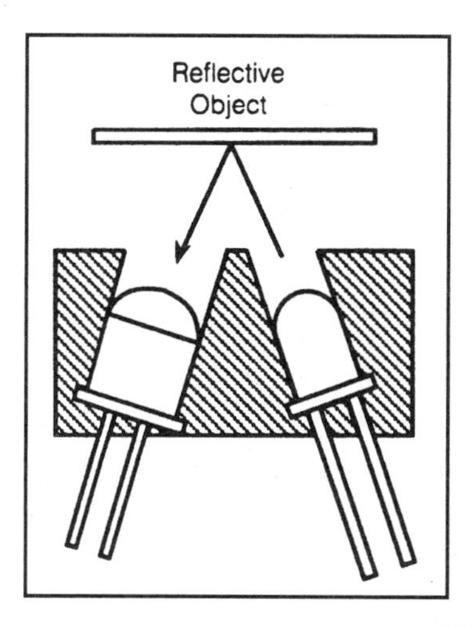

Figure 9-3c. Reflective optocoupler construction.

Figure 9-4 shows an optocoupler for the home electronics hobbyist. A red LED is placed inside a drilled out dowel, opaque plastic rod, piece of insulating "spaghetti" material or heat- shrink tubing with a phototransistor. Seal the ends with opaque silicone seal, if necessary.

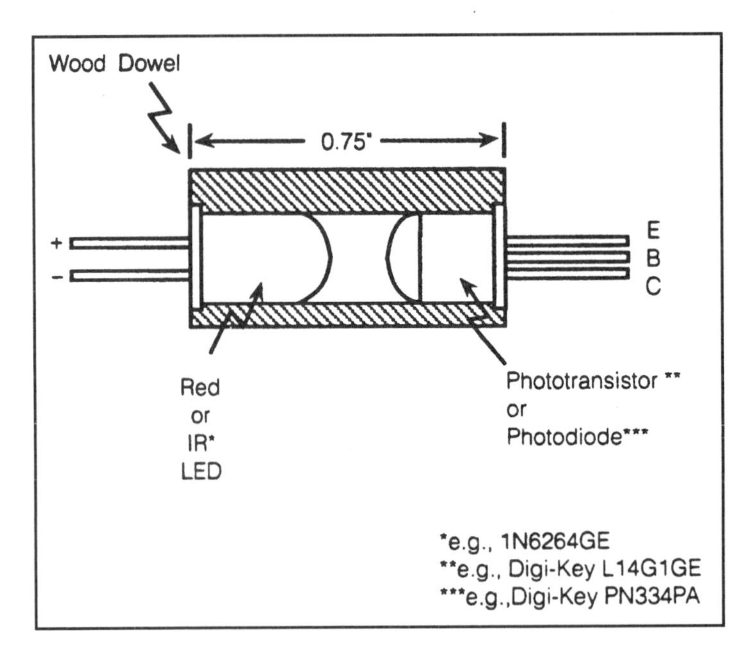

Figure 9-4. Homemade optocoupler.

Optoisolator/Optocoupler Circuits

A variable isolated resistor is shown in **Figure 9-5**. In this application, a device is used that has an LED as the light source (D1) and a photoconductive cell (photoresistor R1) as the sensor. The brightness of the LED determines the resistance between terminals A-B, and is controlled by the current through the LED. The LED current is set by a dc power supply, and a pair of resistors (R1 and R2). The fixed resistor, R1, is used to limit the current to 20 mA when the resistance of the potentiometer, R2, is at zero ohms. Otherwise, the LED might burn out when the current reaches too high a value. The brightness of the LED and the resistance of the photocell are a function of the setting of potentiometer R2.

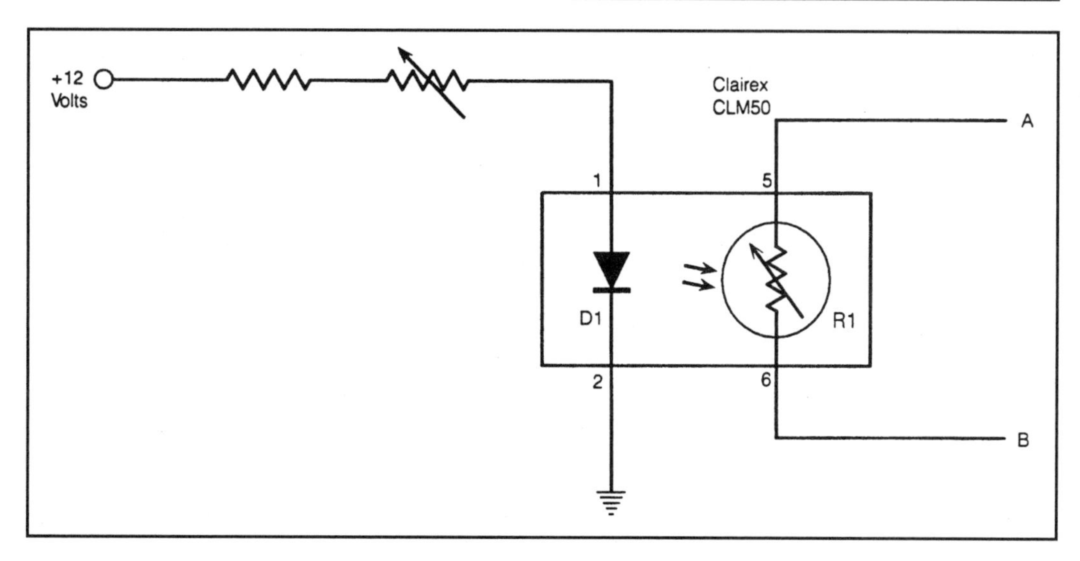

Figure 9-5. Isolated variable resistor.

Figure 9-6 shows two circuits for exciting the LED in any form of optoisolator, regardless of the type of sensor on the other side. In **Figure 9-6a**, the cathode of the LED is grounded, and a positive voltage is applied to the current limiting resistor, R1, connected to the LED anode. The normal voltage drop, V_d, across the LED is about 1.7 volts. The value of R1 is found from:

$$R1 = \frac{(V+) - V_d}{I_L} \quad \text{eq. (9-1)}$$

The current I_L is found from the specifications sheet for the specific device being used, but is typically 10, 15, or 20 mA.

Example 9-1

Find the resistance needed in series with a normal LED of an optocoupler if the maximum allowable current is 15 mA and V+ is 12 volts.

Solution:

- $$R1 = \frac{(V+) - V_d}{I_L}$$

- $$R1 = \frac{(12\ volts) - (1.7\ volts)}{0.015\ Amperes}$$

- $$R1 = \frac{10.3\ volts}{0.015\ amperes} = 687\ ohms$$

The nearest standard resistor value is 680 ohms.

When a digital signal is applied to the LED instead of a dc voltage, the LED is ON when the signal is HIGH (a positive voltage), and OFF when the signal is LOW (near ground potential).

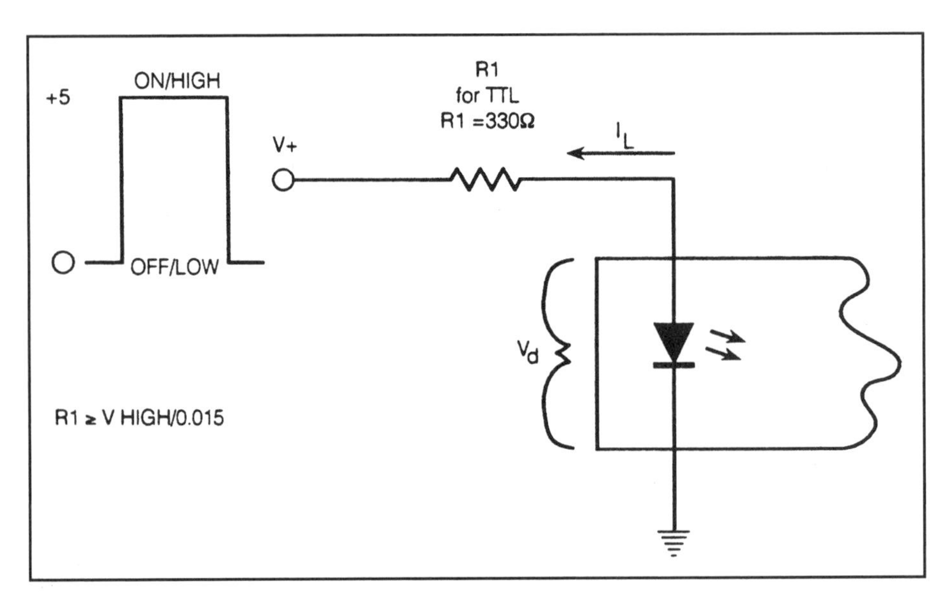

Figure 9-6a. Digital LED driver: ON = HIGH.

A second variation of the circuit is shown in **Figure 9-6b**. In this case, the cathode is not permanently grounded; a fixed V+ potential is applied to the resistor at the anode side of the LED. When the cathode is LOW (grounded), the LED is ON. But when the cathode is HIGH the LED is OFF.

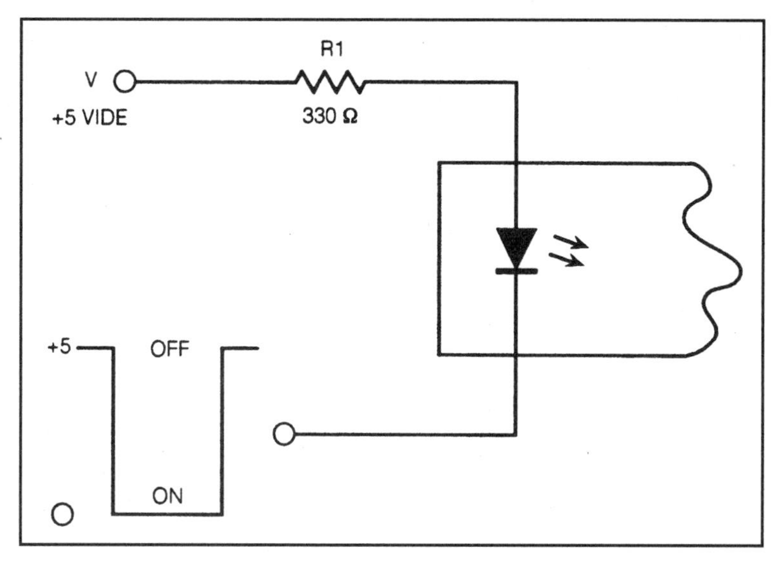

Figure 9-6b. Digital LED driver: ON = LOW.

Figure 9-7 shows the complete circuit for optocouplers and optoisolators. The LED side of the circuit is similar to the previous circuit in both **Figure 9-6a** and **Figure 9-6b**. Note that there are separate dc power supply sources, +V1 and +V2, in each circuit. If D1 and Q1 are connected to the same dc power supply, then the isolation feature is lost.

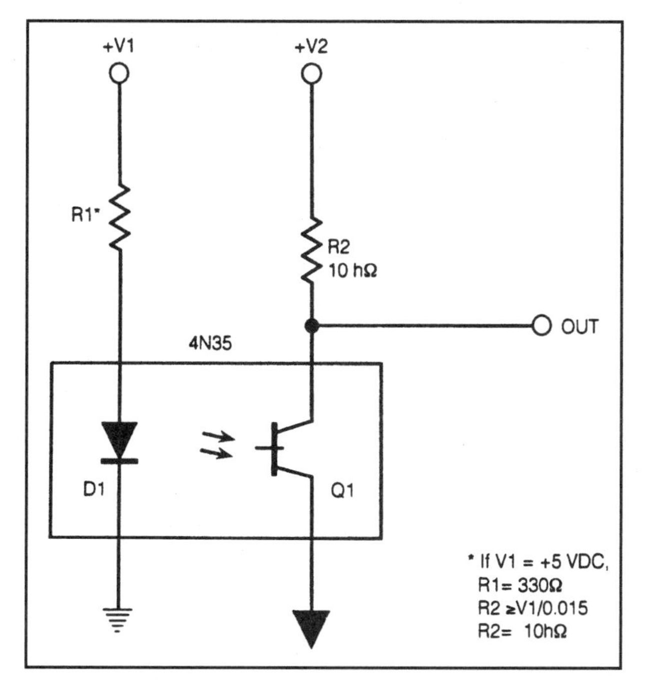

Figure 9-7a. Collector output optocoupler.

In **Figure 9-7a**, the phototransistor, Q1, is connected so that the output signal is taken from the collector of the transistor. The emitter of Q1 is directly grounded, while the collector is connected to a dc power supply, +V2. This configuration is a common *emitter-collector output circuit*.

In **Figure 9-7b**, the output signal is taken from the emitter. Note that a resistor is connected between the emitter and ground. This is the *emitter-follower output* configuration.

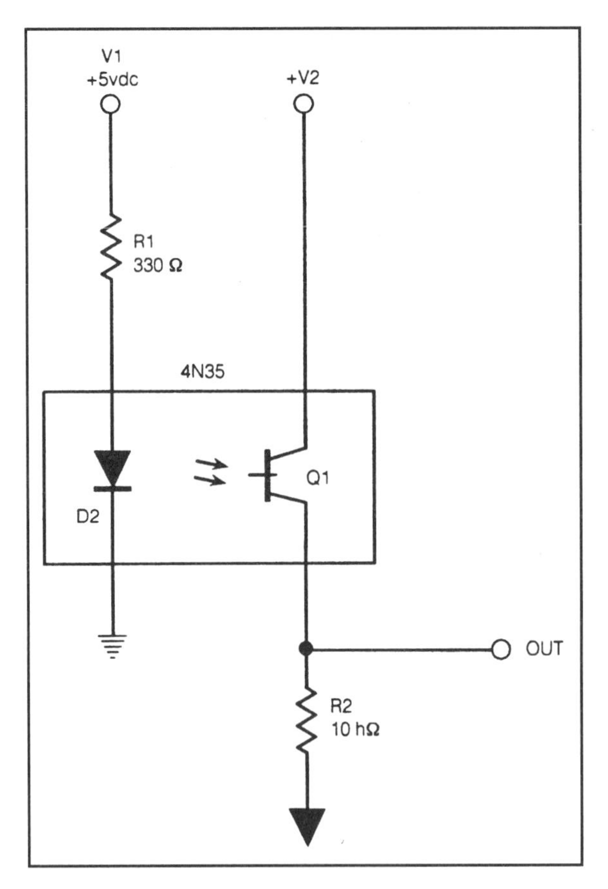

Figure 9-7b. Emitter output optocoupler.

Packaging of Optocouplers and Optoisolators

Over the years these components have had a wide variety of packaging, including a cylinder the size of a broken pencil. The incandescent lamp style heated to a rather warm temperature on the lamp end! Modern devices, however, are typically packaged in integrated circuit *dual inline packages* (DIP), most of which are of the miniDIP variety.

Figure 9-8a shows a typical six-pin miniDIP optocoupler. This device uses an LED juxtaposed with a phototransistor, with the LED situated between pins 1 and 2, and the phototransistor emitter and collector between pins 4 and 5; some models have the transistor base at pin 6. These are "industry standard" locations (pin-outs), but always check the specific component for individual pin-outs.

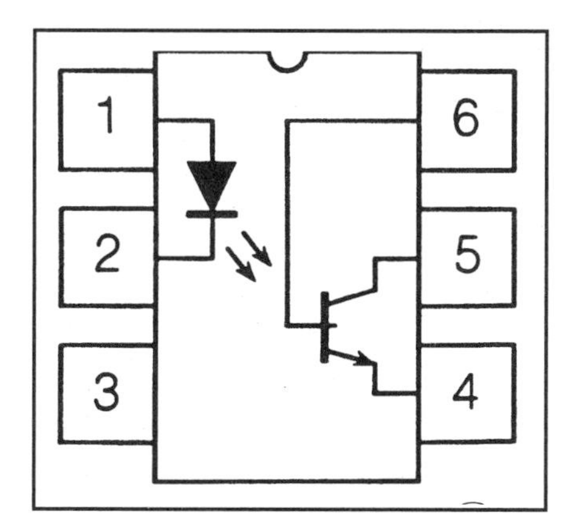

Figure 9-8a. 6-pin miniDIP optocoupler package.

A different sort of package is shown in **Figure 9-8b**. This device is a four-pin miniDIP package, and consists of a phototransistor as the sensor and a pair of LEDs as the source. The two LEDs are connected back-to-back so that voltages of both polarities or an ac current are accommodated.

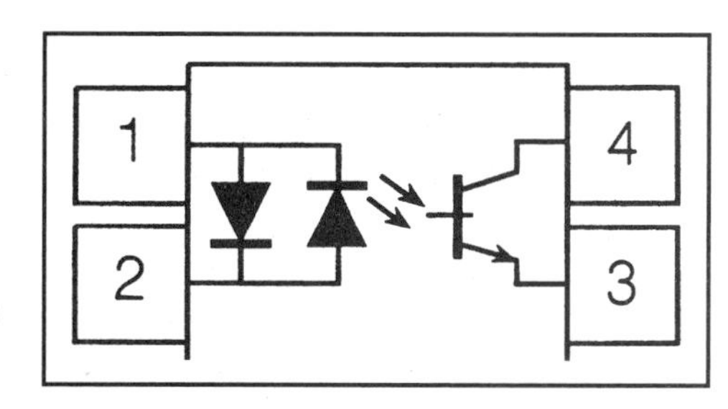

Figure 9-8b. 4-pin miniDIP package.

A multiple device is shown in **Figure 9-8c**. This is similar to the device shown singly in **Figure 9-8a**, but it contains four LED/transistor pairs in a 16-pin DIP package. Versions with up to sixteen pairs are on the market. These devices can be used to control multiple circuits or to isolate multiline data. For example, in an eight bit computer, two of these devices will provide isolation for all eight lines of one bit each.

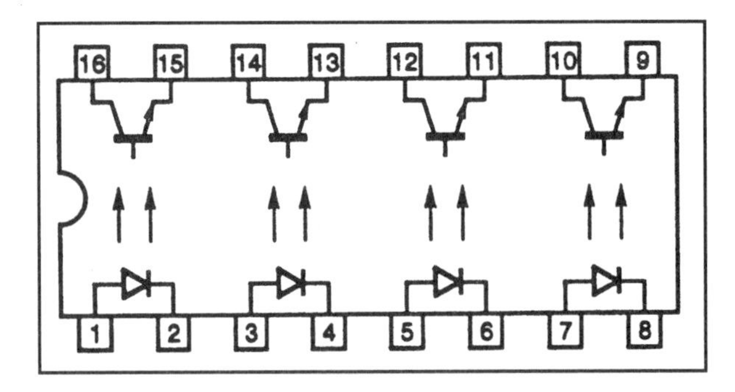

Figure 9-8c. Quad optocoupler package.

Optocoupler Application Circuits

Figure 9-9 shows the circuit for a TTL-to-TTL isolator. In this circuit, a TTL level drives the LED in the optoisolator. The driver circuit is an open-collector TTL inverter (G1). When the input is HIGH, the output of the inverter is LOW, and grounds the cathode end of the LED, causing the LED to turn on. Resistor R1 limits the LED current to the maximum allowable value. The dc potential, +V1, can be +15 or +30 volts or less with some inverters (consult the inverter specifications sheets or Don Lancaster's *TTL Cookbook*[1]).

The output circuit is the collector output form shown earlier, but in this case, the collector is also connected to the input of a TTL inverter, or other TTL digital logic input. TTL inputs are current sources (typically 1.8 mA), and will supply current to the collector when the transistor is LOW (the LED is ON). A pull-up resistor to a +5 volt power supply, +V2, keeps the input of the device HIGH when the transistor is OFF.

The output device shown in **Figure 9-9** is a TTL inverter. It can be any form of TTL device that has a fan-in (input loading) of one.

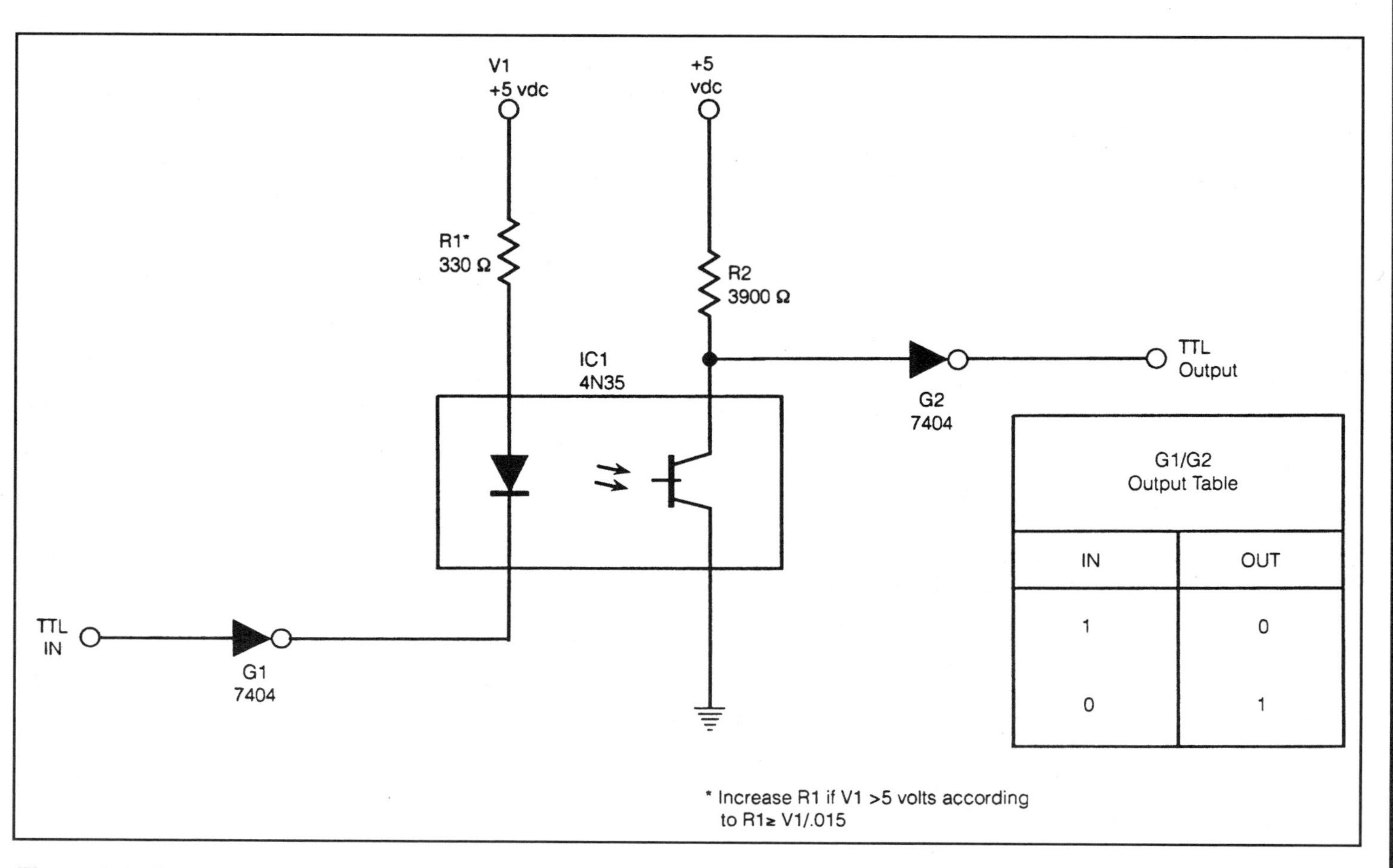

Figure 9-9. TTL-to-TTL isolated circuit.

Figure 9-10 through **Figure 9-12** shows three different communications interface circuits using optocouplers. These devices are often used to isolate the two communications circuits from each other. In **Figure 9-10** the 4N35 optocoupler is driven by a TTL input data signal, and drives a 20 mA current-loop serial communications circuit.

This circuit is commonly used in instruments or in teletypewriter systems. An open-collector TTL inverter is connected to the LED cathode. When the input is HIGH, the TTL inverter output is LOW and the LED is turned ON. When the LED is ON, the phototransistor is illuminated, and is saturated. The loop is closed when the LED is ON and the input is HIGH. On the other hand, when the input of the circuit is LOW, the inverter output is HIGH and the LED is OFF. The transistor is also not conducting, so the loop is open.

The 1N4007 high-voltage diode across the phototransistor of the 4N35 optocoupler is used to suppress high-voltage transients. Teletypewriters used solenoids excited by direct current. When the solenoids are de-energized, there is a counterelectromotive force (CEMF) "inductive kick" spike generated that can destroy or disrupt the circuits connected to it. The diode reduces the spike to about 0.6 volt.

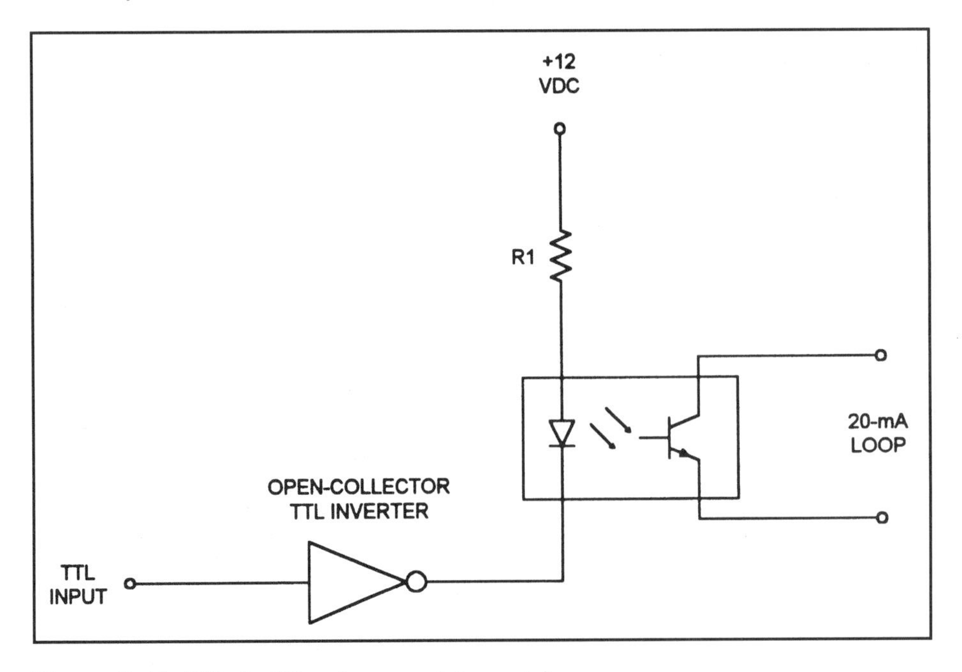

Figure 9-10. TTL to 20-mA current-loop driver.

The opposite case is shown in **Figure 9-11**. In this circuit, the input side is the 20-mA current loop driving the LED. When the loop is closed, the LED is turned on, but when the loop is open the LED is turned off. Note that the input is polarity sensitive because of the LED. The output circuit is the collector output transistor used to drive a TTL inverter. The 0.01-µF capacitor, C1, is used to reduce noise problems. Sometimes a 220 ohm resistor is placed in series with the LED to prevent accidental overload of the LED.

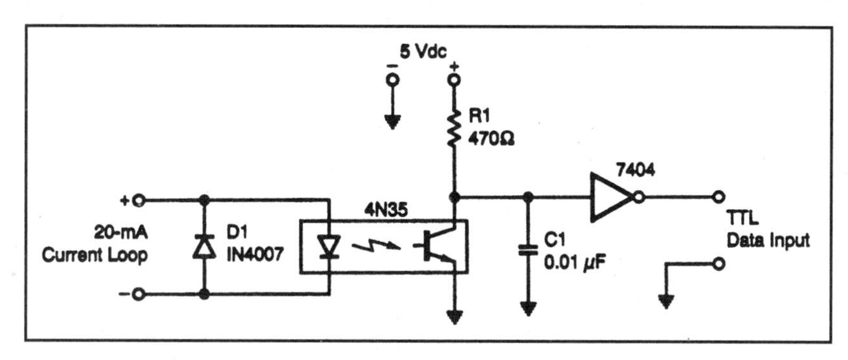

Figure 9-11. 20-mA current loop to TTL driver.

A level translator circuit for driving the standard RS-232C serial communications port used in modern computers is shown in **Figure 9-12**. In this case, the input circuit is similar to the previous 20 mA circuit, but the output circuit is an operational amplifier that supplies the V- and V+ levels that RS-232C defines as HIGH and LOW, respectively. A +5 volt bias applied to the noninverting input of operational amplifier, A1, sets a minimum action threshold.

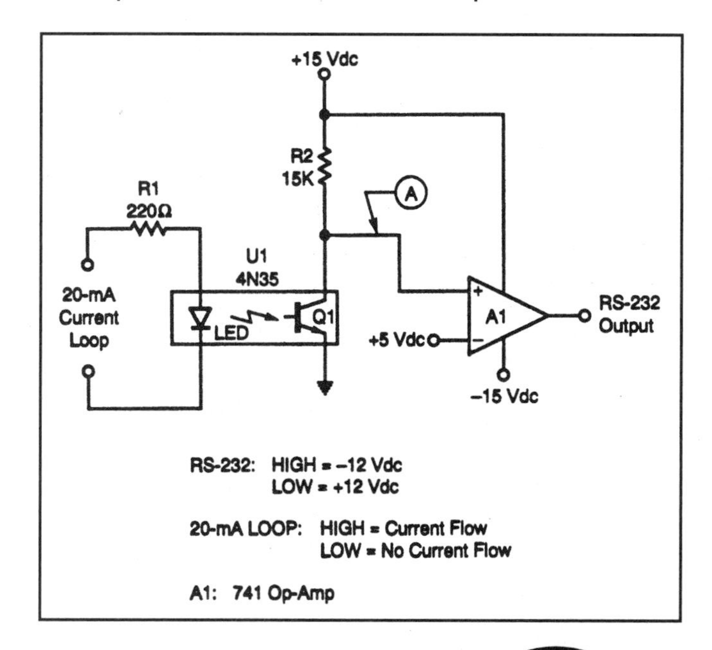

Figure 9-12. 20-mA current loop to RS-232C driver.

An isolated relay circuit for controlling high voltage or high current from a low-voltage electronic circuit or computer is shown in **Figure 9-13**. In **Figure 9-13a**, the output transistor (Q1) is connected to a driver transistor (Q2) such that Q1 operates as an emitter follower. When transistor Q1 is saturated because the LED is turned on, a voltage appears at the base of Q2, turning it on. The load for this transistor is the actuating coil of an electromechanical relay (K1). In the case shown, both normally open and normally closed contacts are used with the common. The diode across the coil is used for spike suppression.

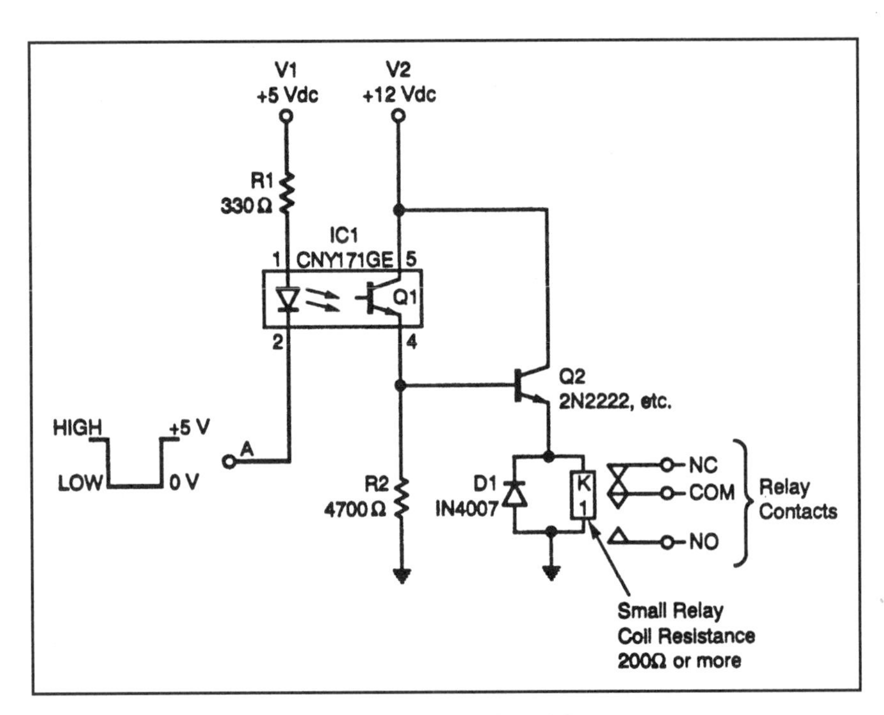

Figure 9-13a. Isolated mechanical relay driver.

An SCR version of the circuit is shown in **Figure 9-13b**. The load is connected in series with the PN junction portion of the internal SCR. When the LED is turned on, the SCR is triggered causing a half-wave rectified current in the load (if ac is needed, then a triac device is used instead of an SCR device). SCRs stay ON once triggered, until the anode-cathode current drops below a pre-set minimum holding value. The normally closed switch (S1) is used to "commutate" the diode (turn it off) when necessary.

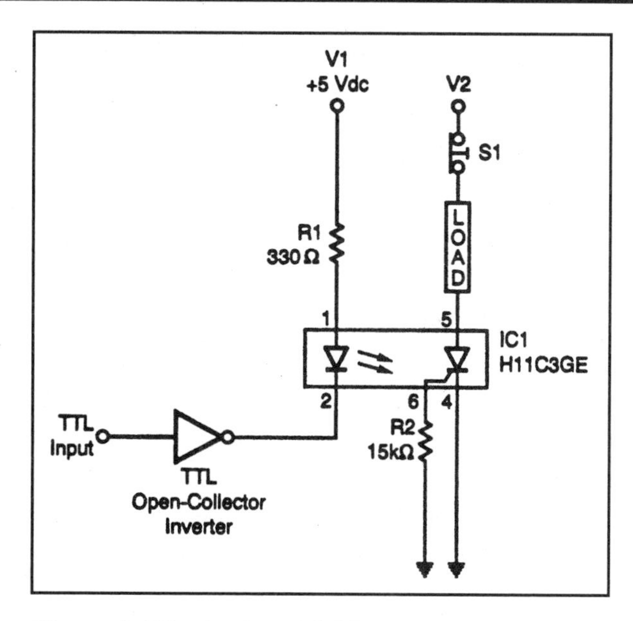

Figure 9-13b. Isolated SCR driver circuit.

Analog Multiplier Circuit

Analog multiplier and divider circuits are available in both monolithic integrated circuit and hybrid circuit forms. Analog multipliers produce an output voltage, V_o, that is the product of two input voltages, V_x and V_y. The general form of the multiplier transfer function is:

$$V_o = K V_x V_y \qquad \text{eq. (9-2)}$$

Where:

V_o is the output potential in volts.

V_x is the potential (in volts) applied to the X-input.

V_y is the potential (in volts) applied to the Y-input.

K is a constant (usually 1/10).

If the proportionality constant K is 1/10, then **Equation 9-2** becomes:

$$V_o = \frac{V_x V_y}{10} \qquad \text{eq. (9-3)}$$

A simple and effective multiplier and divider circuit can be constructed from a pair of operational amplifiers and a special gain-setting block (see **Figure 9-14**), consisting of a pair of matched photoresistors (R1 and R2), both which are illuminated by the same incandescent lamp (IL1). Because R1 and R2 are matched for any given light level, it is true that R1 = R2 = R. Because R1 is in the feedback loop of amplifier A2, nonlinearities in the voltage versus brightness ratio of IL1 are effectively "servoed-out."

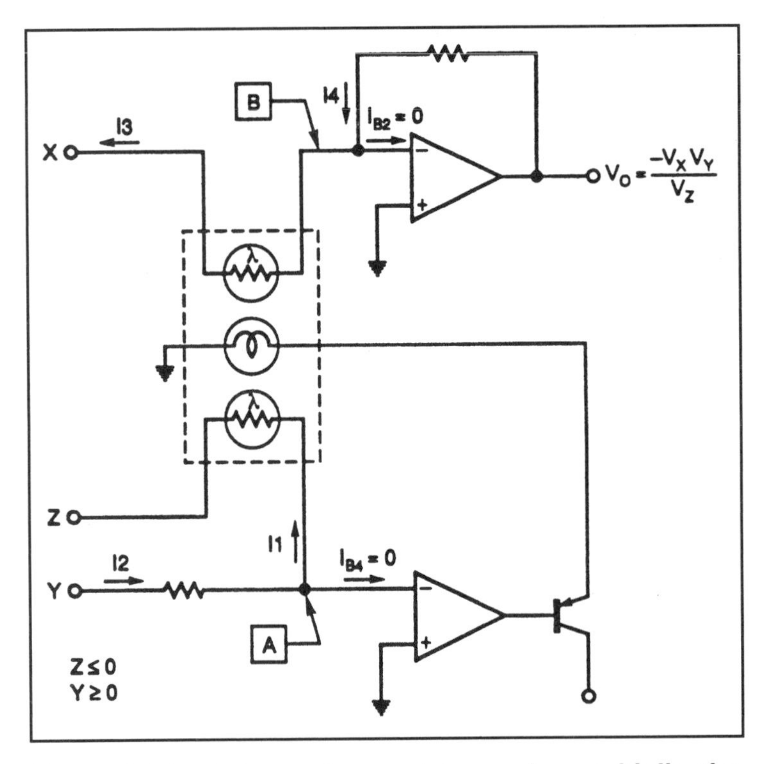

Figure 9-14. Optocoupler used for analog multiplication circuit.

In **Figure 9-14**, the input bias current of each amplifier (I_{b1} and I_{b2}) is zero. In addition, the noninverting inputs of both amplifiers are grounded, so the two summing junctions (A and B) are at zero-volt (ground) potential. By Ohm's law, and considering that R1 = R2 = R:

- $$I1 = \frac{V_z}{R}$$ *eq. (9-4)*

and,

- $$I2 = \frac{V_y}{R3}$$ *eq. (9-5)*

By Kirchoff's current law:

- $$I1 = -I2$$ *eq. (9-6)*

Substituting **Equation 9-4** and **Equation 9-5** into **Equation 9-6**:

- $$\frac{V_z}{R} = \frac{-V_y}{R3}$$ *eq. (9-7)*

and, rearranging terms:

- $$R = \frac{-V_z R3}{V_y}$$ *eq. (9-8)*

Current I3 is, by Ohm's law:

- $$I3 = \frac{V_x}{R}$$ *eq. (9-9)*

So, by substituting **Equation 9-8** into **Equation 9-9**:

$$I3 = \frac{V_x}{\left(\frac{-V_z R3}{V_y}\right)} \qquad \text{eq. (9-10)}$$

and, then rearranging terms:

$$I3 = -\left(\frac{V_x V_y}{V_z R3}\right) \qquad \text{eq. (9-11)}$$

Because $I_{b1} = 0$, and because point B is at ground potential, by Kirchoff's current law it can be stated:

$$I3 = -I4 \qquad \text{eq. (9-12)}$$

From Ohm's law:

$$I4 = \frac{V_o}{R4} \qquad \text{eq. (9-13)}$$

By substituting **Equation 9-11** and **Equation 9-13** into **Equation 9-12**:

$$-\frac{V_o}{R4} = -\frac{V_x V_y}{V_z R3} \qquad \text{eq. (9-14)}$$

or, by rearranging terms:

$$V_o = \frac{V_x V_y R4}{V_z R3} \qquad \text{eq. (9-15)}$$

Accounting for V_z restricted to negative values:

$$V_o = -\frac{V_x V_y R4}{V_z R3} \qquad \text{eq. (9-16)}$$

Placing **Equation 9-16** into the standard form:

$$V_o = -\frac{K V_x V_y}{V_z} \qquad \text{eq. (9-17)}$$

In which K = R4/R3.

From **Equation 9-17** we can see clearly that the circuit is an analog multiplier. It can also function as an analog divider under the correct circumstances. Use of this circuit may not be practical now that analog multipliers are available in integrated circuit form, but the circuit does illustrate the process of analog multiplication and division, and a clever use for optical isolators.

References

Lancaster, Don. *TTL Cookbook.* Carmel, IN, SAMS, Division of Macmillan Publishing, 1974.

Chapter 10

Electronic Amplifiers for Electro-Optical Circuits

Few electro-optical sensors can be used directly in an application without a supporting electronic circuit. In this chapter we will examine amplifier circuits that are suitable for use with E-O sensors. In addition, we will look at certain other circuits used in signal processing of E-O signals, other than amplification. This subject is expanded in Chapter 12 where we will discuss integrators, differentiators, logarithmic amplifiers, and laboratory amplifiers.

The basis for the discussion in this chapter is the *operational amplifier*, or op-amp. These integrated circuit (IC) devices are very popular, and are very easy to apply in both E-O and other electronic circuits. This chapter's material is not specific to E-O circuits, but some examples are provided. We will begin our discussion with the properties of the "ideal" op-amp device.

Properties of the "Ideal" Operational Amplifier

An ideal operational amplifier is an integrated circuit gain block that has the following general properties:

1. Infinite open-loop (no feedback) gain ($A_{vol} = \infty$).
2. Infinite input impedance ($Z_{in} = \infty$).
3. Zero output impedance ($Z_o = 0$).
4. Infinite bandwidth ($f_o = \infty$).
5. Zero noise generation.
6. Differential inputs that follow each other.

Of course, it is not possible to meet these properties—the properties are, after all, ideal—but if we read "infinite" as "very, very high" and "zero" as "very, very low," then the approximations of the ideal situation are close. Working IC operational amplifiers can have an open-loop voltage gain from 50,000 to over 1,000,000, so it can be classed as relatively infinite, and the equations work in most cases.

Differential Inputs

Figure 10-1 shows the basic symbol for the common operational amplifier, including power terminals. In many schematics of operational amplifier circuits, the V_{cc} and V_{EE} power terminals are deleted, so the drawing will be less "busy."

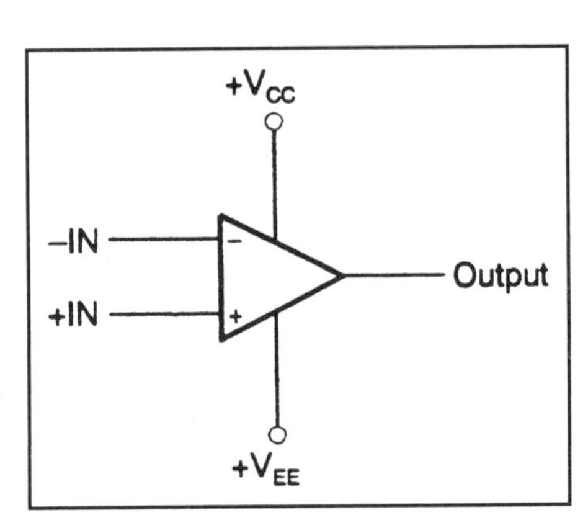

Figure 10-1. Operational amplifier schematic circuit symbol.

Note that there are two input terminals, which are labelled (-) and (+). The terminal labeled (-) is the inverting input. The output signal will be out of phase with signals (there will be a 180-degree phase shift) applied to this input terminal. The terminal labeled (+) is the noninverting input, so output signals will be in phase with signals applied to this input. It is important to remember that these inputs look into equal open-loop gains, so they will have equal but opposite effects on the output voltage.

This characteristic implies that the two inputs will behave as if they were at the same electrical potential, especially under static conditions. In **Figure 10-2** we see an inverting follower circuit in which the noninverting (+) input is grounded. The sixth aforementioned property requires us to treat the inverting (-) input as if it were also grounded. Many textbooks and magazine articles like to call this property a *virtual ground*, but such a term serves only to confuse the reader. It is better to accept as a basic axiom of operational amplifier circuitry that, for purposes of calculation and voltage measurement, the (-) input will act as if grounded whenever the (+) input is actually grounded.

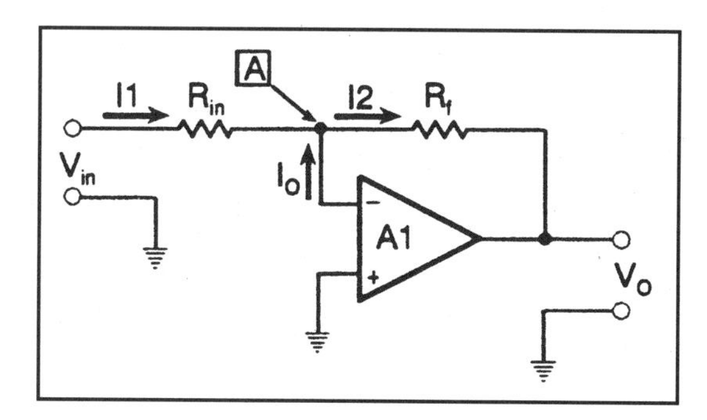

Figure 10-2. Inverting amplifier circuit.

Circuit Analysis Using Ohm's Law and Kirchoff's Current Law

We know from Kirchoff's current law that the algebraic sum of all currents entering and leaving a point in a circuit must be zero. The total current flow into and out of point A in **Figure 10-2**, then, must be zero. Three possible currents exist at this point: input current I2, and any currents flowing into or out of the (-) input terminal of the operational amplifier, I_o. But according to Ideal Property No. 2, the input impedance, Z_{in}, of this device is infinite. Ohm's law tells us that by:

$$I_o = \frac{V}{Z_{in}} \quad \text{eq. (10-1)}$$

Current I_o is zero, because V/Z_{in} is zero. So if current I_o is equal to zero, we conclude that I1 + I2 = 0 (Kirchoff's current law). Since this is true, then:

$$I2 = -I1 \quad \text{eq. (10-2)}$$

We also know that:

$$I1 = \frac{V_{in}}{R_{in}} \quad \text{eq. (10-3)}$$

and,

$$I2 = \frac{V_{out}}{R_f} \quad \text{eq. (10-4)}$$

By substituting **Equation 10-3** and **Equation 10-4** into **Equation 10-2**, we obtain the result:

$$\frac{V_{out}}{R_f} = \frac{-V_{in}}{R_{in}} \quad \text{eq. (10-5)}$$

Solving for V_{out} gives us the transfer function normally given in operational amplifier literature for an inverting amplifier:

$$V_{out} = -V_{in}\frac{R_f}{R_{in}} \qquad \text{eq. (10-6)}$$

The term R_f/R_{in} is the voltage gain factor, and is usually designated by the symbol A_v, which is written as:

$$A_v = \frac{-R_f}{R_{in}} \qquad \text{eq. (10-7)}$$

We sometimes encounter **Equation 10-6** written using the left-hand side of **Equation 10-7**:

$$V_{out} = -A_V V_{in} \qquad \text{eq. (10-8)}$$

When designing simple inverting followers using operational amplifiers, use **Equation 10-7** and **Equation 10-8**. Let us look at a specific example.

Example 10-1

Suppose that we have a requirement for an amplifier with a gain of 50. The amplifier driver has an output impedance of 1000 ohms. A standard rule of thumb for designers to follow is to make the input impedance not less than ten times the source impedance. The amplifier source impedance must be greater than 10 kΩ, so we use a 10-kΩ value for R_{in} in this example:

Solution:

- $A_v = \dfrac{-R_f}{R_{in}}$

- $50 = \dfrac{-R_f}{10{,}000\ ohms}$

- $R_f = 500{,}000\ ohms$

Our gain-of-50 amplifier will look like **Figure 10-3**.

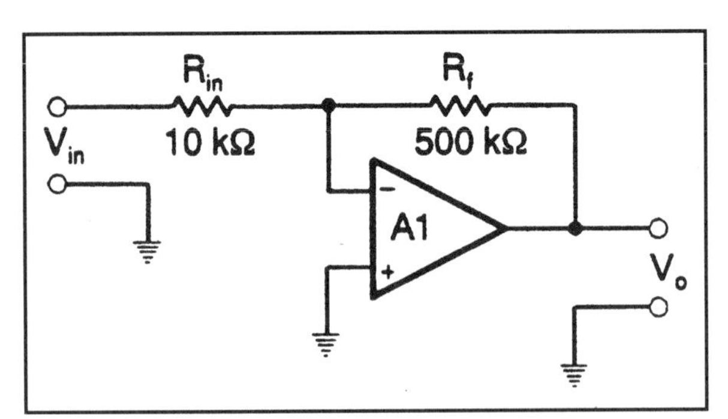

Figure 10-3. Gain-of-50 inverting amplifier.

Noninverting Followers

The inverting follower circuits of **Figure 10-2** and **Figure 10-3** suffer badly from low input impedance, especially at higher gains, because the input impedance is the value of R_{in}. This problem becomes acute when we attempt to obtain even moderately high gain figures from low-cost devices. Although some operational amplifiers allow the use of 500 kΩ to 2 MΩ input resistors, they are costly and often uneconomical. The noninverting follower of **Figure 10-4** solves the input impedance problem very nicely, because the input impedance of the op-amp is typically very high (Ideal Property 2).

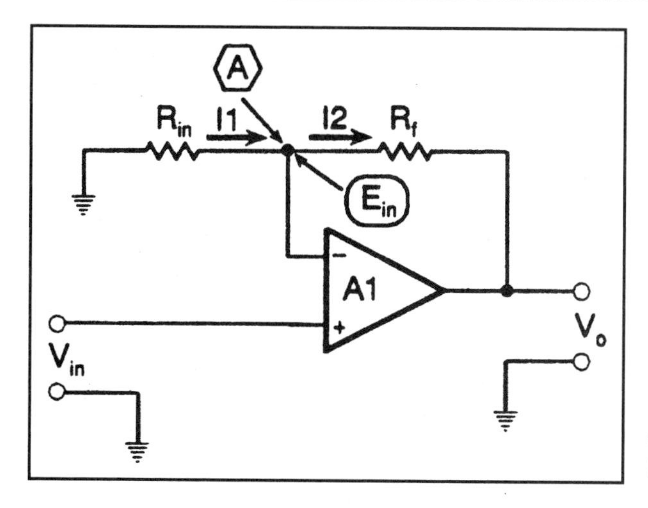

Figure 10-4. Noninverting follower circuit.

We may once again resort to Kirchoff's current law to derive the transfer equation from our basic ideal properties. By property No. 6 we know that the inputs follow each other, so the inverting input can be treated as if it were at the same potential as the noninverting input, which is the input signal voltage, V_{in}. We know that:

- $$I1 = I2 \qquad \text{eq. (10-9)}$$

- $$I1 = \frac{V_{in}}{R_{in}} \qquad \text{eq. (10-10)}$$

- $$I2 = \frac{V_{out} - V_{in}}{R_f} \qquad \text{eq. (10-11)}$$

By substituting **Equation 10-10** and **Equation 10-11** into **Equation 10-9**, we obtain:

- $$\frac{V_{in}}{R_{in}} = \frac{V_{out} - V_{in}}{R_f} \qquad \text{eq. (10-12)}$$

Solving **Equation 10-12** for V_{out} results in the transfer equation for the noninverting follower amplifier circuit.

Multiply both sides by R_f:

$$\frac{R_f V_{in}}{R_{in}} = V_{out} - V_{in} \qquad \text{eq. (10-13)}$$

Add V_{in} to both sides:

$$\frac{R_f V_{in}}{R_{in}} + V_{in} = V_{out} \qquad \text{eq. (10-14)}$$

Factor out V_{in}:

$$V_{in}\left(\frac{R_f}{R_{in}} + 1\right) = V_{out} \qquad \text{eq. (10-15)}$$

This illustrates both transfer functions commonly used in operational amplifier design using only the basic properties of Ohm's law and Kirchoff's current law. We may safely assume that the operational amplifier is a feedback device that generates a current which cancels the input current. **Figure 10-5** gives a synopsis of the characteristics of the most popular operational amplifier configurations. The unity gain noninverting follower of **Figure 10-5c** is a special case of the circuit in **Figure 10-5b**, in which $R_f/R_{in} = 0$. In this case, the transfer equation becomes:

$$V_{out} = V_{in}(0 + 1) \qquad \text{eq. (10-16)}$$

$$V_{out} = V_{in}(1) = V_{in} \qquad \text{eq. (10-17)}$$

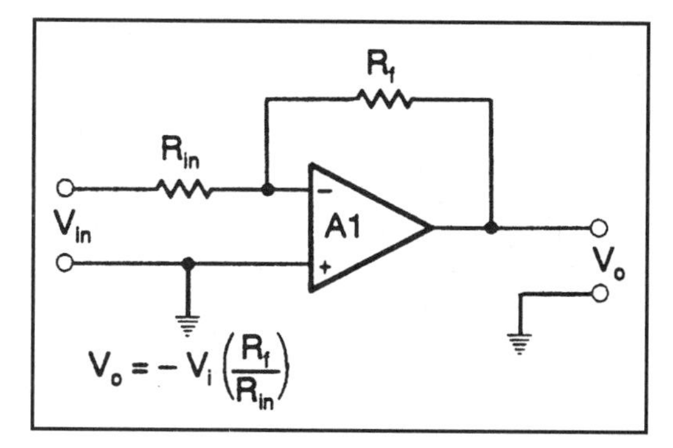

Figure 10-5a. Inverting follower.

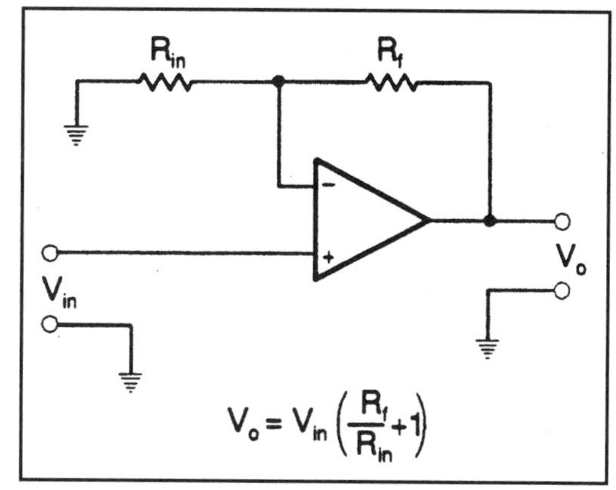

Figure 10-5b. Non-inverting follower with gain.

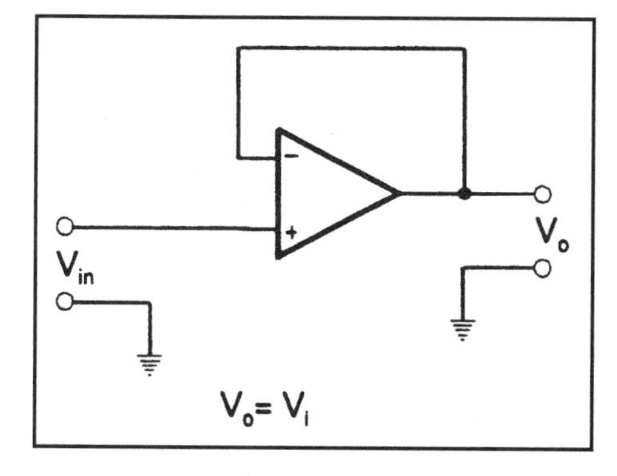

Figure 10-5c. Unity gain noninverting follower.

Operational Amplifier Power Sources

Although most circuits using operational amplifiers use a dual polarity power supply, it is possible to operate the device with a single polarity supply. An example of single supply operation might be in equipment designed for mobile operation or circuits where the majority of circuitry requires only a single polarity supply. Generally speaking, it is better to use the bipolar supplies as intended by the manufacturer.

There are two separate power terminals, V_{cc} and V_{ee}, on the typical operational amplifier device. The V_{cc} supply is positive with respect to ground, while the V_{ee} supply is negative to ground. These supplies are shown in **Figure 10-6**. Although batteries are shown in the example, ac-operated dc power supplies may be used instead. Typical values for V_{cc} and V_{ee} range from ±3 volts dc to ±22 volts dc. The value most often selected for these potentials will be between ±9 volts dc and ±15 volts dc.

One further constraint is placed on the operational amplifier power supply in some cases (such as older devices): V_{cc} - V_{ee} must be less than some specified voltage (usually 30 volts). So if V_{cc} is + 18 volts dc, then V_{ee} must be not greater than (30 - 18), or 12 volts dc.

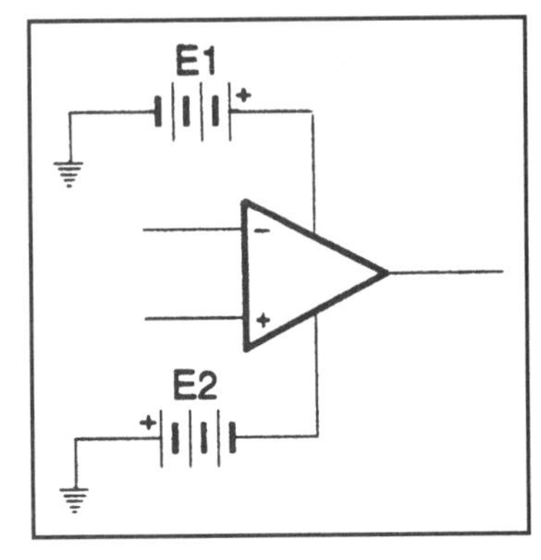

Figure 10-6a. Connection to op-amp.

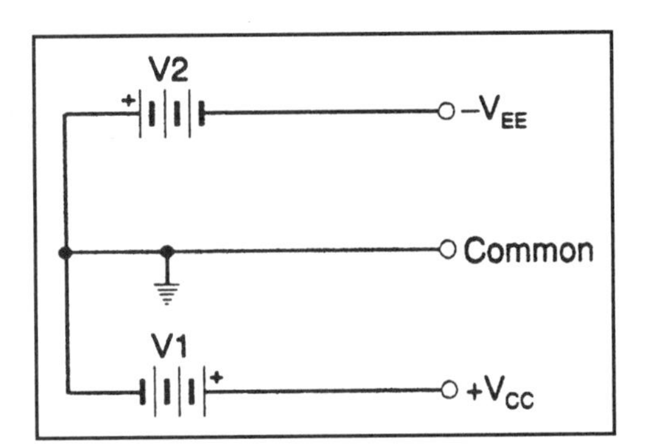

Figure 10-6b. Battery equivalent circuit of op-amp power supply.

Practical Op-Amp Devices: Real-World Problems

Before we can properly apply operational amplifiers in real equipment we must learn some of the limitations of real-world devices. The devices that we have considered up until now have been ideal, so they do not exist. Real IC operational amplifiers range in price from fifty cents to fifty dollars. The lower the cost, generally, the less ideal the device.

Three main problems exist in real operational amplifiers: offset current, offset voltage, and frequency response. (A problem of internal noise generation is of less importance.)

In real operational amplifier devices the input impedance is less than infinite, implying that a small input bias current exists. The input current may flow into or out of the input terminals of the operational amplifier. In other words, current I_o of **Figure 10-2** is not zero, so it will produce an output voltage equal to $-I_oR_f$. The cure for this problem is shown in **Figure 10-7**, and involves placing a compensation resistor between the noninverting input terminal and ground. This tactic works because the currents in the respective inputs are approximately equal. Since resistor R_c is equal to the parallel combination of R_f and R_{in}, it will generate the same voltage drop that appears at the inverting input. The resultant output voltage is zero, since the two inputs have equal but opposite polarity effect on the output.

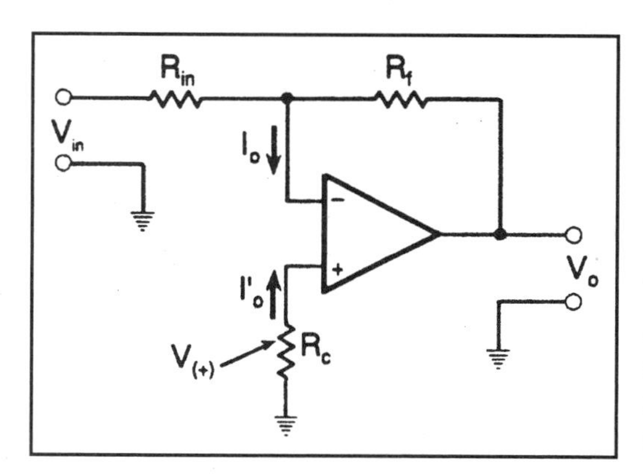

Figure 10-7. Use of a compensation resistor for nulling output offsets that are due to input bias currents.

Output offset voltage is the value of V_{out} that will exist if the input end of the R_{in} is grounded ($V_{in} = 0$). In the ideal device, V_{out} would be zero, but in real devices there may be some offset potential present. This output potential can be forced to zero by any of the circuits in **Figure 10-8**.

The circuit in **Figure 10-8a** uses a pair of offset null terminals found on most operational amplifiers. Although many IC operational amplifiers use this technique, some do not. Alternatively, the offset range may be insufficient in some cases. In either event, we may use the circuit of **Figure 10-8b** to solve the offset potential problem.

The offset null circuit of **Figure 10-8b** creates a current flowing in resistor R1 to the summing junction of the operational amplifier. Since the offset current may flow either into or out of the input terminal, the null control circuit must be able to supply currents of both polarities. Because of this requirement, the ends of the potentiometer, R2, are connected to V_{cc} and V_{ee}.

In many cases, the offset is small compared with normally expected values of input signal voltage. This is especially true in low-gain applications, where the nominal offset current will create such a low output error that no action need be taken. In still other cases, the offset of each stage in a cascade chain of amplifiers may be small, but their cumulative effect may result in a large offset error. In this type of situation, it is usually sufficient to null only one of the stages close to the output stage.

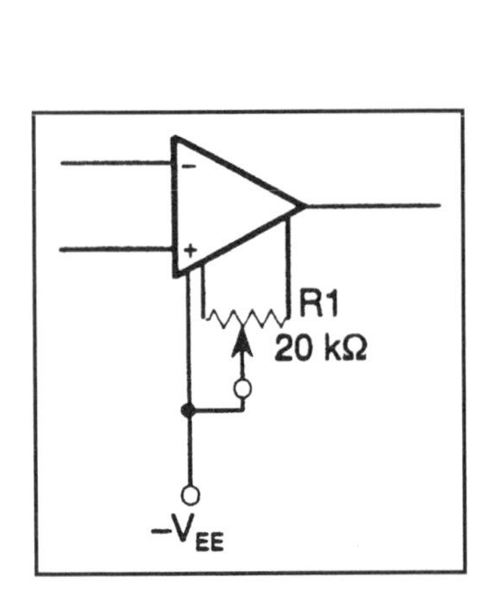

Figure 10-8a. Using the op-amp's own offset null terminals.

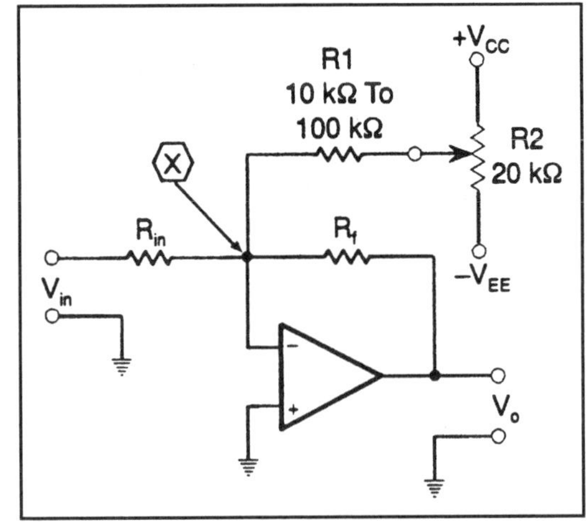

Figure 10-8b. Simple external offset circuit.

In those circuits where the offset is small, but critical, it may be useful to replace R1 and R2 of **Figure 10-8b** with one of the resistor networks of **Figure 10-8c** through **Figure 10-8e**. These perform essentially the same function, but have superior resolution. That is, there is a smaller change in output voltage for a single turn of the potentiometer. This type of circuit will have a superior resolution in any event, but even further improvement is possible if a ten-turn (or more) potentiometer is used.

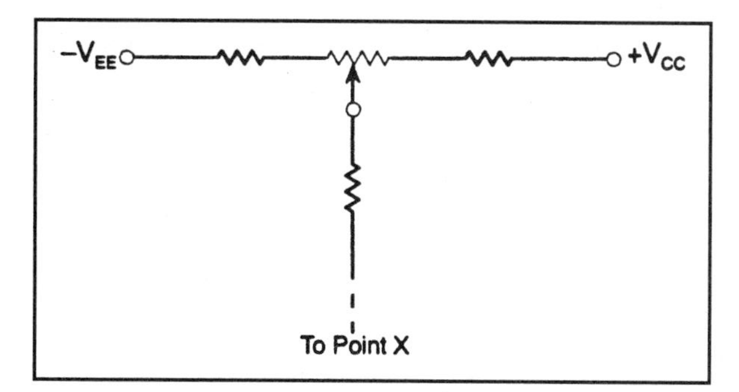

Figure 10-8c. Simple fine resolution offset null circuit.

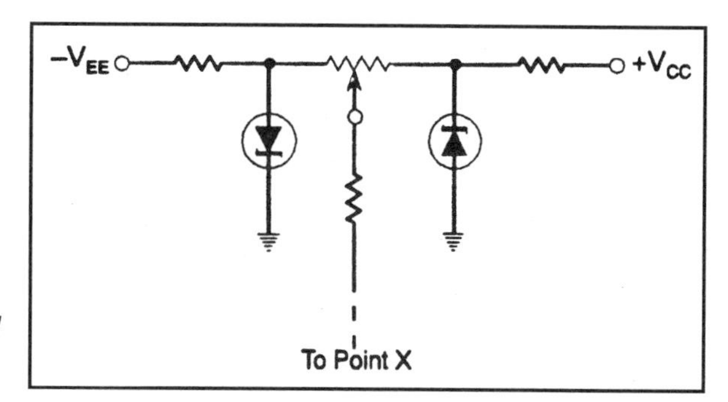

Figure 10-8d. Zener diode-based fine resolution offset null circuit.

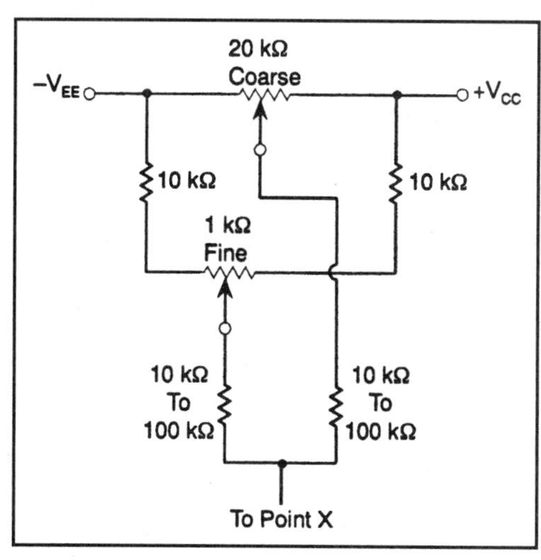

Figure 10-8e. "Coarse" and "Fine Vernier" offset null circuit.

DC Differential Amplifiers

The fact that an IC operational amplifier has two complementary inputs, inverting and noninverting, allows it to be used as a differential amplifier. These circuits produce an output voltage that is proportional to the difference between two ground-referenced input voltages. Recall from our previous discussion that the two inputs of an operational amplifier have equal but opposite effect on the output voltage. If the same voltage, or two equal voltages, are applied to the two inputs (such as a common-mode voltage, V3 in **Figure 10-9**), the output voltage will be zero. The transfer equation for a differential amplifier is:

$$V_{out} = A_v(V1 - V2) \qquad \text{eq. (10-18)}$$

So, if V1 = V2, then $V_{out} = 0$.

The circuit of **Figure 10-9** shows a simple differential amplifier using a single IC operational amplifier. The voltage gain of this circuit is given by:

$$A_v = \frac{R3}{R1} \qquad \text{eq. (10-19)}$$

Provided that R1 = R2 and R3 = R4.

The main appeal of this circuit is that it is economical, as it requires only one IC operational amplifier. It will reject common-mode voltages if the equal value resistors are well matched. A serious problem exists for some applications, however, because the circuit has a relatively low input impedance. Additionally, with the problems of real operational amplifiers, this circuit may be difficult to manage in high-gain applications. As a result, designers frequently use an alternate circuit.

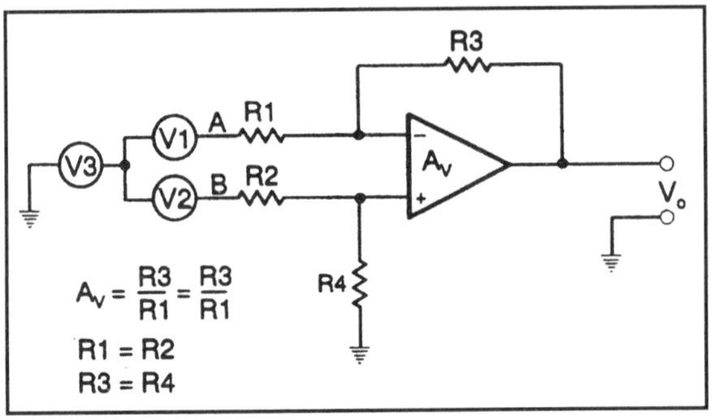

Figure 10-9. Differential amplifier with input and common mode signal voltage sources.

In recent years, the instrumentation amplifier (IA) of **Figure 10-10** has become popular because it alleviates most of the problems associated with the circuit of **Figure 10-9**. The input stages are noninverting followers, so they will have a characteristic high input impedance. Typical values run to as much as 1000 MΩ.

The instrumentation amplifier is relatively tolerant of different resistor ratios used to create voltage gain. In the simplest case, the differential voltage gain is given by:

$$A_v = \frac{2\,R3}{R1} + 1 \qquad \text{eq. (10-20)}$$

provided that R3 = R2 and R4 = R5 = R6 = R7.

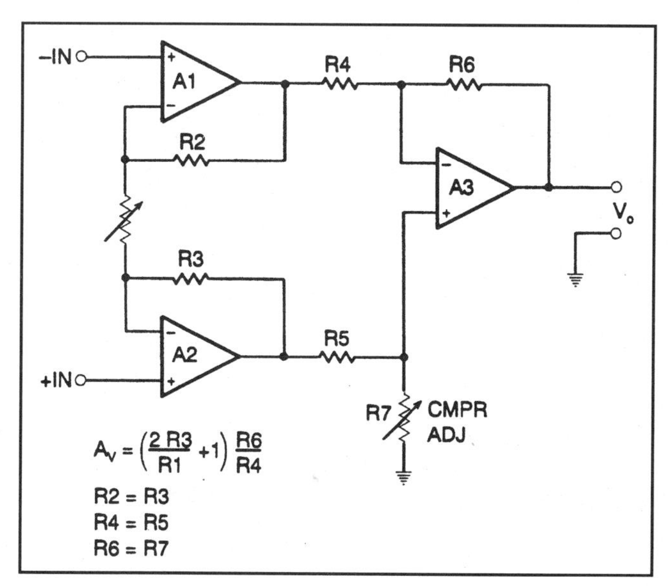

Figure 10-10.
Instrumentation amplifier.

It is interesting to note that the common-mode rejection ratio is not seriously degraded by mismatch of resistors R2 and R3; only the gain is affected. If these resistors are mismatched, a differential voltage gain error will be introduced.

The situation created by **Equation 10-20** results in having the gain of A3 equal to one, which is a waste. If gain in A3 is desired, **Equation 10-20** must be rewritten into the form:

$$A_v = \left(\frac{2\,R3}{R1} + 1\right)\left(\frac{R7}{R5}\right) \qquad \text{eq. (10-21)}$$

One further equation that may be of interest is the general expression from which the other instrumentation amplifier transfer equations are derived:

$$A_v = \frac{R7\,(R1 + R2 + R3)}{R1\,R6} \qquad \text{eq. (10-22)}$$

when the ratio R7/R6 = R5/R4.

Equation 10-22 is especially useful, since you need not be concerned with matched pairs of precision resistors, but only that their ratios be equal.

Practical Circuit

This section will present a practical design example using the instrumentation amplifier circuit. The problem requires a frequency response to 100 kHz and shielded input lines. The latter requirement will deteriorate the signal at high frequencies because of the shunt capacitance of the input cables. To overcome this problem, a high-frequency compensation control is built into the amplifier. Voltage gain is approximately ten.

The circuit to a preamplifier is shown in **Figure 10-11**. It is the instrumentation amplifier of **Figure 10-10** with some modifications. When the frequency response is less than 10 kHz, we may use any of the 741-family devices (741, 747, 1456, and 1458), but premium performance demands a better operational amplifier. In this case, one of the most economical is the RCA CA3140, although an L156 would also suffice.

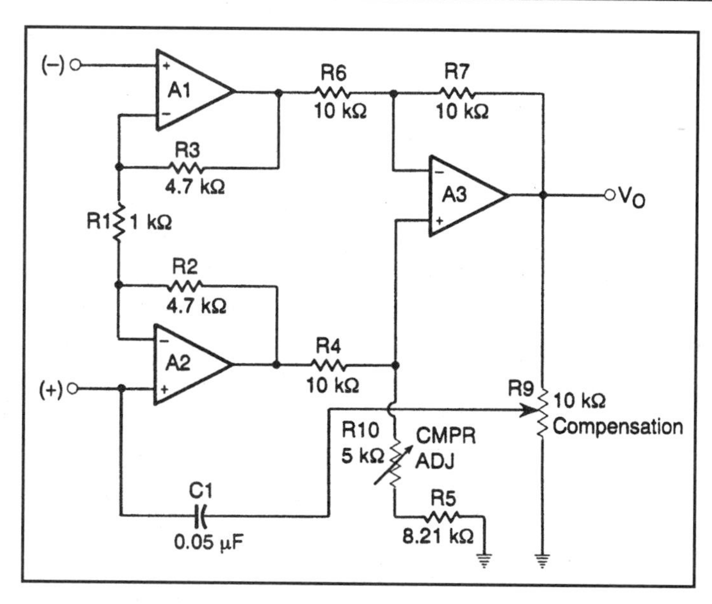

Figure 10-11a. Instrumentation amplifier with built-in capacitance compensation network.

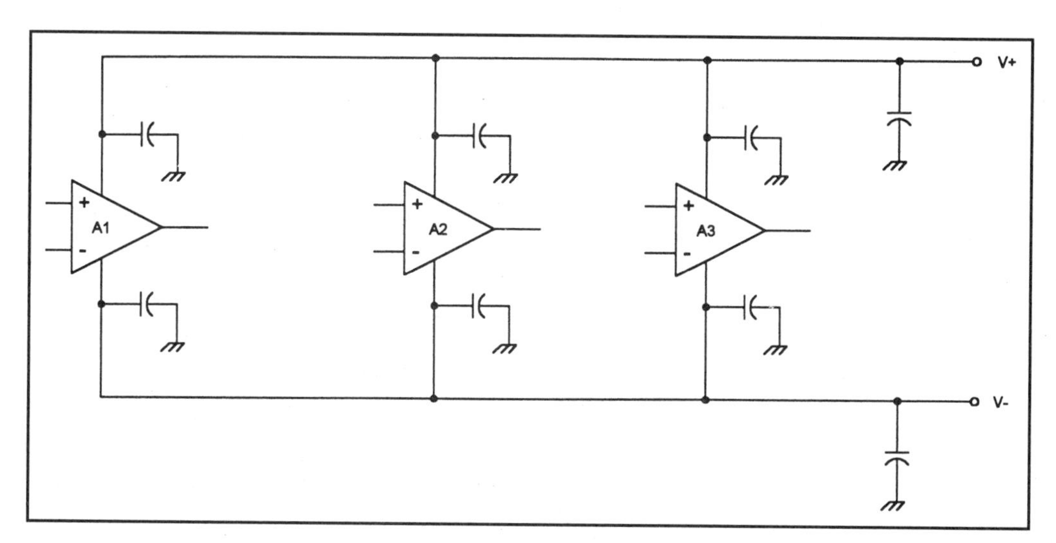

Figure 10-11b. dc power supply arrangement for the circuit.

Common-mode rejection can be adjusted to compensate for a mismatch in the resistors or IC devices by adjusting R10. This potentiometer is adjusted to zero output when the same signal is applied simultaneously to both inputs.

The frequency response characteristics of this preamplifier are shown in **Figure 10-12** through **Figure 10-16**. The input in each case is a 1000-Hz square wave from a function generator. The waveform in **Figure 10-12** shows the output signal when resistor R9 is set with its wiper closest to ground. Note that it is essentially square and shows only a small amount of roll-off of high frequencies. The waveform in **Figure 10-13** is the same signal when R9 is at maximum resistance. This creates a small amount of regenerative feedback; although it is not sufficient to start oscillation, it will enhance amplification of high frequencies. The frequency response is shown in **Figure 10-14.** To obtain a particular response curve, modify the values of C1 and R9 accordingly.

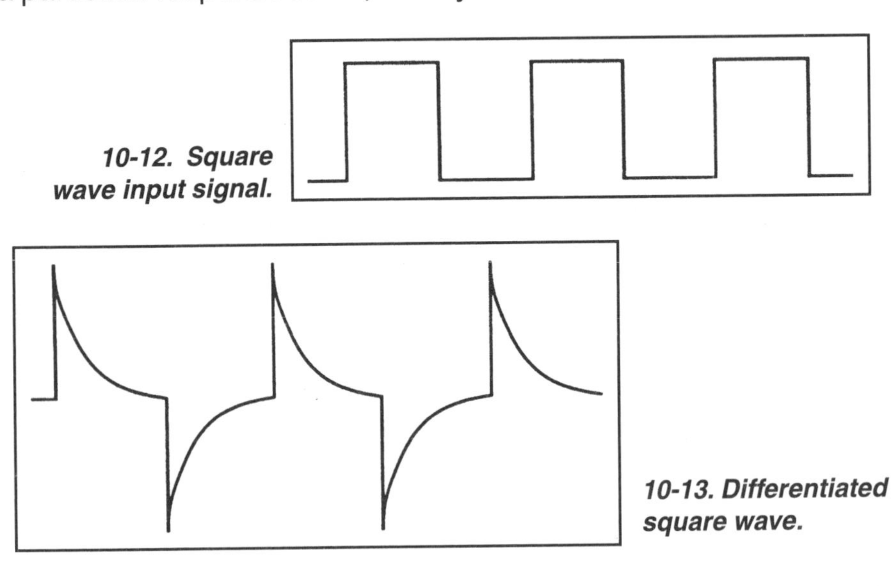

10-12. Square wave input signal.

10-13. Differentiated square wave.

35
30
RC = Maximum
25
20
15
RC = Minimum
10
5
F
10^1 10^2 10^3 10^4 10^5 10^6
Frequency In hertz

10-14. Frequency response of circuit under conditions of RC value.

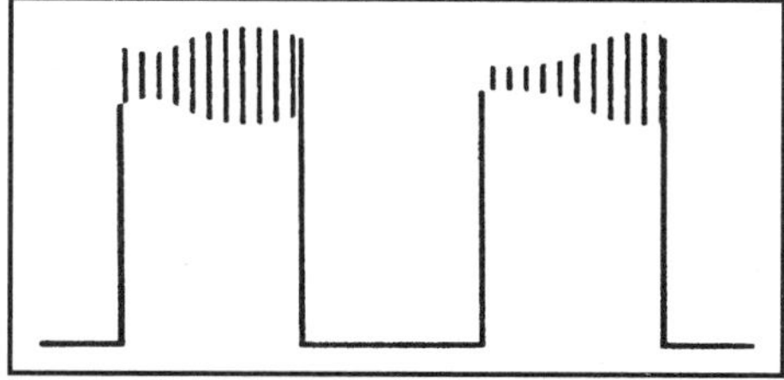

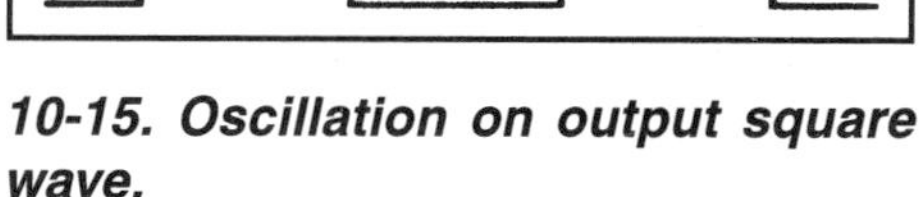

10-15. Oscillation on output square wave.

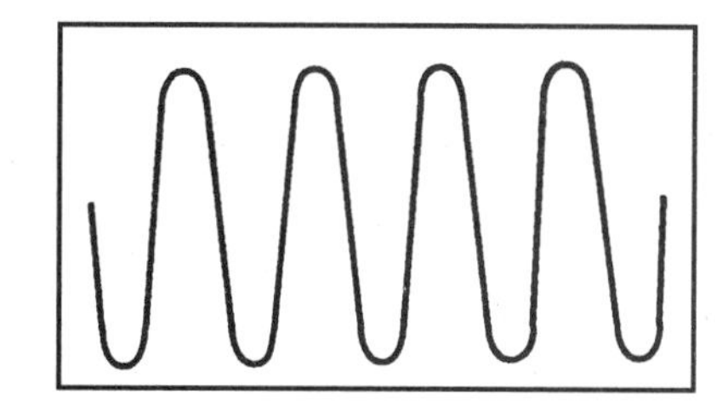

10-16. Oscillation waveform expanded.

Oscillation can be a quite serious problem (**Figure 10-15**). Certain precautions, most of which involve limiting the amplitude of the feedback signal, should be taken. A 2200 ohm resistor in series with the potentiometer is adequate.

Oscillation may also come from capacitor C1. When a 0.001 μF capacitor is used at C1, an 80 kHz oscillation is created (see **Figure 10-16**).

Differential Amplifier Applications

Differential amplifiers find application in many instrumentation situations. Of course, it should be realized that they are required wherever a differential signal voltage is used. They are also used to acquire signals or to operate in control systems in the presence of large noise signals. Many medical applications use the differential amplifier because they look for minute bipotentials in the presence of strong common-mode 60-Hz fields from the ac power mains.

Another application is the amplification of output signal from a Wheatstone bridge, as shown in **Figure 10-17**. If one side of the bridge's excitation potential is grounded, the output voltage is a differential signal voltage. This signal can be applied to the inputs of a differential amplifier or instrumentation amplifier to create an amplified, single-ended output voltage.

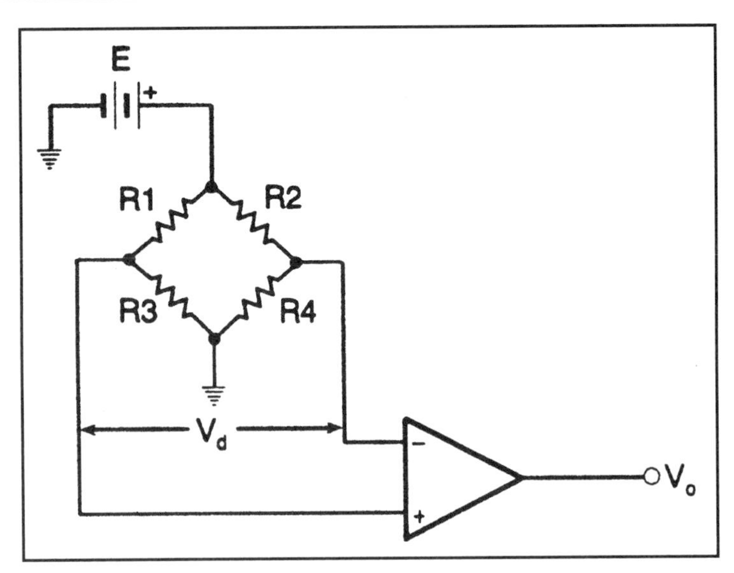

10-17. Differential amplifier used as a Wheatstone bridge amplifier.

A "rear end" stage suitable for many operational amplifier instrumentation projects is shown in **Figure 10-18**. This circuit consists of three low-cost operational amplifier ICs. Since they follow most of the circuit gain, we may use low-cost devices such as the 741 in this circuit. The gain of this circuit is given by $R2/10^4$.

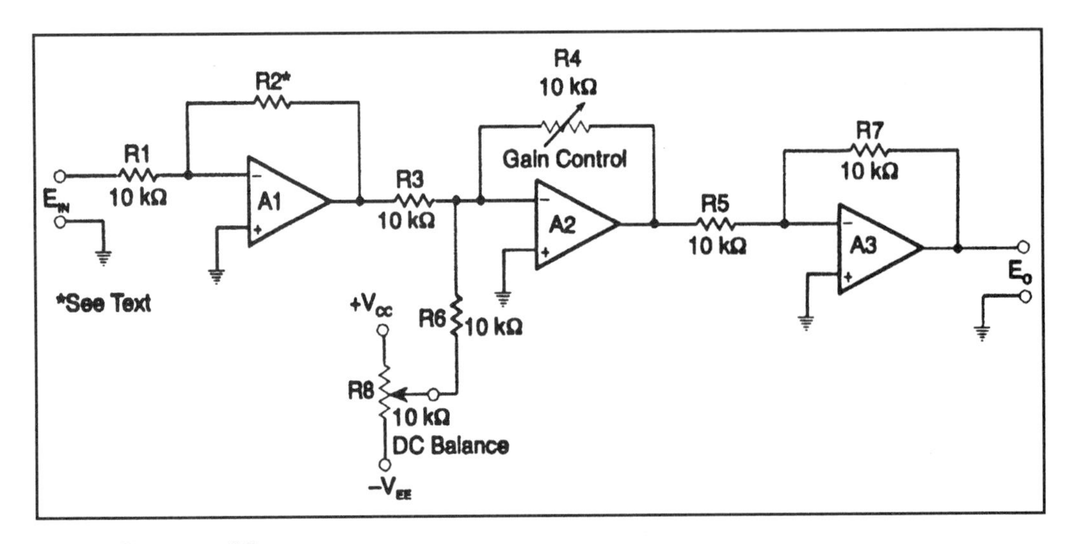

10-18. Postamplifier.

Chapter 11

Electro-Optical Medical Instruments

Electro-optic devices have long been used in medical instruments, and will continue to be for a long time to come. In this chapter, we will take a look at several different electro-optical medical instruments, including one (a pulsometer circuit) that you can build and use at home, using devices and circuits discussed in the last two chapters.

Photocolorimetry

One of the most basic forms of instrument circuitry is the oldest and most commonly used: *photometry* (also called *photocolorimetry*). These photometry circuits are used to measure oxygen (O_2) content of blood, carbon dioxide (CO_2) content of air, water vapor content in a gas, electrolyte levels (sodium, Na, and potassium, K) of blood, among other measurements. The basis for these measurements is that different wavelengths of light can be absorbed when passed through a substance, reflected when the substance is illuminated or emitted when it is burned. For example, when infrared light passes through a gas containing CO_2, certain IR wavelengths are heavily absorbed. By measuring the relative absorption of these wavelengths, one can infer how much carbon dioxide is present in the mixture.

Photocolorimetry is basically a measurement technique in which infrared, visible, or ultraviolet light absorption, reflection, or transmission is compared with a fixed set of standards. **Figure 11-1a** shows the basic circuit of the most elementary form of visible light colorimeter. The sensors are photoresistors, such as cadmium sulphide (CdS) cells. Although the circuit is very basic, it is the one actually used in a number of instruments.

The circuit contains a Wheatstone bridge, and uses a pair of photoresistor cells (R2 and R4) as the light sensors. Potentiometer R5 in is a bridge balance control, and is adjusted for zero output ($V_o = 0$) when the same light shines equally on both photoresistors. The output voltage from the bridge (V_o) will be zero when the two legs of the bridge are balanced. In other words, V_o is zero when R1/R2 = R3/R4. It is not necessary for the resistor elements to be equal (although that is often the case), only that their resistance ratios be equal. Thus, a 500k/50k ratio for R1/R2 will produce zero output voltage when R3/R4 = 100k/10k.

The photoresistors are arranged such that light from a calibrated source illuminates both resistors equally and fully, except when an intervening filter or sample is present in one or both pathways. In most instruments based on this

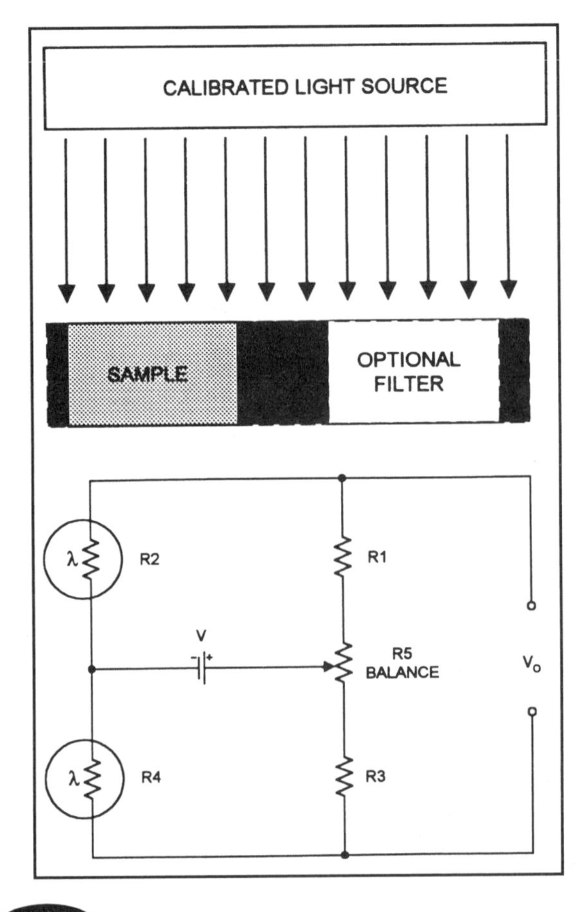

Figure 11-1a. Photocolorimeter circuit.

principle, a translucent sample is placed between the light source and one of the photocells. The amount of light transmission allowed by the sample is a measure of its optical density, and is thus a transducible property.

A variation on this principle is shown in **Figure 11-1b**. This circuit uses either silicon solar cells or silicon phototransistors (a dc power supply is required for the latter) as the sensors. These sensors have a wider spectral response than the CdS cells, and will work into the IR region. The two matched sensors (PC1 and PC2) are used to drive a differential amplifier, A1. The output, V_o, of the amplifier is the difference between the output of PC1 (V1) and PC2 (V2). This voltage, V_o, is proportional to the difference in the relative IR levels applied to the two sensors.

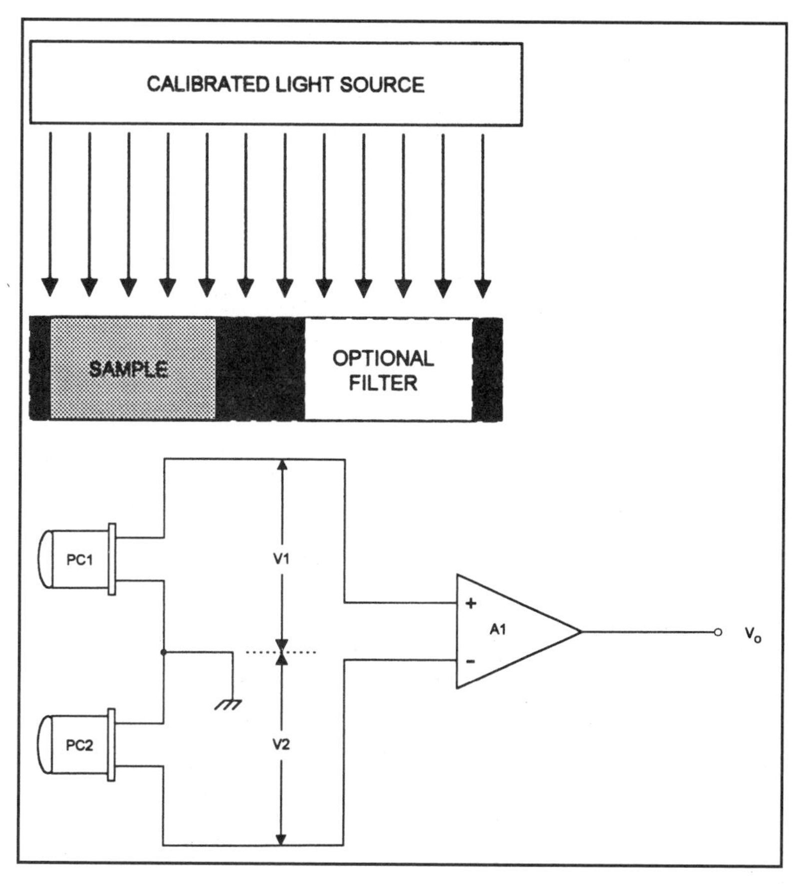

Figure 11-1b. Photovoltaic cell version of the photocolorimeter.

Blood O_2 Level

A classical and still widely used method for measuring blood oxygen level is similar to the basic colorimeter. It works because the intensity of the color red in the blood is a measure of its oxygenation: oxygenated arterial blood is more red than deoxygenated venous blood. **Figure 11-2** shows the relative optical absorption of blood according to its oxygen content. At a light wavelength of approximately 800 nanometers the two types of blood show the same absorption. This point is called the *isoabsorption point*, or more commonly the *isobestic point*. By comparing the light absorption properties of a blood sample with the absorption at 800 nm, one can determine the oxygen content.

An instrument such as **Figure 11-1b** can be built to read blood O_2 content. An 800 nm standard filter cell is introduced between the light source and one sensor, and a blood sample is placed between the light source and the other sensor. The degree of blood O_2 saturation in the sample is thus reflected by the difference in the bridge reading between the sample path and filter path.

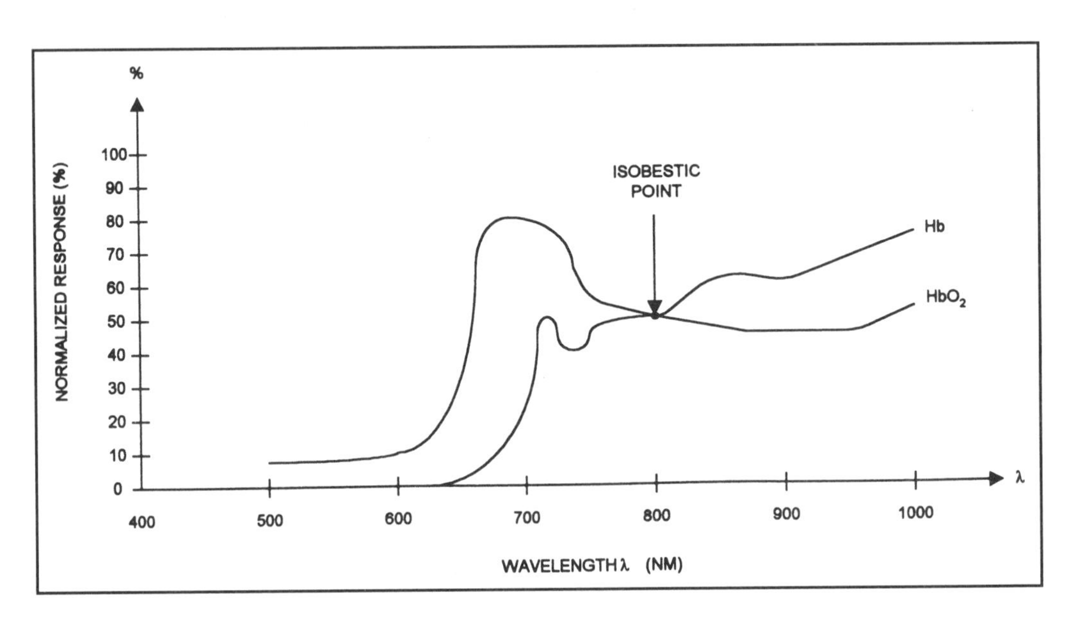

Figure 11-2. Change of absorption spectrum in blood allows transduction of oxygen level.

Respiratory CO_2 Level

Human breath is approximately 2 to 5 percent carbon dioxide (CO_2), while the percentage of CO_2 in normal room air is negligible. A popular form of "End Tidal CO_2 Meter" is based on the fact that CO_2 absorbs infrared waves at several discrete wavelengths (**Figure 11-3**). The "light source" in many of these instruments is actually nothing more than a *Cal-Rod* identical to the one that heats your coffee pot! In this instrument, ambient air is passed through a glass cuvette placed between one sensor and the heat source, while patient expiratory air is passed through the same type of cuvette placed between the heat source and other sensor. The difference in the IR transmission across the two paths is a function of the percentage of CO_2 in the sample circuit.

Water vapor also absorbs IR energy, but with a different wavelength curve than CO_2 (see **Figure 11-3**). It is not easy to fine-tune the IR emitter to produce only those wavelengths that are absorbed by CO_2, but it is possible to filter out the unwanted wavelengths that do not easily provide comparative data. The filter is placed between the source and the sample in order to block the wavelengths strongly absorbed by water.

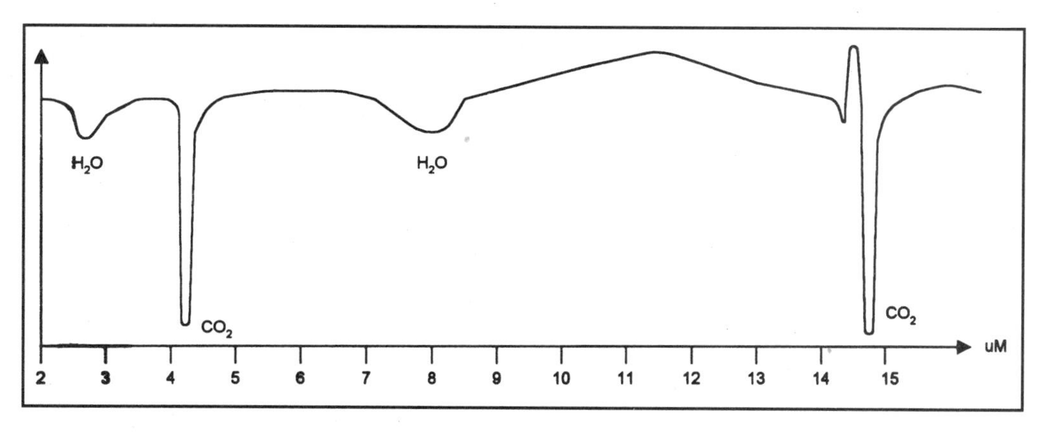

Figure 11-3. Absorption of infrared radiation by CO2.

The associated components (not shown) for the circuit in **Figure 11-1b** will allow zero and maximum gain adjustment. The zero point is adjusted with ambient air in both cuvettes, while the maximum scale is adjusted with a sample cuvette of calibration gas (usually 5-percent CO_2, 95-percent nitrogen). This calibration gas must be obtained from a local scientific laboratory supplies outlet, and be specified for use as a calibration gas.

Blood Electrolytes

Blood chemistry tests often test the levels of sodium (Na) and potassium (K). An instrument commonly used for these measurements is the flame photometer (**Figure 11-4**). This instrument replaces the light source with a flame produced by burning gas in a carburetor. The sample is injected into the carburetor, and burned along with the gas/air mixture. The colors emitted by burning are proportional to the concentrations of Na and K ions in the sample. A special gas is used to burn with a clean blue flame when no sample is present.

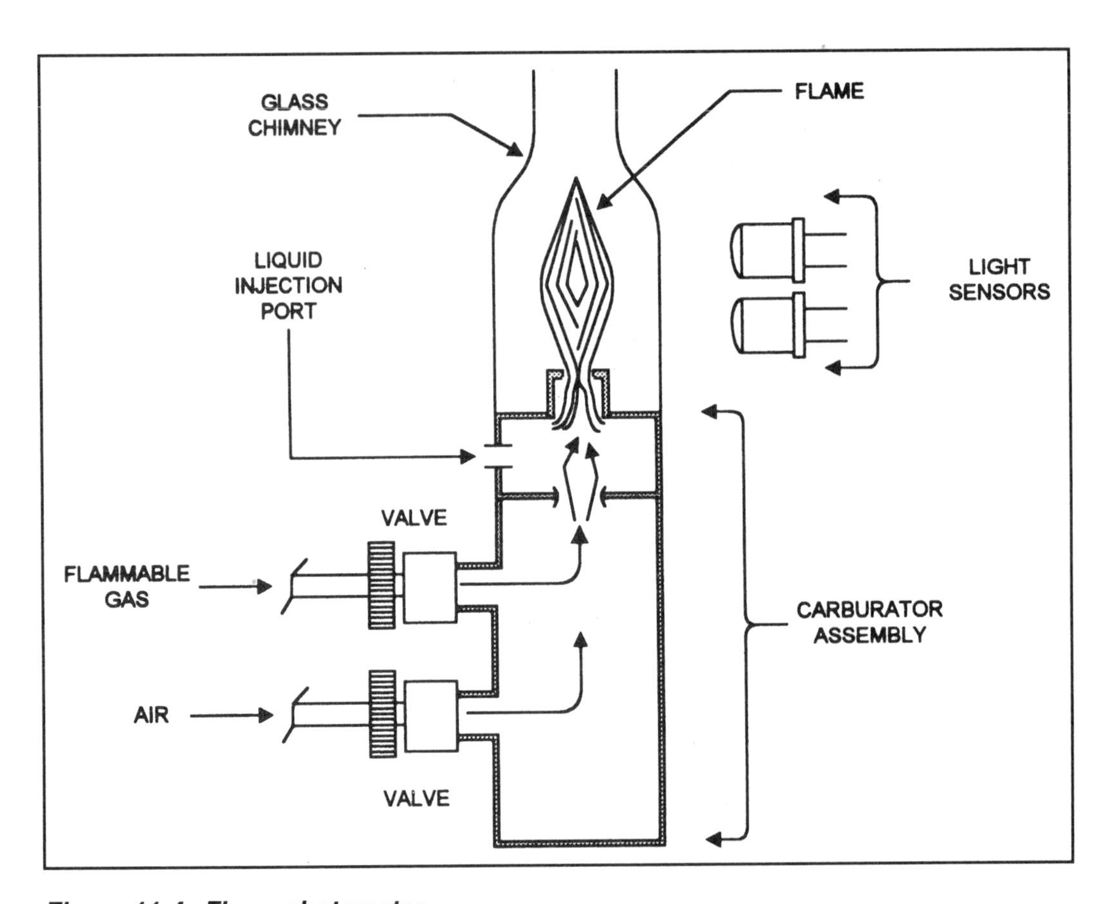

Figure 11-4. Flame photometer.

In medical applications, a specified volume of the patient's blood is mixed with a lithium calibrating solution (also a precise, predetermined amount). The solution is well-mixed, and then applied to the carburetor. The instrument can determine the concentration of the two elements by comparing the color intensities of the burned Na and K ions to the intensity of the calibration solution.

Flame photometers can give terribly flawed readings if improperly maintained. The carburetor and surrounding glass structures must be cleaned frequently, or the build-up of material from past tests will bias the results. In addition, soot (carbon) on the associated glass windows will obscure the flame, and may produce a lack of sensitivity and an erroneous reading (especially if the windows are not uniformly carboned).

Photoplethysmograph: The Sensor You Can Build

The devices already discussed are not easily built or used by the amateur scientist, but there is one sensor that is easily assembled from materials available to amateurs. A *photoplethysmograph* (PPG) is a device that uses a light source and sensor in order to produce a waveform similar to the blood pressure pulse (**Figure 11-5**).

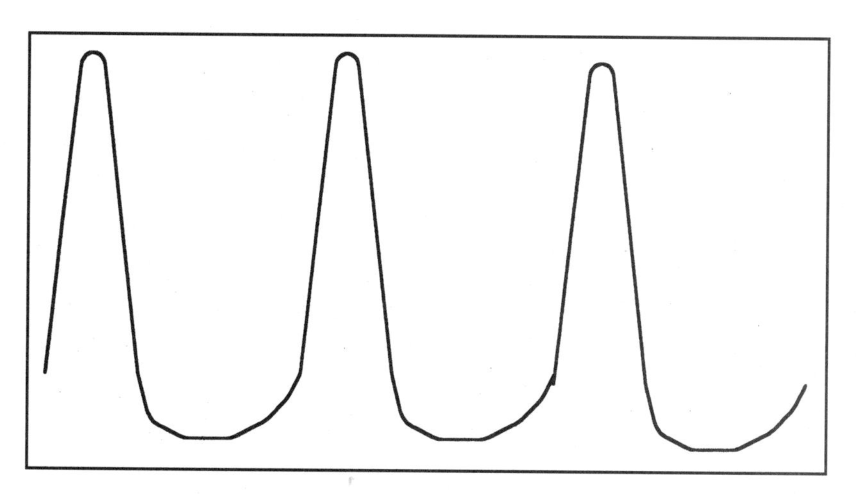

Figure 11-5. Output pulse from photoplethysmograph (PPG).

Figure 11-6 shows two basic forms of PPG: *reflective* and *direct*. Both work because blood pulses through tissue as the heart beats. The optical density of the tissue in which the blood flows changes as the pulse passes. The PPG is a heart rate meter often attached to exercisers. It can be clipped to an earlobe, or placed over a finger or thumb in order to acquire the pulse signal.

The reflective PPG shown in **Figure 11-6a** works when light from the source is scattered from bone in the finger or thumb. The direct type of PPG (**Figure 11-6b**) transmits the light through the tissue to a sensor on the other side. Of these two methods, the direct form is easiest to make. The reflective form requires a very high degree of isolation between the source and sensor, and good alignment of the two.

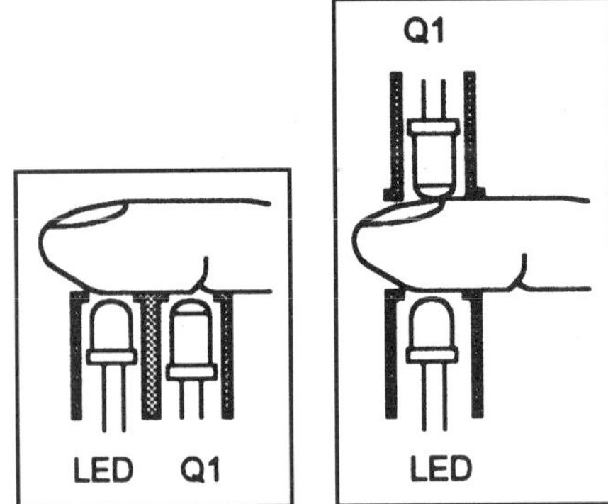

Figure 11-6a. Basic structure of the PPG reflective type sensor.

Figure 11-1b. Transmission-type PPG sensor.

The circuit for a basic PPG sensor is shown in **Figure 11-7**. This circuit uses a high-intensity red LED as the emitter, and a high-gain silicon photodarlington transistor (such as *Digi-Key*[1] L14R1GE) as the sensor. The tissue, usually a finger or thumb, is placed between the emitter and the sensor. Signal from the collector of the phototransistor, Q1, is a low-frequency pulse, so a relatively high-value coupling capacitor is needed. The amplifier is an operational amplifier connected in the noninverting follower configuration. The op-amp selected must have a high input impedance (the CA-3140 BiMOS device or one of the BiFET op-amps is preferred, see Chapter 10). The voltage gain, A_v, of the amplifier is given by:

$$A_v = R\frac{5}{R}4 + 1 \qquad \text{eq. (11-1)}$$

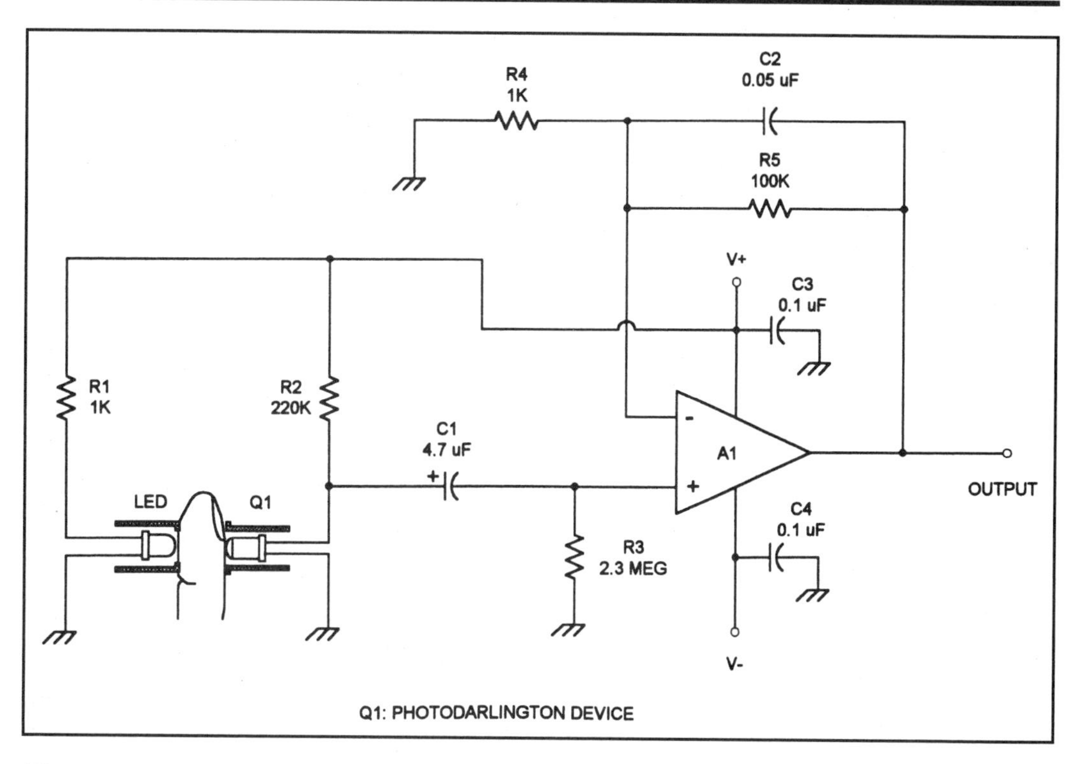

Figure 11-7. PPG circuit.

The output of the op-amp is a voltage pulse waveform with an amplitude between 200 mV and 1 volt. It will contain a fair amount of 60-Hz and high-frequency noise on the baseline. The capacitor (C2) shunting the op-amp feedback resistor (R5) will help cut some of the noise by limiting the frequency response of the circuit. If additional amplitude is required, then follow the circuit of **Figure 11-7** with another amplifier.

The PPG preamplifier can be used to drive a pulsometer circuit. A voltage comparator (A1) is used to clean up the output signal (**Figure 11-8**). The comparator is used to produce an abrupt output level shift when the pulse voltage rises above voltage V1 at the output of voltage divider R1/R2. Voltage V1 is set so that noise signals do not cross the threshold. This level can then be used to trigger a monostable multivibrator (one-shot) circuit, if desired. Counting the one-shot output pulses will yield a dc voltage that is proportional to heart rate.

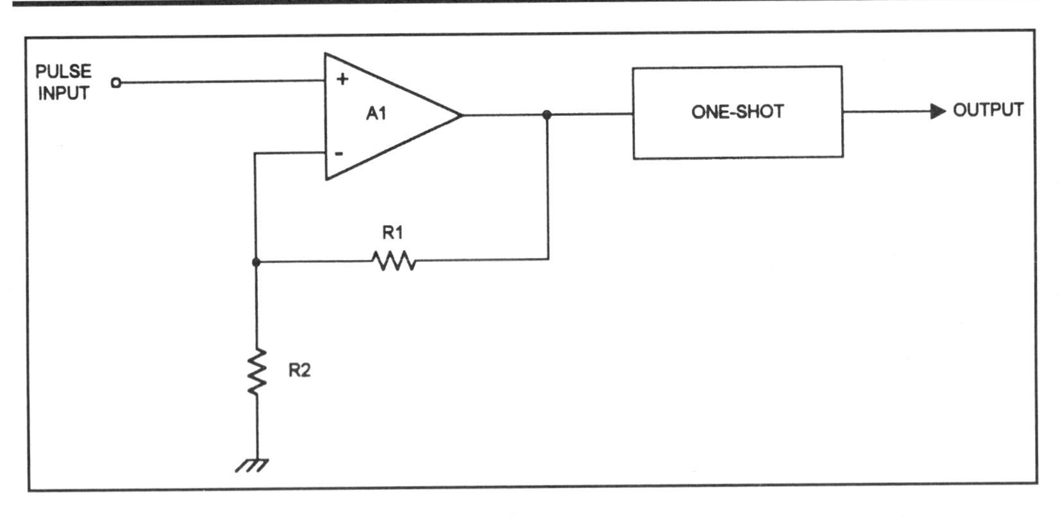

Figure 11-8. Circuit to clean up waveform.

The construction of the pulsometer sensor is shown in **Figure 11-9**. The sensor housing is a large rubber boot normally used for insulating jumbo alligator clips. These can be purchased at most retail electronics parts stores. Select a size that fits over either the thumb, index finger, or middle finger. Two holes are cut into the rubber boot to accommodate the sensor and LED emitter. Once these are wired into a circuit such as **Figure 11-7**, the whole assembly is covered with black insulating tape. It is very important to prevent ambient light from affecting the sensor transistor.

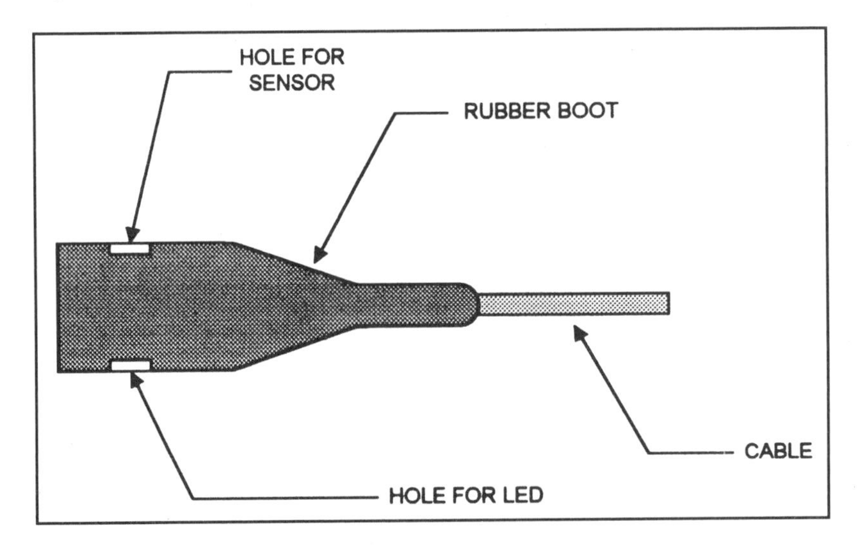

Figure 11-9. Pulsometer.

Blood Oxygen Measurement by PPG

One of the big problems in surgery is ensuring that the patient receives an adequate oxygen supply. One popular blood oximetry method is based on the PPG principle. **Figure 11-10** shows the basic PPG sensors required for O_2 measurements. A single wide-spectrum sensor is placed opposite a pair of LEDs. One LED emits visible red light near 660 nm, and the other emits IR light near 920 nm; the difference between the two resistors is near the isobestic. Coupled to a microprocessor which executes a complex algorithm, this instrument is able to produce an output that measures blood O_2 level.

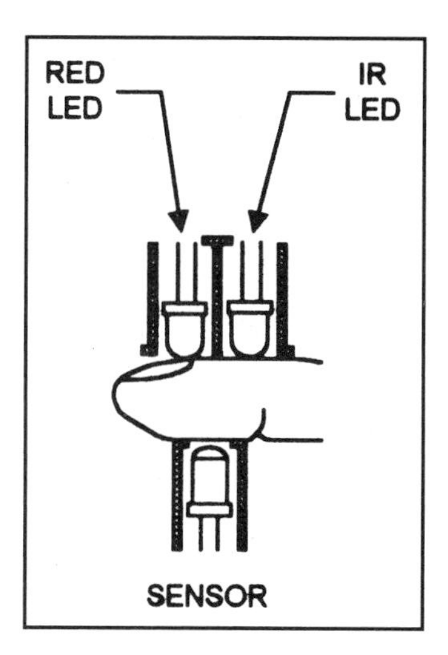

Figure 11-10.* *Photopletheysmography can be used for measuring blood oxygen level by comparing the light absorption at two wavelengths, one of which is infrared.

Summary

Availability of parts and simple construction techniques make the PPG instruments practical for the amateur scientist to pursue. These simple electro-optical instruments, as well as the more complex, play an important role in modern diagnostic medicine.

References

Digi-Key Corporation, P.O. Box 677, Thief River Falls, MN 56701-0677. 1-800-344-4539.

Chapter 12

Integrated Circuit Electro-Optical Sensors

There are a number of different types of light sensor on the market: photoresistors, photovoltaic cells, and photodiodes or phototransistors are among the most common. In this chapter we will look at a type of sensor that builds a photodiode or transistor into a package with an operational amplifier.

Figure 12-1 shows several different package styles for electro-optical sensors. **Figure 12-1a** and **Figure 12-1b** are eight-pin miniDIP integrated circuit packages. The version in **Figure 12-1a** is the standard black or dark gray package, but with a clear window over the die containing the photodiode and op-amp circuitry. In **Figure 12-1b** a different tactic is used. In this type of sensor the entire body of the IC is made from clear plastic. The variant shown in **Figure 12-1c** is a side-mounting type. The five pins needed for connection of the amplifier come out the side of the package. The package itself is approximately the size of the eight-pin miniDIP package. This "sideways" design solves some practical mounting problems sometimes encountered in electro-optical sensor projects. The final form of package (**Figure 12-1d**) is a

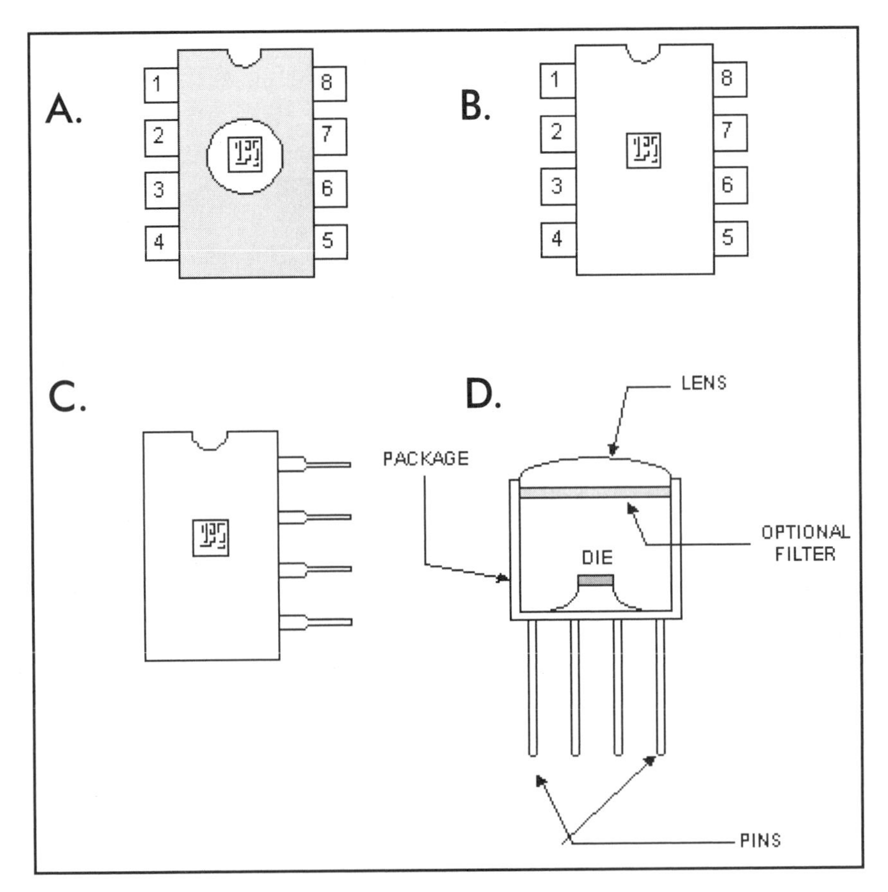

Figure 12-1. Assorted IC photosensor packages.

standard TO-5 transistor or TO-39 IC metal can, but with the top removed and replaced with either a clear glass windows to admit light, or a lens. The package is handled in the same manner as any other TO-5/TO-39 package.

In some cases, electro-optical IC sensors will include a filter as part of the light window to admit some wavelengths and reject others. Forrest Mims used a special electro-optic IC sensors that has an ultraviolet filter incorporated into the window of the sensor in his famous UV detector. It was detailed in *Science Probe!* magazine. Other projects using this type of sensor were published in *Scientific American* when Forrest was the editor of the Amateur Scientist column.

Although there are a number of different sensor circuits, **Figure 12-2** is sufficiently representative to cover the class. This circuit is derived from the internal circuitry of the Burr-Brown OPT202 product. The OPT202 includes a photodiode and transimpedance amplifier in an IC package (Figures 12-1a, b and c are available for the OPT202 device).

Like most sensors of this class, the peak response is in the near-infrared region, around wavelengths of 750 nanometers. Other models also peak in the IR region, but at slightly different wavelengths. Significant output can be obtained at wavelengths as long as 1000 nm, or as short as 350 nm (in the UV range). If the peak response output of the sensor is 500 mV/μW, then "significant output" at the ends is defined (by me) as 100 mV/μW. Some response with lower output levels is also available.

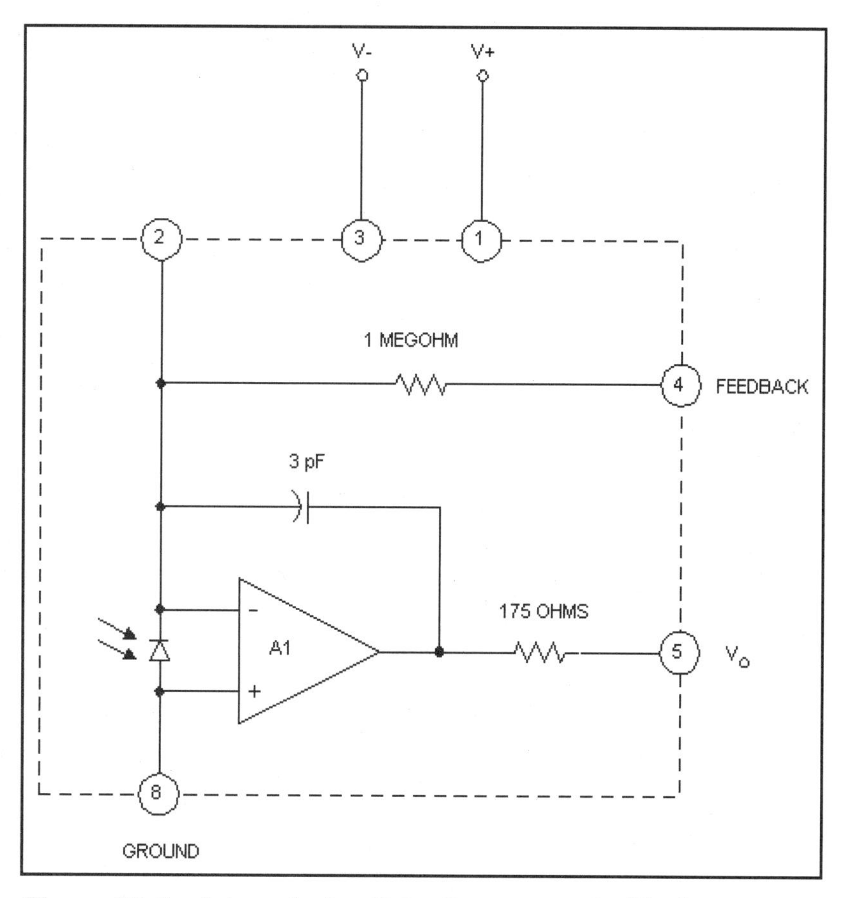

Figure 12-2. Internal circuitry of an example IC photosensor (OPT202).

If you need an ultraviolet sensor, then the Burr-Brown OPT301 device may suit your needs. It has a double-humped spectral response. One peak is about 750 nm in the infrared region, and the other peak is at 300 nm in the ultraviolet region. The response shown in the data sheet for the IR peak is 500 mV/μW, while for the UV it is about 180 mV/μW or so. The response between peaks, at least over about 450 nm to 650 nm (visible light), shows excellent linearity.

Figure 12-3 shows the basic connection for the device. The feedback terminal (pin no. 4) and the output terminal (pin no. 5) are strapped together, and the ground terminal (pin no. 8) is, well, grounded. The DC power supply is bipolar, with a voltage range of ±2.25 to ±18 volts dc. The output of the sensor is $V_o = I_D R_F$, where I_D is the current in the photodiode and R_F is the 1 megohm internal resistor.

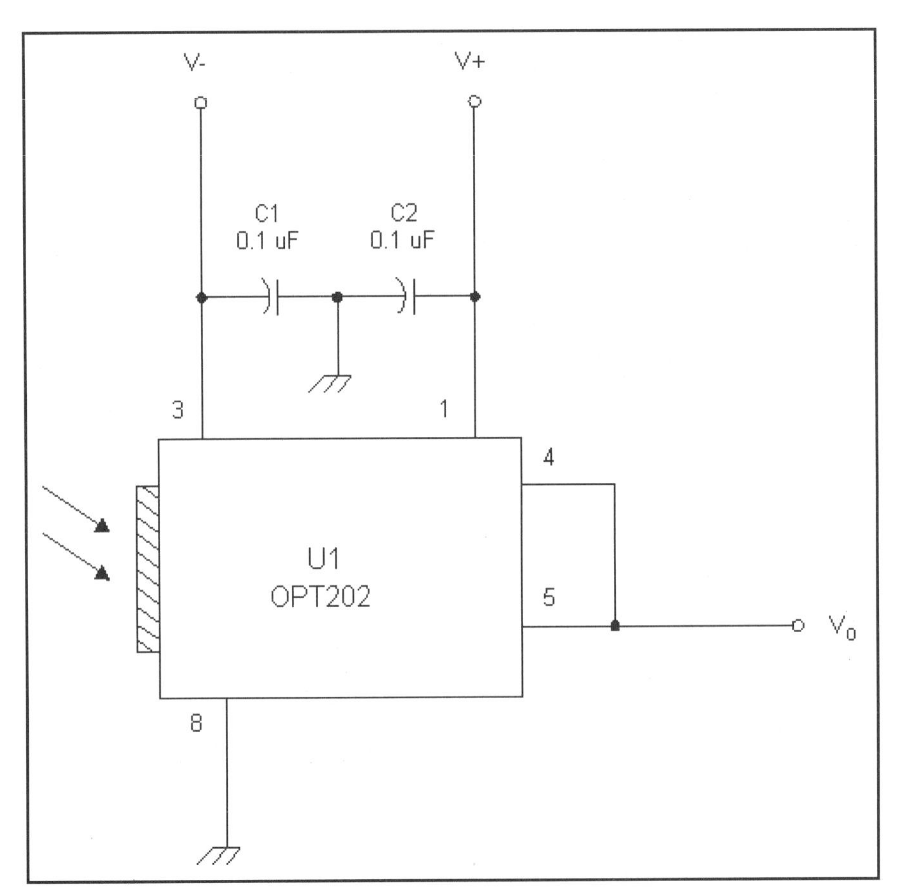

Figure 12-3. Basic connection of the photosensor.

If an external resistance is used in series with the feedback path (i.e., connected between pins 4 and 5 as shown in **Figure 12-4a**), then the total value of R_F is R_{EXT} + 1 megohm. Similarly, if the external resistor is connected in parallel with the internal resistor (**Figure 12-4b**), then the total value of R_F is the parallel combination:

$$R_F = \frac{R_{EXT} \times 1\,megohm}{R_{EXT} + 1\,megohm} \qquad \text{eq. (12-1)}$$

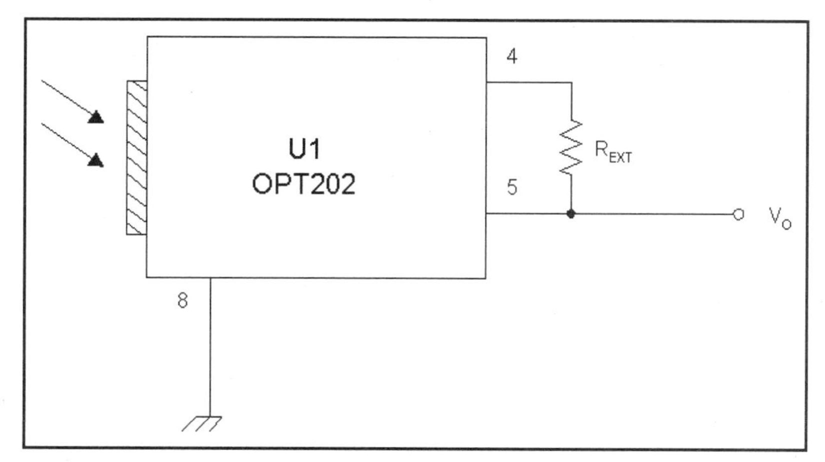

Figure 12-4a. Adding gain by using external feedback resistor.

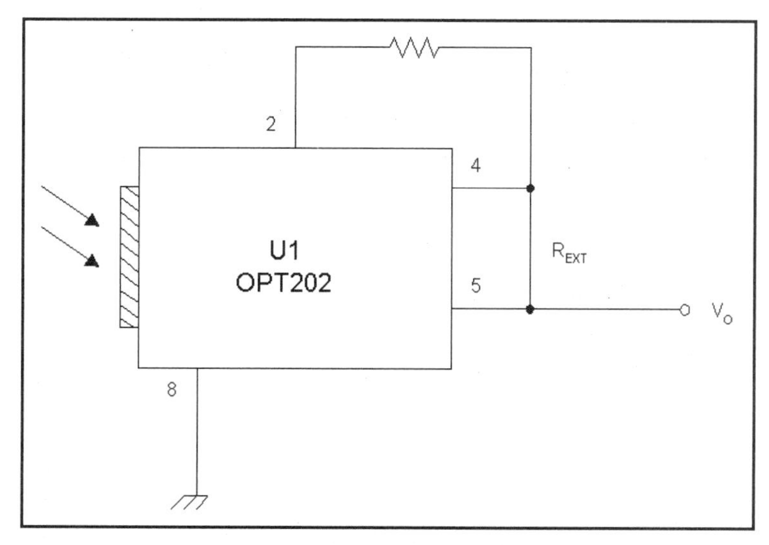

Figure 12-4b. Reducing gain with parallel external feedback resistor.

In **Figure 12-4** the dc power supply connections are deleted for sake of simplicity. If R_{EXT} is 1 megohm or higher, or ≤100K, then no external shunting capacitor is needed, otherwise use 2 pF.

A higher gain circuit is shown in **Figure 12-5**. The feedback signal is provided by a voltage divider consisting of resistors R1 and R2. This type of circuit is sometimes called a "super-gain" circuit in operational amplifier books. The value of the output voltage, V_O, is:

$$V_O = \frac{(R1 + R2)\, I_D\, R_F}{R2} \qquad \text{eq. (12-2)}$$

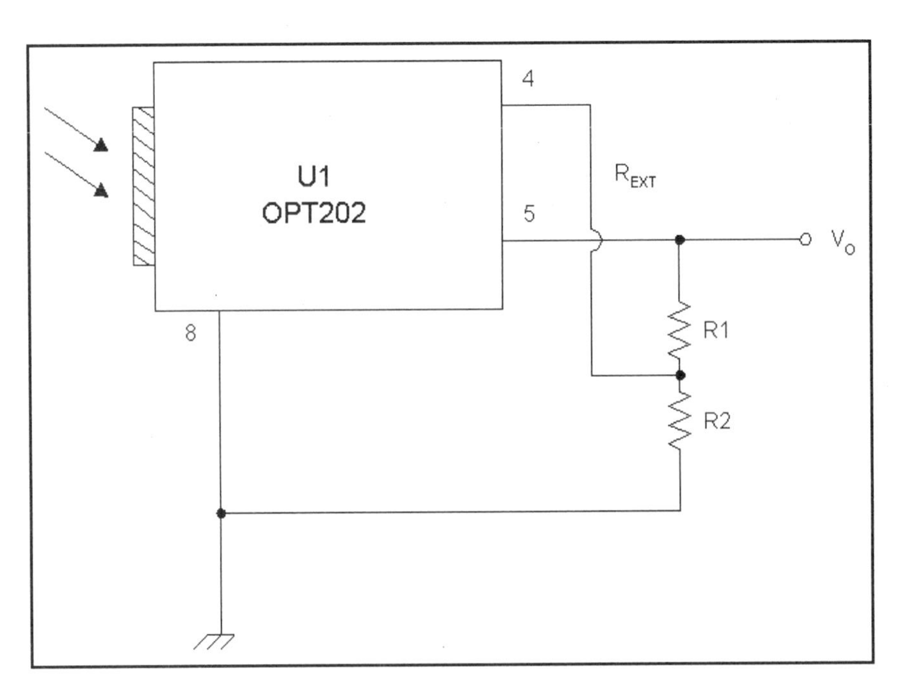

Figure 12-5. High gain circuit uses external voltage divider.

On the down side, this circuit also increases the offset voltage error and the noise contribution of the sensor. These problems are largely due to the gain. However, on the plus side, in many applications further amplification is not needed. Also, printed circuit board leakage, which can be a problem in any circuit where high value resistors are used, is minimized. It is possible to obtain the required gain with series connected external resistors, but this circuit uses lower value resistors to achieve the same end.

One of the problems with any light sensor circuit is that we often are looking for light changes, or small point variations in a larger ambient light field. What is needed in those cases is the ability to remove the effect of ambient light on the overall output voltage. **Figure 12-6** shows a method for doing that job. This circuit uses the feedback principle to adapt the circuit to ambient light, and is called a dc restoration circuit.

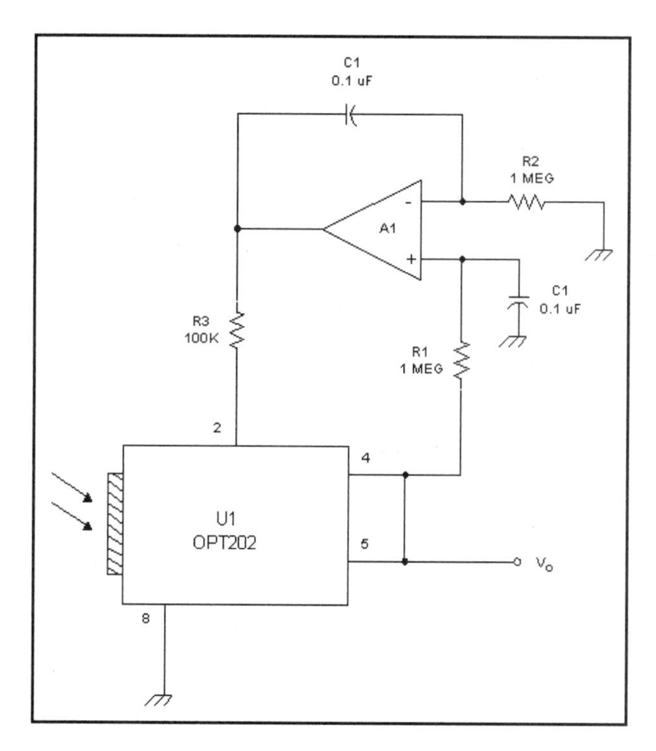

Figure 12-6. dc restoration circuit to keep ambient light from affecting output signal.

The operational amplifier (A1) in the feedback path between pins 2 and 4 is a Miller integrator. This circuit acts like a high-pass filter with a 16 Hz cut-off frequency. As a result, slowly changing or steady state light is below the cut-off frequency, so does not affect the output voltage. For other cut-off frequencies:

$$F_{-3dB} = \frac{R1}{2\pi\, R2\, R3\, C2} \qquad \text{eq. (12-3)}$$

Where:

F is the frequency in hertz.

R1, R2 and R3 are in ohms.

C2 is in farads.

Differential Light Measurement

A differential light measurement circuit is shown in **Figure 12-7**. In this case, the entire circuitry of U1 is used in the normal manner. The output terminal (pin no. 5) and the feedback terminal (pin no. 4) are strapped together in the manner of **Figure 12-3**. The only part of U2 that is used is the photodiode. It is connected in series with the photodiode of U1 in a manner that the overall current produced by the pair is the difference of the currents in the two diodes. Hence, the overall output voltage is proportional to the difference in light intensities applied to the two sensors.

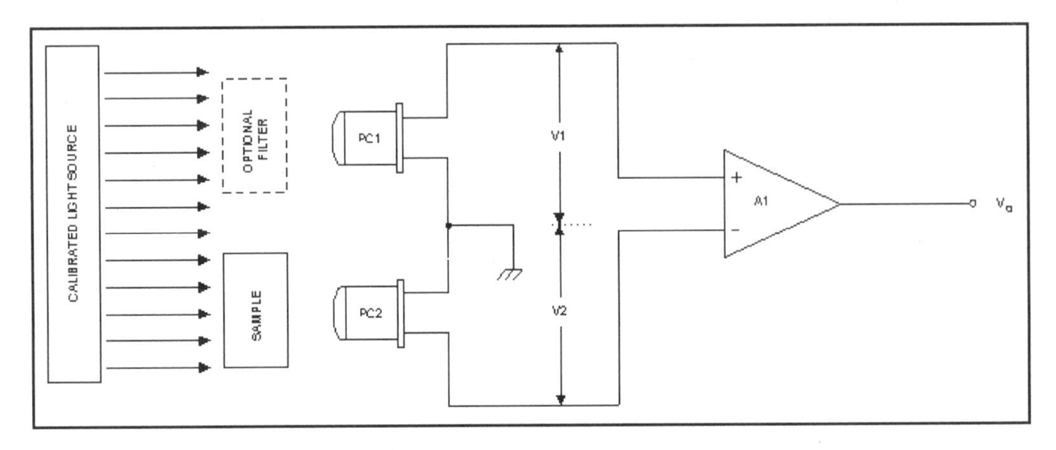

Figure 12-7. Differential light sensor.

Chapter 13

X-ray Sources and Radiation Detectors

The term "radiation" is used to refer to several different physical phenomena. From the atom come three forms of radiation: *alpha*, *beta*, and *gamma*. Alpha particles are positively charged helium nuclei with two neutrons and two protons. Beta particles are electrons ejected from the atomic nucleus, and are negatively charged. Gamma rays, like X-rays, are high energy electromagnetic waves, differing from them primarily in their origin. Like alpha and beta rays, gamma rays are from nuclear sources, while X-rays originate outside the atomic nucleus.

X-rays, a form of electromagnetic energy, are very similar to light and ultraviolet, except that the wavelength of X-rays is shorter (0.1 nm or less). Indeed, many spectrum charts (**Figure 13-1**) show the ultraviolet and X-ray bands overlapping.

X-rays are most familiar for their medical uses, but other uses are also known: industrial inspection, airport security, physics experiments, and even gemstone identification. In the familiar medical context, X-rays are a means of taking either film-emulsion photographs or video pictures of internal organs,

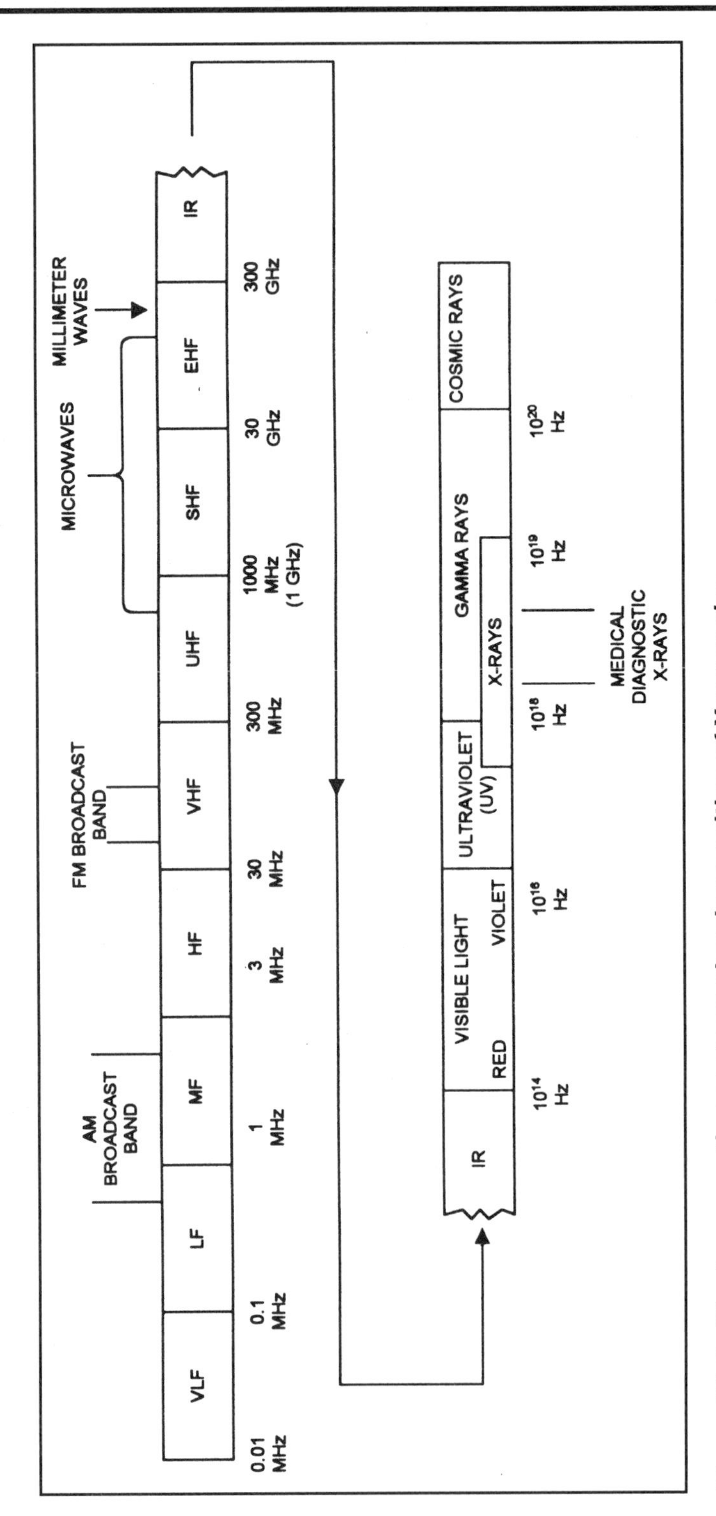

Figure 13-1. Electromagnetic spectrum showing position of X-ray region.

bones or foreign objects inside the human body. Another medical application is in cancer treatment, where X-rays are one source of the radiation in radiation therapy. Industry uses X-rays in a variety of inspection and testing modes, while scientists often use them to investigate nature.

Although X-rays are commonplace today, they were once viewed as almost magical. X-rays were first discovered in 1895 by the German physicist Wilhelm Roentgen. Using a modified Crookes tube apparatus, Roentgen passed X-rays through his wife's hand to a photographic plate. When the plate was developed, the plate showed her bones and wedding ring. Although Roentgen proposed the term "X-ray" because the rays were of unknown origin, some of his contemporaries insisted on using the term "Roentgen rays" in honor of the discoverer. Both terms are used today, but in the USA X-ray is the more generally accepted term.

The Nature of X-Rays

X-rays are a form of electromagnetic radiation similar to radio waves and light waves, except that the frequency is much higher and the wavelength is much shorter. One should know the terminology of waves, of which electromagnetic waves are but one form, so a quick review of the fundamentals is provided. Wavelength is denoted by the Greek letter lambda (λ), and is measured in meters (m), micrometers (μm), nanometers (nm) or Angstroms (Å).

A radiating wave does not consist of a single cycle, but rather a train of cycles proceeding one after another. The frequency of the wave is the number of cycles that occur in one second. The unit of frequency is thus cycles per second, also designated hertz (Hz). Wavelength and frequency of an electromagnetic wave are related by the expression:

$$c = \lambda f \qquad \text{eq. (13-1)}$$

Where:

c is the speed of electromagnetic wave propagation (3×10^{10} m/s).

λ is the wavelength in meters (m).

f is the frequency in hertz (Hz).

X-rays, like light and radio waves, are transverse electromagnetic waves; that is, their electric and magnetic fields oscillate orthogonal to each other and to the direction of travel. X-rays wavelengths are shorter than those of infrared, visible light or ultraviolet waves. **Figure 13-1** shows a partial spectrum chart that puts some of the various forms of electromagnetic radiation into perspective. Note that the X-ray band overlaps the ultraviolet and gamma ray bands of the spectrum. (The band of the diagnostic medical X-rays is smaller than the total X-ray band.) It is artificial, although commonplace, to show gamma rays and X-rays as separate entities because the principal difference is their respective origins. The practice is to designate as X-rays those waves that are generated outside of the nucleus by the interaction of an energetic kinetic electron with a heavy nucleus; gamma rays, on the other hand, originate within the nucleus. But a spectrum chart, that considers only wavelength and frequency, shows that X-rays are a subset of the gamma ray spectrum.

Like other forms of electromagnetic radiation, X-rays can be viewed both as particles and waves. In other words, the X-ray has a complementary nature. This theory, proposed by Niels Bohr, was discussed in the first chapter of this book.

A discovery about electromagnetic radiation was made in 1900 by Max Planck. While working on certain experimental anomalies in the thermal radiation of blackbodies he observed that the wavelengths did not meet the pattern that classical (Newtonian) physics predicted. Planck proposed a solution that required electromagnetic radiation to exist only in discrete bundles, later called *quanta*. Thus, radiation can be regarded either as a wave or a particle with an energy that is proportional to the frequency by the relationship:

$$E = h f \qquad \text{eq. (13-2)}$$

Where:

E is the energy level in joules (J).

h is Planck's constant (6.626×10^{-34} J-s).

f is the frequency in hertz (Hz).

Example 13-1

Calculate the energy of an X-ray that has a frequency of 10^{19} Hz.

Solution:

- $E = h f$

- $E = (6.626 \times 10^{-34} J - s)(10^{19} Hz) = 6.626 \times 10^{-15}$ joules

As the frequency increases into the X-ray/gamma-ray region, it is commonplace to express energy not in joules, but rather in electron-volts (eV), such that 1 eV = 1.602×10^{-19} joules. The energy level in Example 13-1, 6.626×10^{-15} joules, is:

- $$6.626 \times 10^{-15} \text{ joules} \times \left(\frac{1 \text{ eV}}{1.602 \times 10^{-19} \text{ joules}} \right) = 41{,}361 \text{ eV}$$

Early in the investigation of X-rays it was discovered that they were dangerous; sadly, it was only after a number of scientists and physicians had died of leukemia and other blood or lymphatic system disorders that the extent of the danger was recognized. The high energy of the X-ray can penetrate deep into organs (which is why they are medically useful), and can damage cells, turning them cancerous in some cases. Many people, including health care professionals, fail to appreciate the danger that X-radiation can pose. A letter was once published in a local newspaper from a dentist in response to an article criticizing the overuse of dental X-rays. The dentist retorted that he was taught in dental school that a dental X-ray contained about the same dosage as the ultraviolet radiation received from the sun during a summer afternoon on the beach. Implicit in his argument is that UV is not terribly dangerous, so makes a favorable point of comparison. There were a couple of problems with the dentist's argument:

First, anyone who has experienced a sunburn from UV after an afternoon on the beach is likely to doubt that UV is safe; anyone who has melanoma (a frequently fatal form of skin cancer) will have no doubt about the danger of UV.

Second, the dentist fallaciously transferred the X-ray problem to the UV problem by a logically illicit equation of UV and X-rays. The dosage is the area under the energy versus time curve. The "afternoon on the beach" is low intensity over a long time, while the X-ray is a very high intensity "ionizing" radiation for a short time (10 milliseconds). The implication here is that the X-ray can: a) damage different structures, b) do more serious damage, and c) damage deeper (hidden) organs that are more difficult to diagnose than skin. Logically, this is called the false comparison fallacy, and in this case it is quite dangerous to accept. The X-ray and UV from the sun may give the same dosage, but the fact is that neither are "safe" and the X-ray is considerably more potent.

Lower-frequency waves are called *nonionizing radiation*, and, until recently, were considered safe at all intensities less than that required to severely heat bodily tissues (microwave ovens cook by heating food internally with very short-wavelength radio waves). Recently, mechanisms other than ionization and heating are suspected of causing some forms of leukemia and non-Hodgkin's lymphomas. The issue is not settled at this writing, although research continues. Therefore, it is prudent to limit exposure to all sources of electromagnetic radiation to the minimum possible level.

Generating X-rays

Medical X-rays are usually generated using a phenomenon called *Bremstrahlung* (**Figure 13-2**). An electron with energy E_i collides with a target containing heavy nuclei. As the electron is deflected around the nucleus, it loses some energy, and assumes a new energy level, E_d. The difference between the incident and deflected electron energy levels, E_i and E_d, must be conserved, so it becomes a photon of X-ray energy. The energy level of the X-ray photon is given by:

$$E_x = E_i - E_d \quad \text{eq. (13-3)}$$

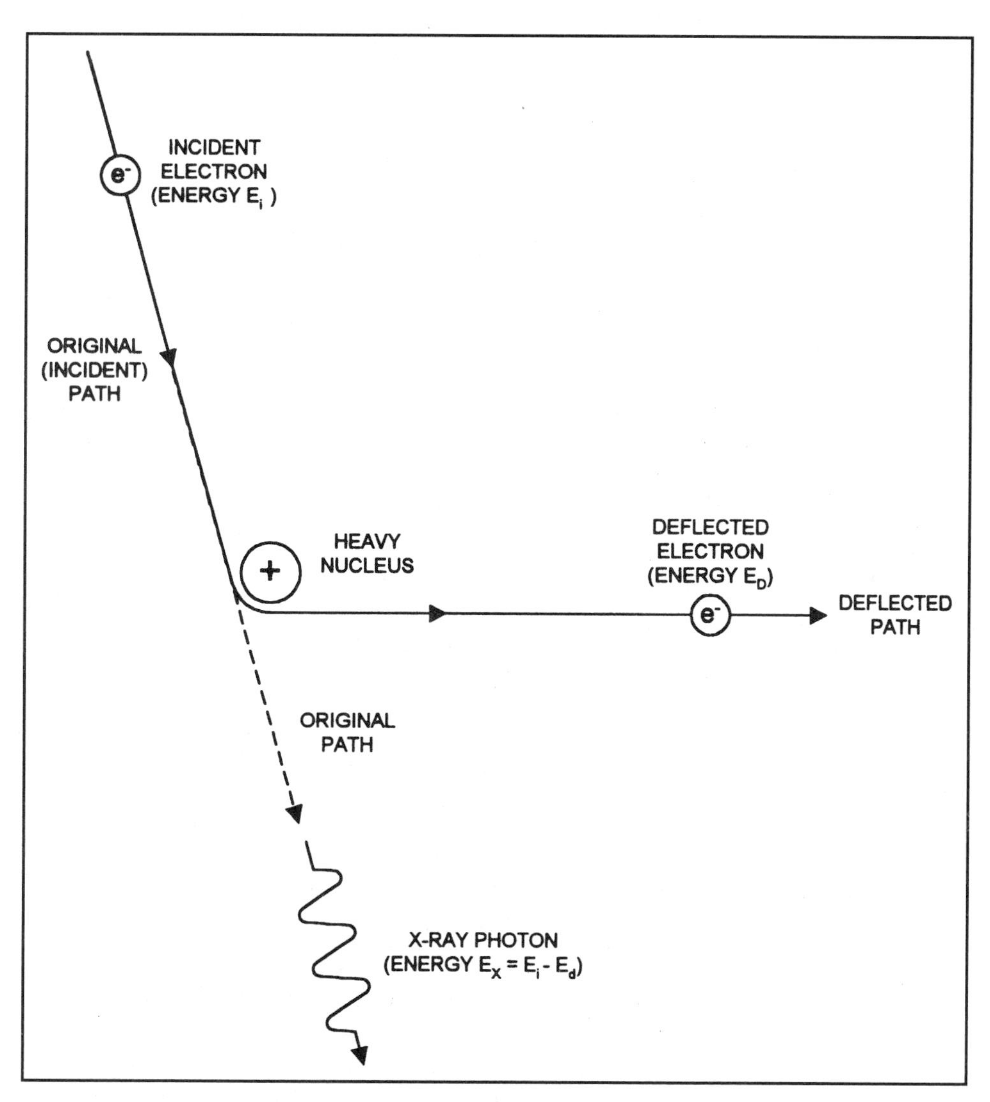

Figure 13-2. Bremstrahlung effect produces X-ray when an electron deflected by a heavy nucleus gives up energy in the form of an X-ray photon.

Figure 13-3 shows an early X-ray generator tube based on the *Bremstrahlung* phenomenon. It is a vacuum tube containing an electron emitting cathode and a target anode. Electrons from the negative cathode are accelerated by the positive charge on the anode and anti-anode, to a kinetic energy state proportional to the electrical potential difference between cathode and anode. When the accelerated electrons strike the anti-anode ("target"), the kinetic energy is given up. Because of the conservation of energy law, the kinetic energy of the electrons is converted to heat (infrared radiation) and X-rays. The wavelength (and therefore frequency) of the radiation is proportional to the kinetic energy of the electrons.

The electrons are generated by a phenomenon known as *thermionic emission*. When a metallic surface is heated to incandescence, a cloud of electrons is emitted, or "boiled off" from the surface. These electrons form a *space charge* around the surface. In X-ray tubes and radio vacuum tubes, the metallic surface is a wire filament similar to that inside an ordinary light bulb.

Thomas Edison could have begun the radio era twenty or thirty years earlier than actually happened because of his efforts to reduce the blackening of the inside of his early light bulbs. One of his early experiments used a positively charged anode inside the tube. Edison noted that an electrical current flowed between the filament and the anode, a phenomenon now called the *Edison effect*. It is this effect that produces the stream of accelerating electrons that strike the anode/target in an X-ray tube.

The materials of the target anode are selected to make X-ray generation easier. The applied high-voltage potential (V+) also helps determine the kinetic energy of the accelerated electrons, and hence the frequency and wavelength of the emitted X-rays. In a few very old television sets, there was a concern over X-ray emissions caused by an HV regulator tube that operated at higher than usual potentials. Modern TV sets do not seem to exhibit this. Medical, scientific and industrial Bremstrahlung generators are more rigorously designed than TV tubes, so present no particular danger when used in accordance with the manufacturer's instructions and X-ray safety guidelines.

Figure 13-4 shows the basic form of a more modern, but still very simple, X-ray tube. The anode is a beveled assembly of copper or molybdenum alloy inset with a tungsten alloy target region. The tungsten is better for X-ray gen-

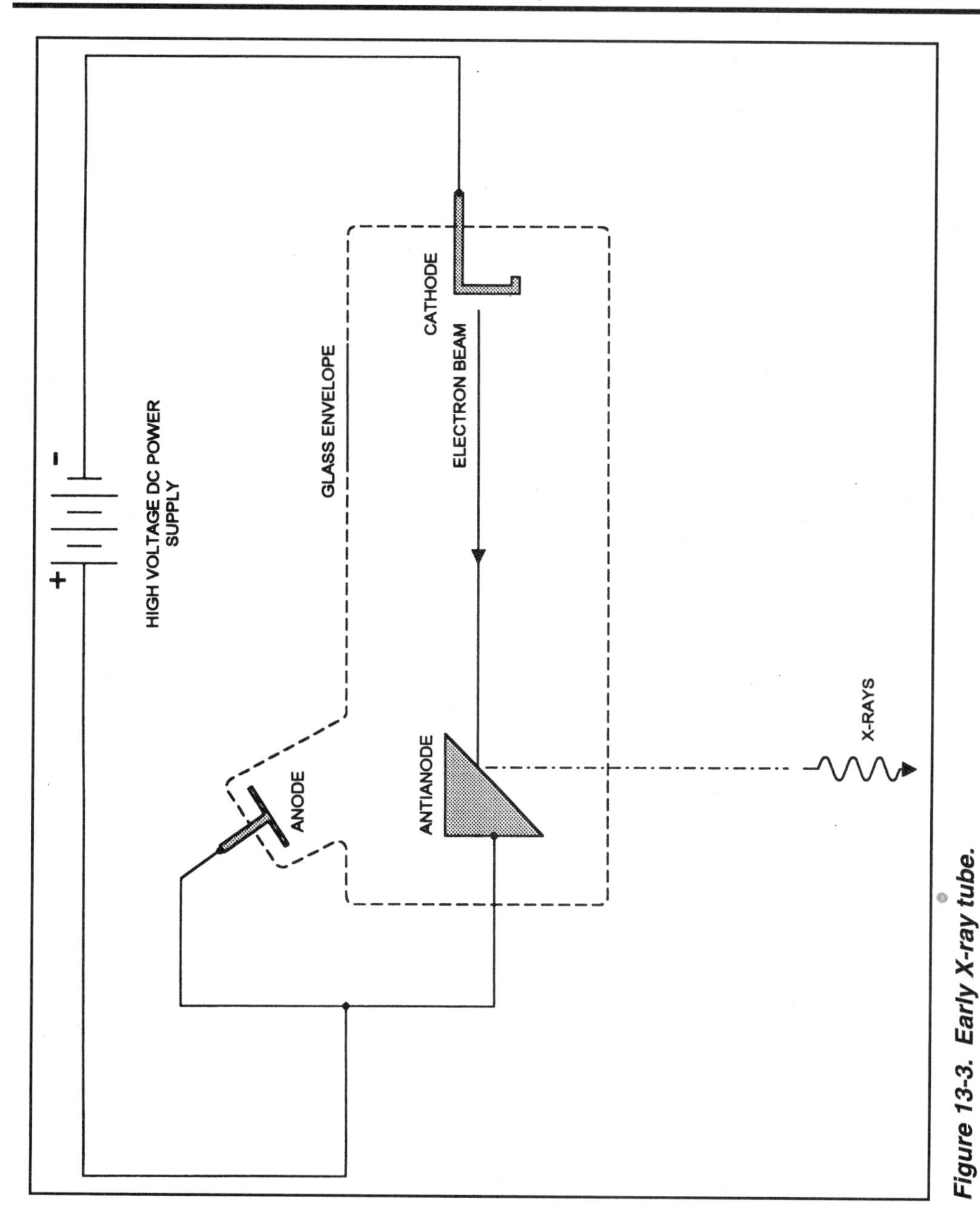

Figure 13-3. Early X-ray tube.

eration, but copper or molybdenum is a better heat sink. The cathode in this case is an incandescent filament. Electrons boil off of the surface of the filament, and then accelerate through a high-voltage electrical potential with an increasing velocity v, and hence increasing kinetic energy toward the target/

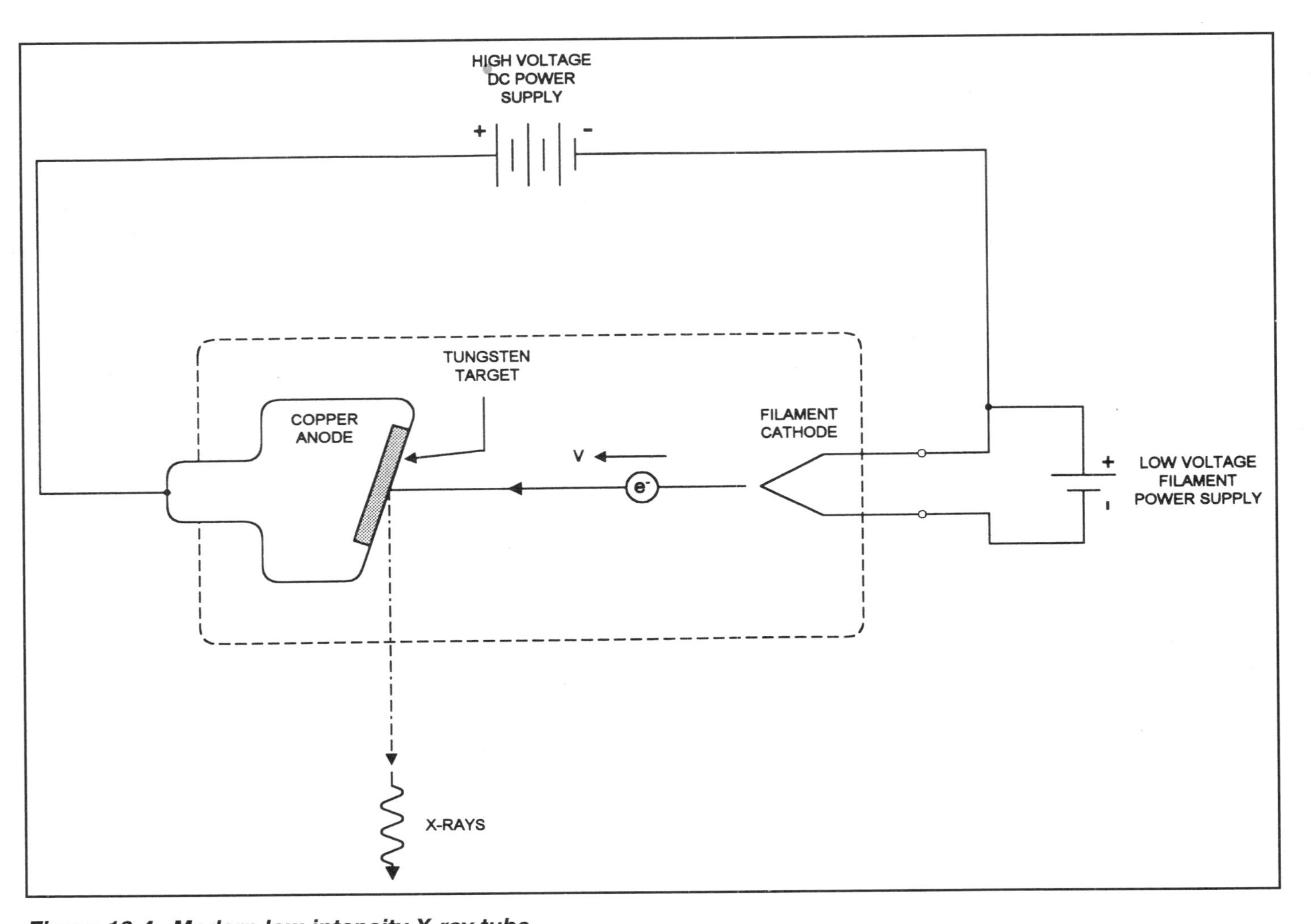

Figure 13-4. Modern low-intensity X-ray tube.

anode. When the electron strikes the target, its kinetic energy is given up as heat and X-radiation. The target is beveled in order to direct the X-rays in the desired direction.

A variant on the basic X-ray tube uses a rotating anode (**Figure 13-5**). The heat created by the kinetic energy of electrons colliding with the target is tremendous, and overheating is a major cause of X-ray tube failures. The rotating anode spreads the heat energy over a larger volume of metal, and incidentally produces a more narrowly focused beam, and hence sharper images on the film or video camera.

Many of the photosensors used in light-operated instruments will also work at X-ray wavelengths. Certain phototubes, photomultipliers, photodiodes, and phototransistors will also work for making X-ray measurements. In this chapter we will look at certain sensors that are used to make X-ray measurements: the Geiger-Mueller tube, the scintillation cell, the PN junction diode and the photomultiplier tube.

Geiger-Mueller Tubes

The Geiger-Mueller (G-M) tube is a glass or metal cylinder that is filled to a less-than-atmospheric pressure with an ionizable gas (e.g., argon with a trace of bromine, at 100 Torr pressure). When radiation impinges on such a gas, its energy forces the gas into ionization, thereby altering the electrical characteristics of the G-M tube. Those changes in electrical characteristics are the measurable event in this instrument.

The general circuit for the G-M tube is shown in **Figure 13-6**. In most applications, the external circuitry consists of a power supply and a series current-limiting (or load) resistor. The signal of interest can be either the current flowing in the external resistor or the voltage spikes appearing across it. Different types of radiation instruments use these signals in different ways.

There are three modes of operation for the G-M tube, and these modes affect both the subsequent circuitry and the voltage level required. These regions are shown in the curves in **Figure 13-7**, and are designated as follows: a) ionization chamber mode, b) proportional counter mode, and c) Geiger counter mode.

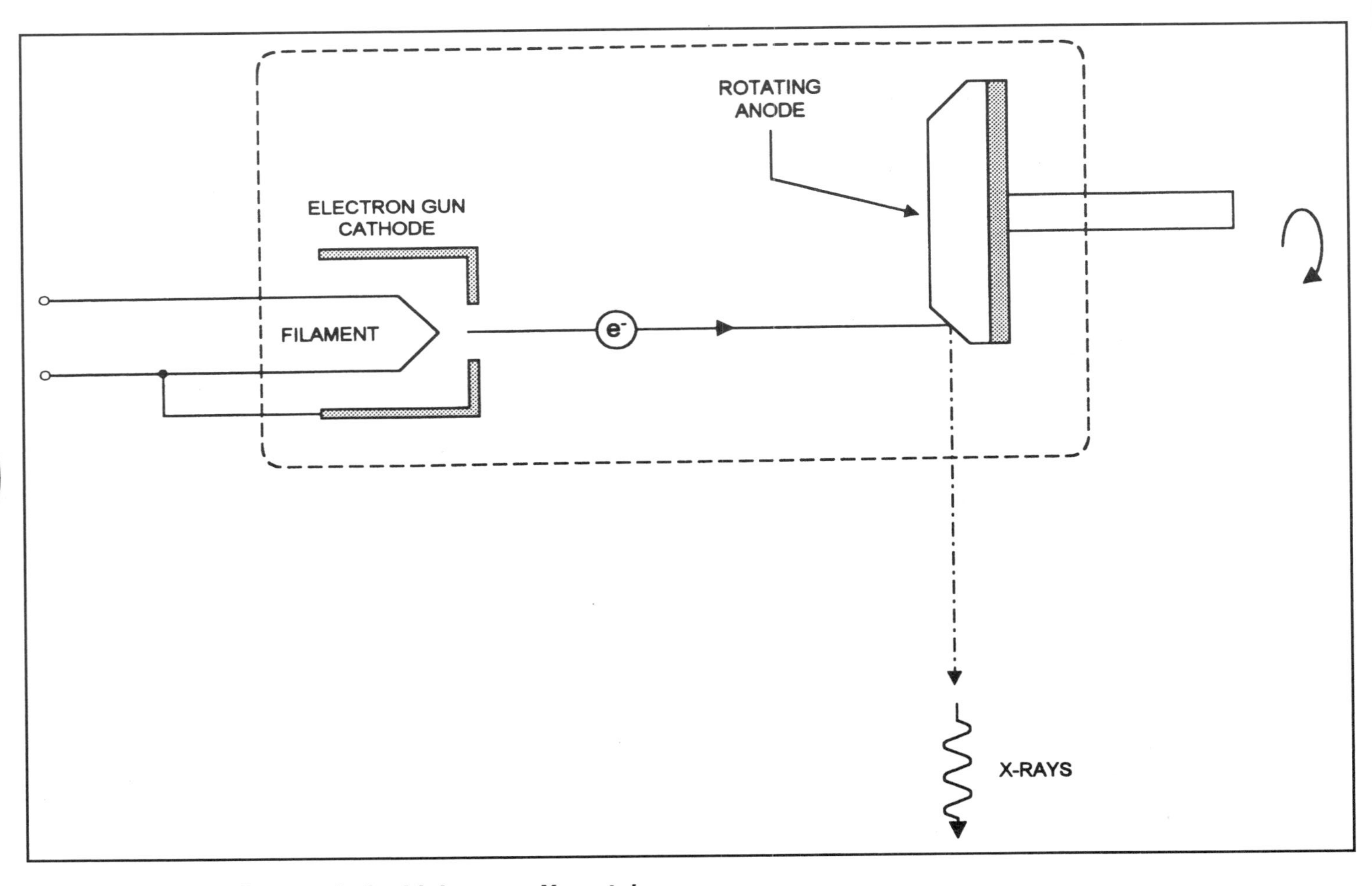

Figure 13-5. Rotating anode for high-power X-ray tube.

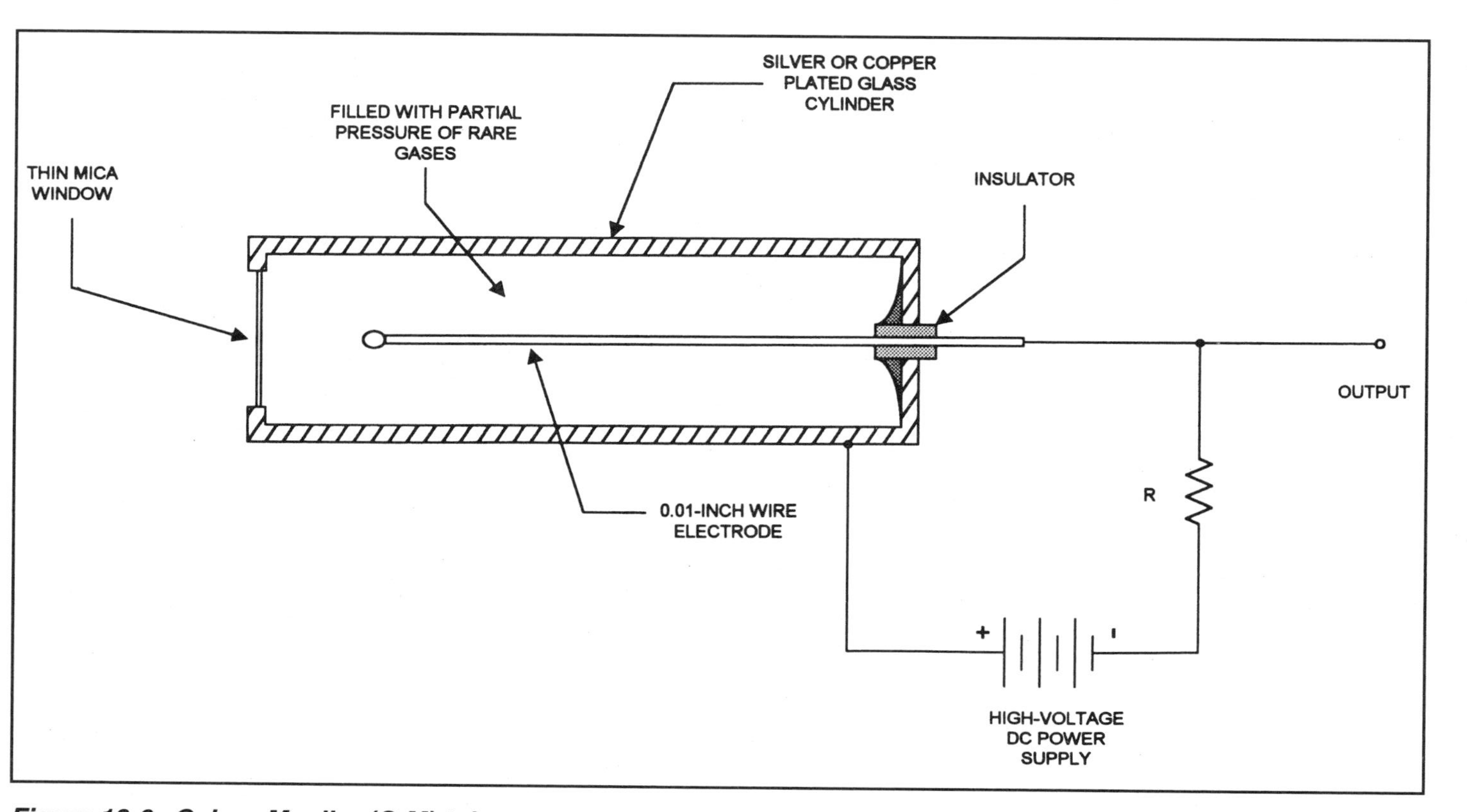

Figure 13-6. Geiger-Mueller (G-M) tube.

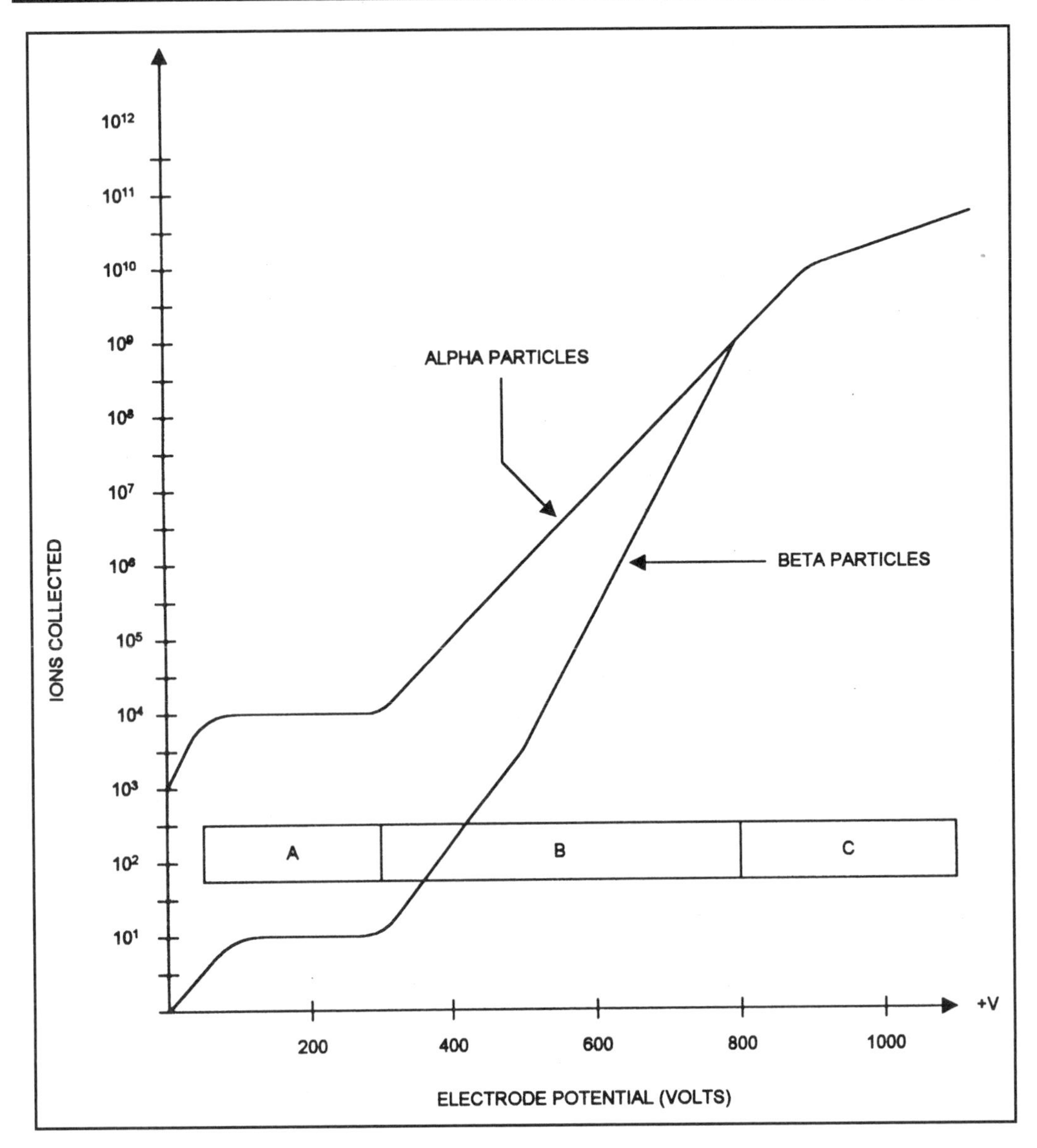

Figure 13-7. Regions of operation of the G-M tube.

In the ionization chamber mode, a weak electric potential is applied across the G-M tube, so only a few electrons are generated by the impinging radiation. Nearly all of these electrons are collected by the electrode. The current flowing in the external load resistor is proportional to, and therefore a measure of, the amount of impinging radiation.

In the proportional counter mode, the electric field is stronger, so the electrons generated by the ionizing gas reach sufficient kinetic energy to release additional (or secondary) electrons by kinetic collisions with gas molecules. The output across the load resistor will be a voltage spike with an amplitude that is proportional to the kinetic energy of the ionizing radiation particle. In this mode, the G-M tube can count ionizing particles on a one-for-one basis.

The Geiger counter mode uses very-high voltage potentials of 800 to 2000 volts. The operation is similar to the proportional counter mode, except that the kinetic energies are so high that a large electron multiplication effect, or avalanche, takes place. The output pulses are approximately the same amplitude every time the tube fires. An external counter circuit will measure the number of pulses per unit of time, indicating the level of radiation.

Photomultiplier Tubes

The photomultiplier tube (PMT) discussed in Chapter 4 is an amplifying device that uses the photoelectric effect to generate a current proportional to the impinging radiation. The photoelectric effect calls for the emission of electrons when a light wave (or higher-frequency UV or X-radiation) strikes a photoemissive metallic surface (the "screen" in **Figure 13-8**).

In a PMT, photons strike the screen and cause electrons to be emitted. The electrons are attracted to a series of anodes (called *dynodes*), each of which has a slightly higher positive potential. When electrons strike each dynode they knock loose other electrons. Each electron is multiplied as the electron stream accelerates towards the final collector anode. Thus, a large current is produced from the collector in response to a relatively low level of radiation applied to the screen.

The photomultiplier tube is used in medical X-rays for at least two applications. One is in fluoroscopic X-ray systems, in which the output of the PMT is applied to the video input of a television monitor system. Thus, the X-ray picture is projected onto a TV screen. Another application is in computer-assisted tomography (CAT) scanners. The PMT is considerably more sensitive than ordinary X-ray film, so one can take a multiple-position X-ray, as needed for a CAT scan, using relatively small doses of radiation.

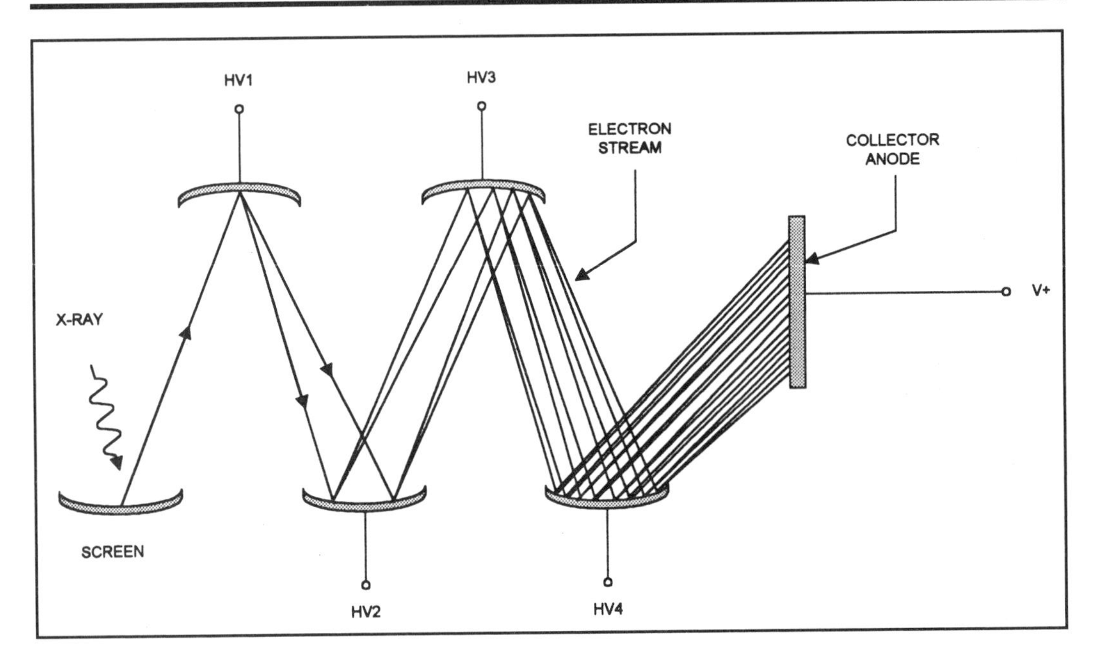

Figure 13-8. Structure of the PMT X-ray detector.

Scintillation Counters

An example of a scintillation cell (counter) is shown in **Figure 13-9**. The word scintillation is used to denote a process similar to that which generates light on the screen of a cathode ray tube. When a radiation particle strikes an atom of certain phosphorous materials, its kinetic energy may be added to the energy of the orbital electrons. When the electrons are thus excited, they jump to a higher, unstable energy state. When they fall back to their ground state, the energy absorbed is emitted as a photon of light. Certain crystal materials possess this property.

Figure 13-9 shows a scintillation device (D1) that has a scintillation crystal window that is attached to the light input window of a photomultiplier tube. Radiation causes the crystal to scintillate, and the light thus produced is received and amplified by the PMT. The current at the output of the PMT is proportional to the light, hence the radiation level.

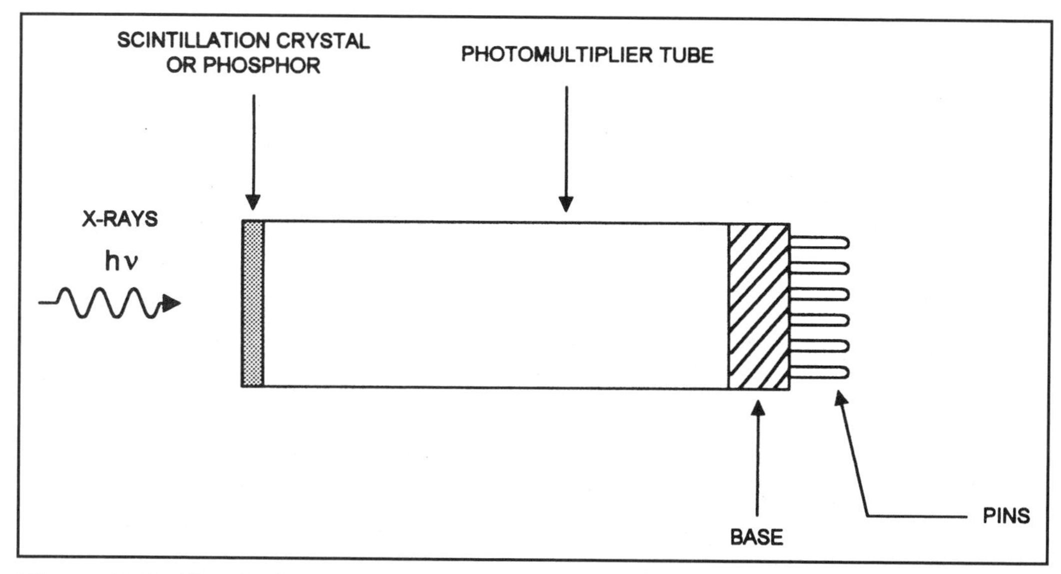

Figure 13-9. Physical arrangement of PMT X-ray detector with a scintillation cell.

PN Junction Diodes

The PN junction diode is used extensively in electronics, and was the first commercially available solid-state device. Its use as a photosensor is well known. The PN junction diode is formed by placing a region of N-type semiconductor material in contact with a region of P-type semiconductor material. When a PN junction diode is forward biased as in **Figure 13-10**, a forward current (I_o) will flow. But if the diode is reverse biased (**Figure 13-11a**), no current should flow. However, as we learned in Chapter 8, there will be a tiny *leakage current* (I_L) across the junction. This negative current can be increased by impinging X-rays (**Figure 13-11b**).

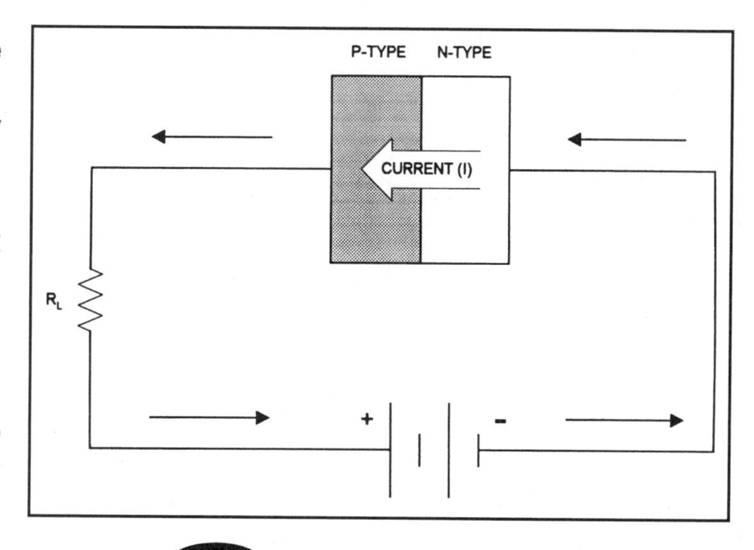

Figure 13-10. PN diode will pass current Io when forward biased.

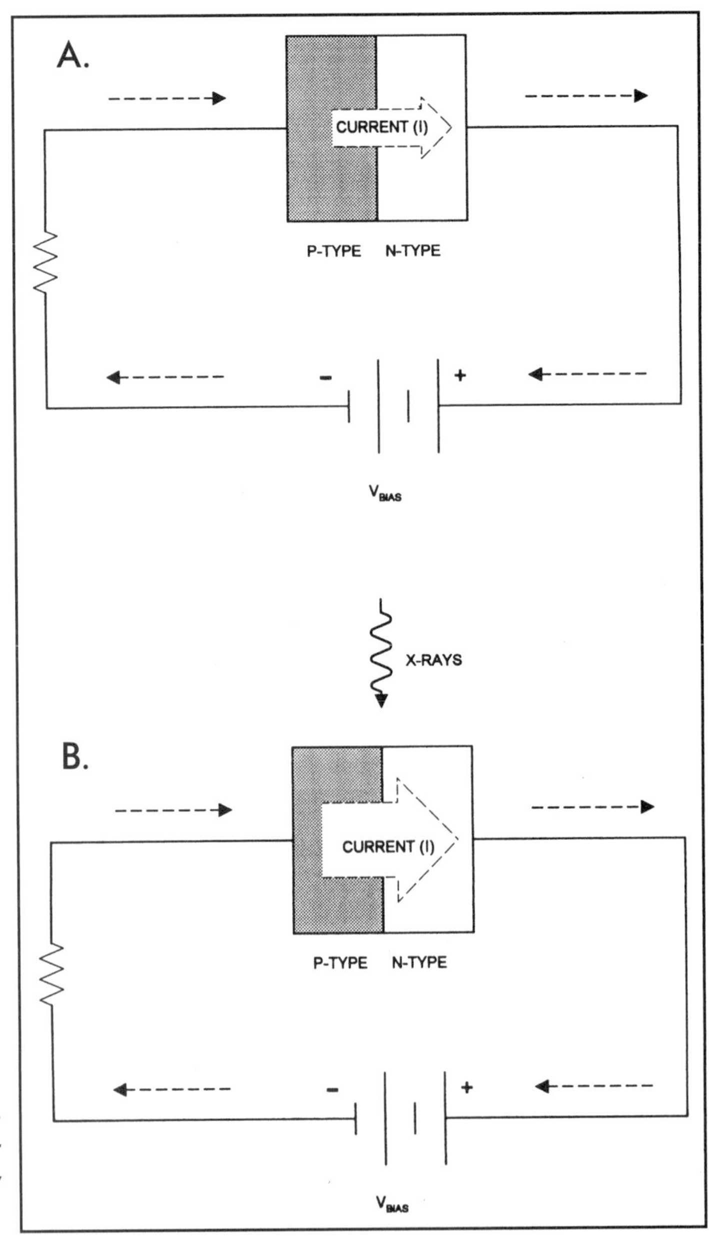

Figure 13-11. A reverse biased PN diode will pass only a small leakage current, I_L, but this current can be increased by impinging X-rays.

Figure 13-12 shows a PN junction diode connected as a radiation detector. A reverse bias is supplied by V_{bias}, and the current is limited by a series resistor, R1. The voltage drop across R1 is proportional to I_L. If radiation causes an increase in leakage current, then the voltage drop (V_o) across R1 will vary in proportion to the applied radiation levels.

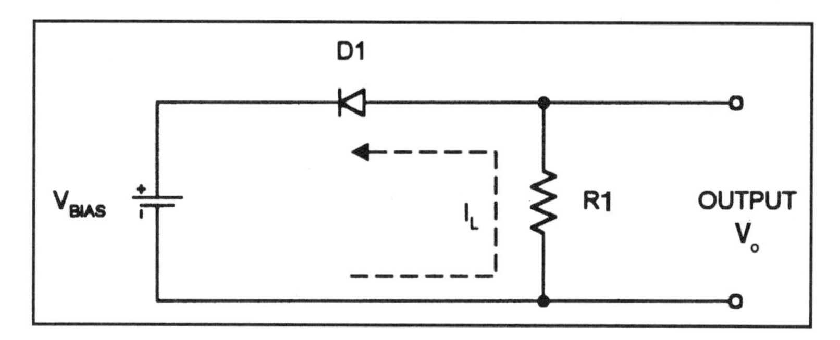

Figure 13-12. Practical circuit using a PN junction diode as a radiation sensor.

Alpha Particle Detector

Figure 13-13 shows a simple but effective alpha particle detector. The sensor (Q1) is a P-channel JFET. The type selected should be in a TO-5 metal can package (e.g., 2N2386). The top of the TO-5 can is removed (see inset to **Figure 13-13**) by sawing or grinding. Be very careful in this operation. The cutting produces small metal particles that could fall into the package. The transistor die is exposed by this operation.

The circuit is a common-source amplifier in which the gate is tied to the +22.5 volt source by a very high value resistor R2, which is 10 megohms. Every time an alpha particle impinges the transistor die gate region, the transistor is forced into conduction, causing the drain-to-source voltage to drop markedly. This action produces a negative-going output pulse, which is coupled to the output through capacitor C1.

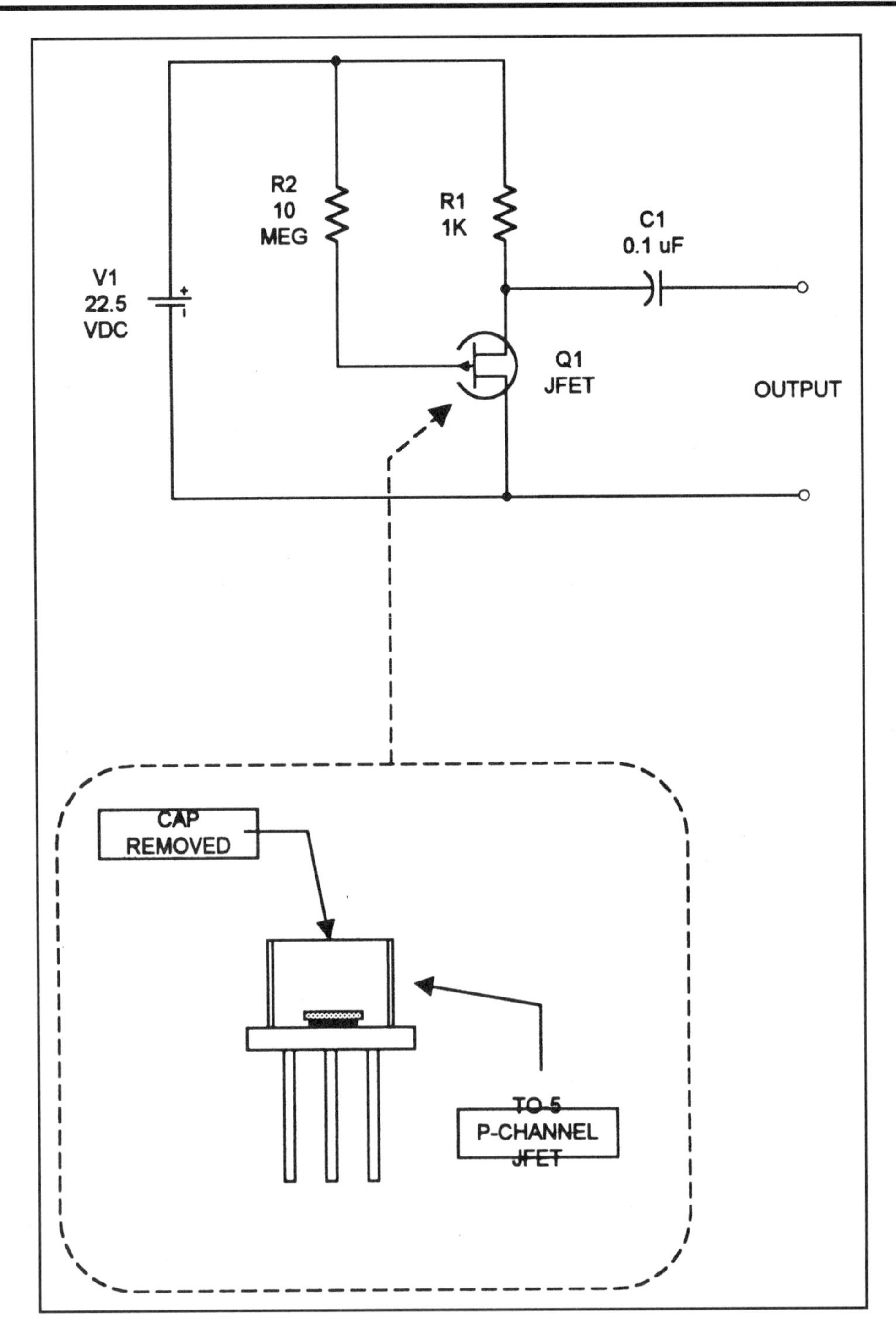

Figure 13-13. Alpha particle detector.

Chapter 14

Lasers

WARNING!

Lasers can be extremely dangerous. The light produced by some lasers can cause severe or permanent damage to eyes and skin. Even low power lasers are dangerous and must be handled with care. Never look into a laser. Reflections can also be dangerous. Become familiar with the laser safety standard set by the American National Standards Institute (ANSI), 1430 Broadway, New York, NY 10018. Always follow the manufacturer's safety guidelines on laser equipment.

Lasers are optical versions of Masers, devices that are used in the amplification of microwave radio signals. Masers (*M*icrowave *A*mplification by *S*timulated *E*mission of *R*adiation) were developed for amplification of very low level radio signals that would normally be obscured by noise. Lasers (*L*ight *A*mplification by *S*timulated *E*mission of *R*adiation) are simply optical-wavelength Masers. Laser light differs from other light in several important ways. The principal characteristics of laser light are: *coherence*, *monochromaticity*, and *low dispersion*. We will deal with these factors shortly.

Laser Classification

Lasers can be classified according to different schemes. A common system is to use the form and material of the laser: gas, solid-state or PN junction, for example. Another method of classifying lasers is according to the safety categories published by the ANSI, and the Laser Institute of America, in ANSI Standard Z136.1-1986.

ANSI classifies lenses as Class I, II, III and IV. A Class I laser is one that is not capable of producing biological damage to the eye or skin during intended uses, by either direct or reflected exposure. These lasers may emit light in the 400 to 1,400 nm wavelength region. Class II lasers emit visible light in the 400 to 700 nm region. Normal aversion responses such as blinking are believed to be sufficient to afford protection against biological damage from direct or reflected light. For Class II CW (continuous wave) lasers, the point source power used in exposure calculations is 2.5 mW/cm^2. According to ANSI, Class III lasers "...may be hazardous under direct and specular reflection viewing conditions, but diffuse reflection is usually not a hazard." Class III lasers typically do not present a fire hazard. A Class IV laser is a dangerous, high-powered device, that may be both a fire hazard and hazard to eyes and skin from both direct and specular reflected beams, and sometimes also from a diffuse reflected beam.

Laser classes are determined by authorized Laser Safety Officers (LSOs), according to criteria established by ANSI standards. Consult the latest edition of the ANSI standards for current guidelines on laser safety.

Basic Concepts

The concept of coherence is very important in understanding lasers, and is responsible for much of their observed behavior. Coherent light is light in which points on two or more waves traveling together maintain their relationships regarding amplitude, polarity and phase. In a non-coherent light source, the various emitted waves have a random relationship, so their mutual interference is random. Lasers exhibit two types of coherence: *longitudinal* and *transverse*.

Longitudinal coherence is shown in **Figure 14-1**, in which two waves of identical frequency and wavelength are propagated along parallel paths. The time phasing of the two waves is the same; that is, they intercept the axis in the same direction and at the same time. The phasing of such waves in laser light is so precise that a laser beam split into two paths can still exhibit constructive interference when the two portions differ in overall path length by as much as 250 to 300 kilometers. The comparable limit for non-coherent light sources is a difference in path length of only 50 to 100 millimeters.

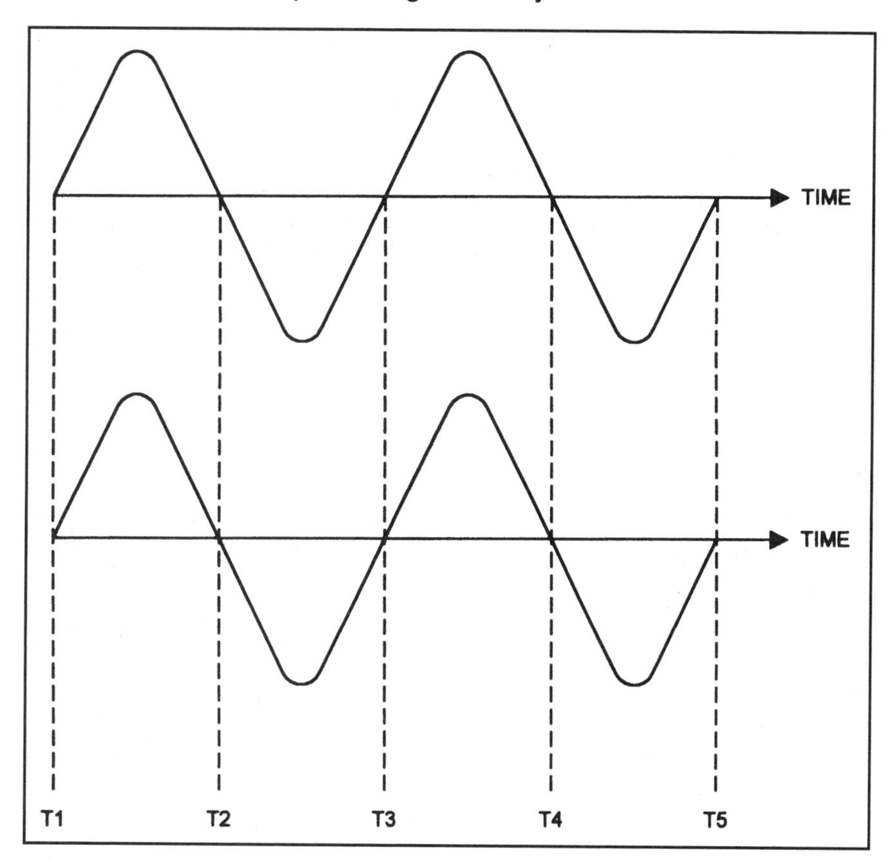

Figure 14-1. Longitudinal coherence is consistency in time phasing.

Transverse coherence is the tendency of adjacent waves to retain their definite relationship when viewed at right angles (orthogonally) to the direction of travel. For electromagnetic waves, a transverse cut would reveal similar amplitudes and alignments of the electric (E) and magnetic (H) fields (**Figure 14-2**).

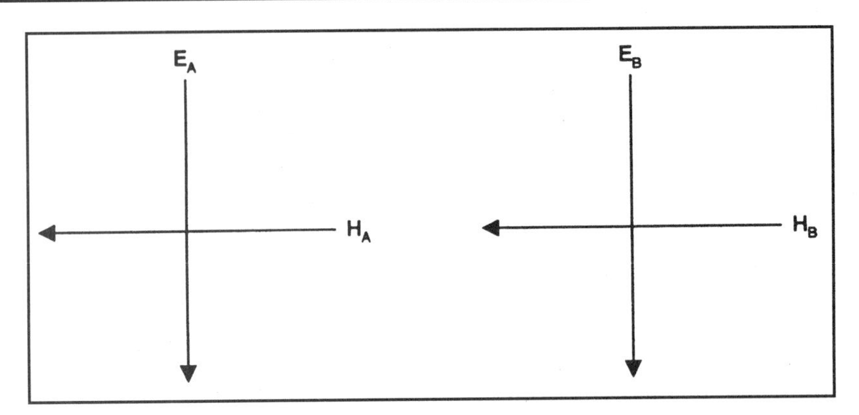

Figure 14-2. Transverse coherence is electromagnetic consistency across the two ways orthogonal to the line of travel.

As a result of transverse coherence, the laser beam is very narrow, and does not disperse as randomly phased light, such as a flashlight. Dispersion of less than 0.1 milliradian (mrad) is typical in lasers. The laser light beam remains very narrow over very long distances. Low dispersion is still another of the principal characteristics of laser light.

The basis of laser operation is the law of conservation of energy applied to quantum systems. When an external source of energy impinges on an atom it will raise the energy level of the associated electrons, causing them to rise to a higher, but unstable, orbit. When electrons decay back to their ground energy state, the acquired energy is given up as photons. The amount of change in energy state is a function of Planck's constant and the frequency of the emitted photon:

$$E = nhf = \frac{nhc}{\lambda} \qquad \text{eq. (14-1)}$$

Where:

E is energy (joules).

n is a positive integer.

c is the speed of light (3×10^8 meters per second).

f is the frequency of the emitted photon in hertz.

λ is the wavelength of the emitted photon in meters.

h is Planck's constant (6.626×10^{-34} joules per second).

Since energy level (E) and wavelength (λ) are the only variables in this equation, for any given energy level there can be only one wavelength. For any given laser material, there is only a very narrow band of possible wavelengths. For example, the common helium-neon (HeNe) gas laser produces a light with a bandwidth only a few kilohertz wide in the frequency domain, even though the center frequency (analogous to the "carrier" frequency in RF systems) is 5×10^{14} hertz (or about 638 nm in wavelength). Monochromaticity is one of the primary characteristics of laser light.

Stimulated emission of radiation was predicted as early as 1917 by Einstein as part of his explanation of Planck's theory of energy quanta. According to Einstein's prediction, an atom can be stimulated into producing radiation when it is excited by an electromagnetic field that has the frequency that would normally be emitted if the atom collapsed from the excited state to a lower state (usually the ground state). In this way, a small excitation light source can create a larger stimulated light emission.

Laser action requires more atoms in the excited upper state than in lower energy states to increase the probability of stimulated emission. In normal situations, at what is called thermodynamic equilibrium, only a small portion of the atoms are in an excited state at any given time, the majority being in a lower or ground state. In good laser materials, it is possible to produce a *population inversion* in which more atoms are excited than are not. Such materials tend to have long-lasting excited-energy states.

The quantum level action in typical lasers is a three-stage (sometimes a four-stage) transition. In **Figure 14-3a**, an atom at ground state (energy level U1) is excited by an external energy source, raising it - or "pumping" it - to a "pumping level", or upper state (U3). It undergoes fast spontaneous decay to a metastable state (U2), and then a slower decay back to the ground state (U1). In this second decay, a photon is emitted - so-called a stimulated emission. The frequency of the photon emitted is a function of the upper energy level (**Equation 14-1**). In four stage emission (**Figure 14-3b**), there are two spontaneous decay regions (U_{4-3} and U_{2-1}), separated by the stimulated emission region (U_{3-2}).

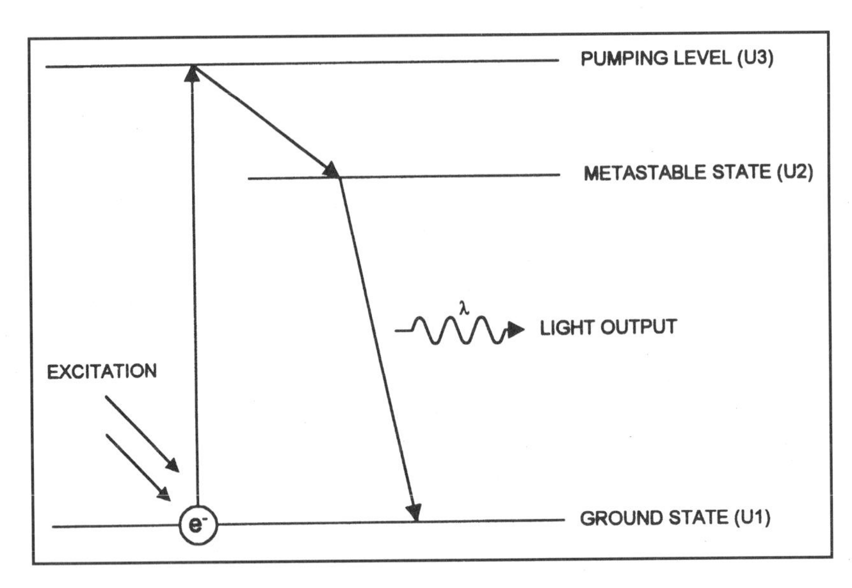

Figure 14-3a. Three-step emission laser operation.

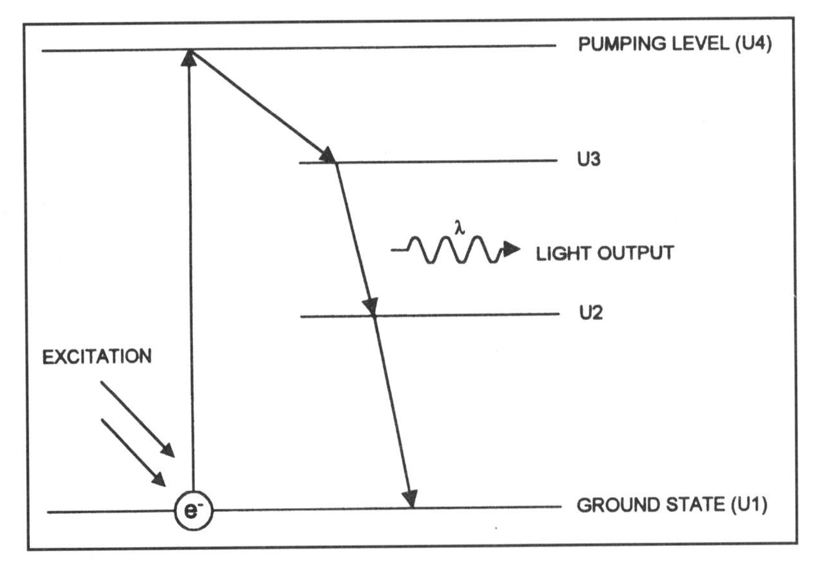

Figure 14-3b. Four-step emission laser operation.

Types of Lasers

There are several different types of laser: insulating solid-state, insulating crystal, excimer, gas, injection (PN junction solid-state), chemical and liquid (organic dye) lasers.

Insulating Solid-State Lasers

The insulating solid-state laser consists of a semitransparent solid, such as ruby, that is optically pumped in an excited state by a pulse of light from a xenon (or similar) flash tube or flashlamp (**Figure 14-4**). The ruby laser is perhaps the most commonly known laser of the insulating solid-state type. It was the first laser demonstrated by Theodore Maiman of Hughes Laboratories in July 1960 (patent preeminence is the subject of a dispute, though legally many of them, covering a wide range of laser science and applications, are held by one man, Gordon Gould).

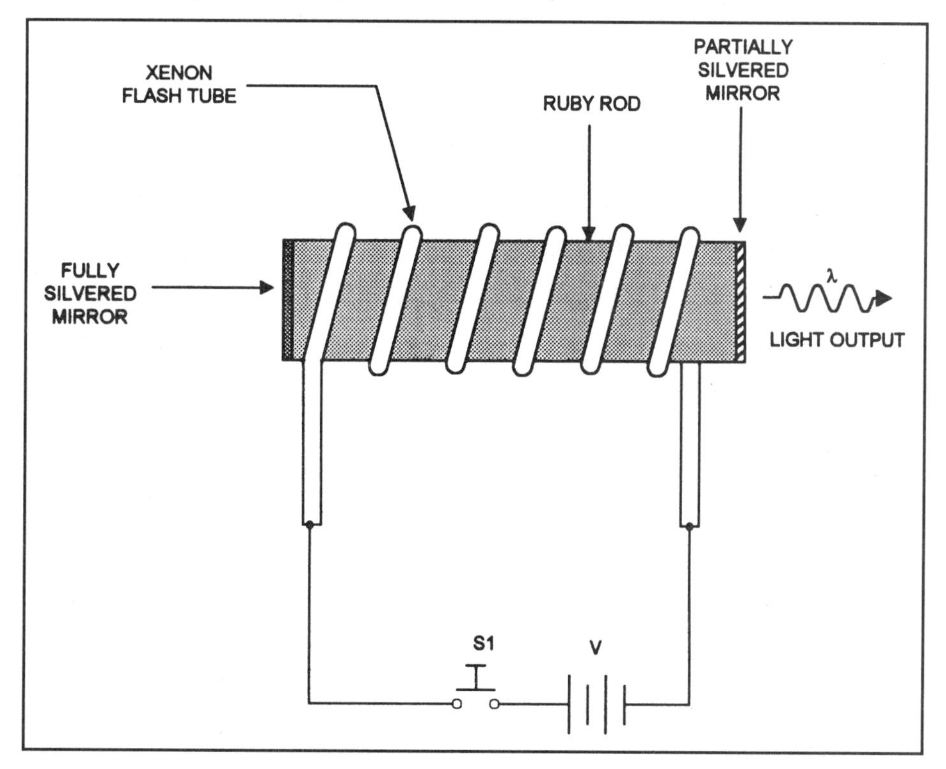

Figure 14-4. Ruby laser structure.

In the ruby laser, a rod of ruby material is fashioned with a fully silvered mirror at one end, and a partially silvered mirror at the other. Surrounding the ruby rod is a xenon high-energy flash tube, similar to those used in photographic flash guns. Because of the this excitation system, ruby lasers fall into the broad category of *optically pumped, solid crystal lasers*. Other pumping mechanisms include krypton-filled arcs and tungsten-iodine lamps.

Rubies are a variety of the material corundum (Al_2O_3), which occurs naturally, but also can be produced synthetically. The red coloring is produced by a small amount of chromium ions (Cr^{3+}) that replace some of the aluminum ions. When the xenon tube fires, it produces a short duration high intensity blast of white light. The chromium ions absorb violet, blue and green wavelengths from the white light, which causes them to become pumped up to a higher energy state. The chromium ions decay to a metastable state, where they remain for a few milliseconds, and then undergo decay to a lower energy level, emitting a photon of red light (wavelength 6,934 A or 693.4 nm) in the process.

The initial decay is omnidirectional, so a small fraction of the photons are directed toward the mirror at the rear of the assembly, where they are reflected toward the other mirror. Some photons escape from the partially silvered end, but others are reflected toward the fully silvered end. In a short time, a number of wavefronts reflect back and forth between the mirrors, a process called oscillation, building energy by mutual reinforcement; some waves escape at the partially silvered end on each reflection, forming an in-phase wavefront.

Because the waves escape from the ruby rod in phase with each other, they are coherent. Transverse coherence of the ruby laser results in a very narrow laser beam that has a dispersion of less than 0.1 milliradians. When the waves are emitted in random phase, there is a great deal of cancellation due to mutual interference, so much of the light energy is lost.

Insulating Crystal Lasers

This category of laser is very similar to the foregoing type: both are examples of optically pumped solid-state lasers and many texts place them in the same category. The insulating crystal devices differ from ruby lasers in that they use a rare earth element, such as neodymium ions (Nd^{3+}), as a dopant for a material such as yttrium aluminum garnet (YAG)—the widespread Nd:YAG—or certain glasses-—Nd:Glass.

The Nd:YAG laser typically emits a beam at infrared wavelengths (1,064 nm and an important secondary line at 1,320 nm), but it can be "doubled" to emit in the visible region. Energy outputs of 100 joules for several milliseconds are easily obtained. Typical power levels on the order of tens to hundreds of watts multimode are achieved using a flashlamp or a krypton arc lamp for pumping. One multistage Nd:YAG experimental laser produced 1,000 watts when pumped with a tungsten-iodine lamp; practical kilowatt Nd:YAG lasers are now making their appearance. In the Nd:Glass configuration, up to 10 kilojoules have been produced in nanosecond pulses. Efficiencies are typically 1 to 3 percent. The beam can be either a pulse or continuous wave.

Excimer Lasers

An *excimer* is a molecule that does not normally exist except in the state in which its constituent atoms are excited to high energy states. When the extra energy of the excitation state is given up as laser output, the de-excited molecules will revert at ground state to their original constituent atoms. Typical excimer lasers mix a noble gas (such as argon, krypton or xenon) with active elements such as chlorine, fluorine, iodine or bromine.

Gas Lasers

Another common form of laser is the *gas laser*; included in this class are neutral atom lasers, ionic gas lasers and molecular glass lasers. Gas lasers consist of a glass tube filled with a gas such as helium-neon (HeNe), carbon dioxide (CO_2) or argon at a low pressure (~0.3 Torr) (**Figure 14-5**). The end mirrors needed for oscillator behavior can be internal, but in most cases they are external, as shown in **Figure 14-5**, for ease of manufacture and adjustment. When external mirrors are used, the ends of the glass envelope of the tube are canted at a critical angle (α), called *Brewster's angle*. At this angle, light waves that are polarized in the correct manner will suffer no reflection losses at the two ends. Thus, reflection can take place between the two ends

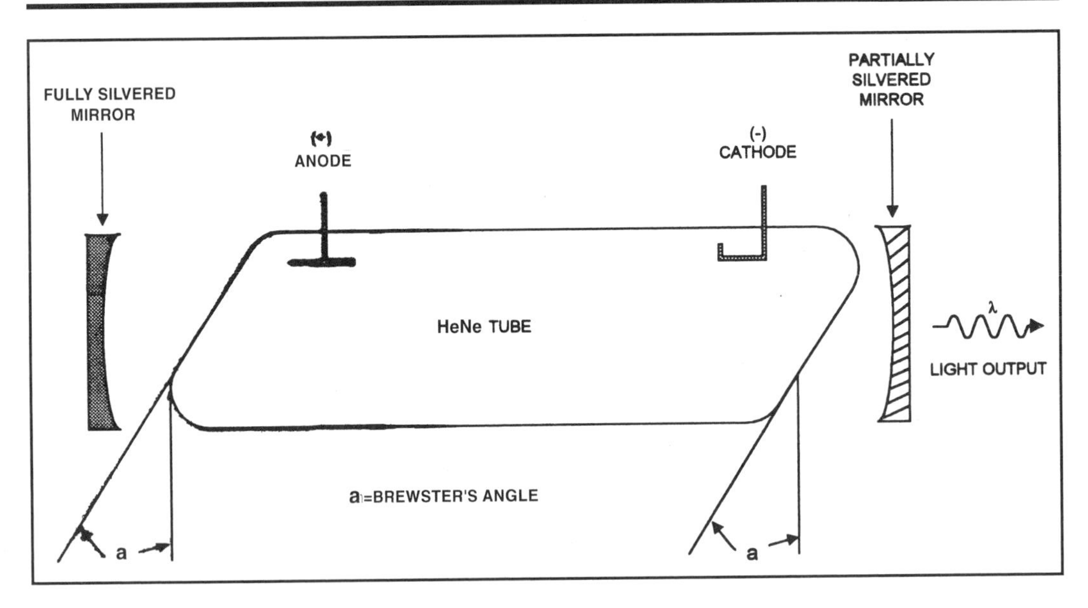

Figure 14-5. Gas laser structure.

at a specific wavelength and phase relationship, forming the lasing action and retaining coherence and monochromaticity. In most gas lasers, the end mirrors are spherical or parabolic so that proper oscillation will take place.

The gas laser works because electrons are injected into the gas chamber from a cathode, and are accelerated toward the positively charged anode. Along the way, some of the electrons collide with gas ions (**Figure 14-6**), causing them to increase their energy states by absorbing the electrons' kinetic energy. The excited state is not stable, so decay can be expected through several different mechanisms: collisions between the excited atom and another free electron, collisions between excited and unexcited atoms, collisions between excited atoms and the glass walls or spontaneous emission.

Helium-neon (HeNe) gas lasers are usually low-powered (0.1 to 10 milliwatts of output power), typically drawing ≤10 mA, at a potential of several kilovolts, from the dc power supply. The principal wavelength produced by the HeNe laser is 632.8 nm (red region). The HeNe laser can also produce light at wavelengths of 594, 604, 612, 1150 and 3390 nm by correct design of the end mirrors. HeNe lasers are frequently used for classroom or hobbyist science demonstration projects. Some are in the very low milliwatt region; these are the types usually recommended for amateur experimenter use. (Note:

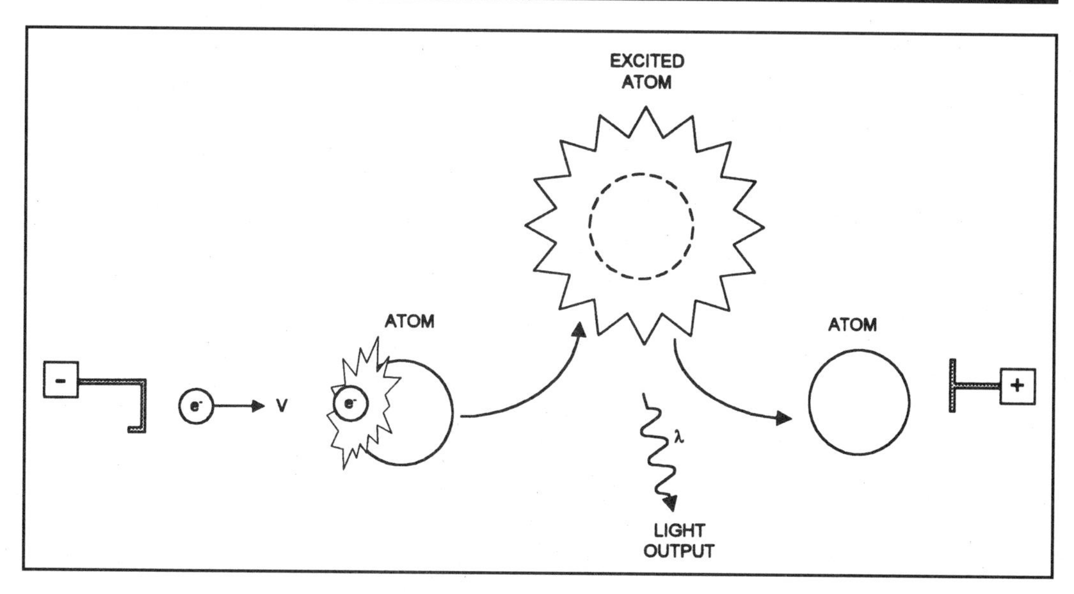

Figure 14-6. Mechanism for gas laser action: kinetic electron excites an atom. When the atom loses its excited state, it gives up a light photon.

even very low-powered lasers can be dangerous to eyesight.) Divergence of the HeNe laser is typically ≤0.1-1 mrad. The glass tube on HeNe lasers is typically less than 100 cm long, and has a bore diameter of 12 mm. The HeNe laser is very inexpensive compared with other lasers.

Argon gas lasers produce up to 20 watts of light power at wavelengths of approximately 488 and 515 nm, in the blue and green portions of the spectrum. Argon lasers have more gain than similar HeNe lasers, but operate at efficiencies of less than 0.1 percent. They often require very high current densities (~100 A/cm^2) for proper operation. Special glass is needed for the outer envelope because of heating problems. Argon and krypton gas lasers produce light in the blue and green regions. These lasers find substantial application in the medical and scientific areas. Blue lasers, for example, are used for ophthalmic applications such as "spot welding" detached retinas in the human eye.

Carbon dioxide (CO_2) lasers are higher-powered, reaching a power level of tens of kilowatts. These lasers can therefore be used for metal cutting, welding, drilling and other industrial chores. The principal output of the CO_2 laser has a wavelength of 10.6 μm, or 106,000 A, in the far-infrared region.

Solid-State PN Junction Lasers

The final laser discussed here is the solid-state PN junction diode laser, or the *injection laser* (**Figure 14-7**). The laser diode is very similar to ordinary PN diodes and LEDs, except that the heterojunction material forming the PN junction is sandwiched between AlGaAs sections that serve as internal resonating mirrors. Population inversion, which is necessary for laser action, occurs when holes from the P-side, and electrons from the N-side, are forced by the applied electrical field into the junction region. As we have seen, the conduction band thus produced becomes the upper laser energy level, and the valence band becomes the lower energy level.

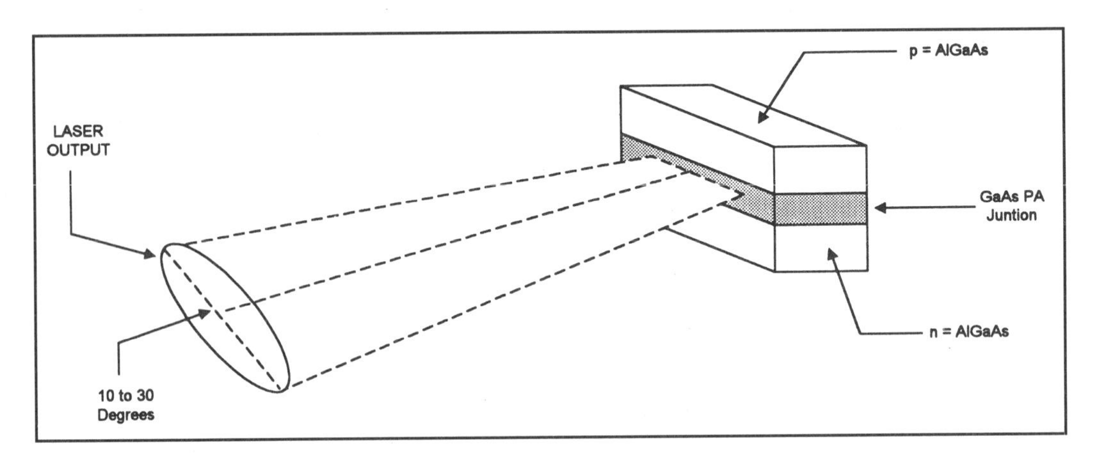

Figure 14-7. PN diode laser structure.

The most common low-cost laser diodes use GaAs material sandwiched between AlGaAs, and produce laser light in the 760 to 905 nm region, depending on the proportions of the materials, and whether a single or double heterojunction is used. Other material combinations include InGaAsP (indium gallium arsenide phosphorous), which emits in the 1200 to 1550 nm range.

In both types of injection laser, room temperature lasing is possible, although only at low power levels and low efficiencies. For example, one AlGaAs laser diode produces 0.02 watt (20 mW) CW, at 7 percent efficiency, at room temperature. Cooling can produce a tremendous increase in both power level and efficiency. A GaAs laser diode cooled to -253°C can have an efficiency as high as 50 percent. In the pulsed mode, commonly available laser diodes can produce 3 to 10 watts, drawing about 10 amperes from the dc power supply. Noncooled GaAs devices produce 3 to 5 mW in some configurations, and 500 mW in others.

At low current levels, the diode acts like other PN junction diodes, and at higher currents will emit a broad spectrum of noncoherent light much like that from other LEDs. If the threshold current (I_{th}) is exceeded, the laser diode will begin to "lase", emitting coherent light. Threshold currents of 1000 or more amperes per square centimeter (1000 A/cm^2) are possible, even at moderate average forward current levels, because of the very small area of the laser diode die (typically a few square millimeters).

An example of the laser diode is the *Panasonic* LN-9705[1] shown in **Figure 14-8**. This device contains two diodes: a laser diode (D1) and a PIN photodiode (D2) to monitor laser light level output. The PIN diode can be used either in a negative feedback circuit to automatically set the laser diode current, or to provide a metered light level output to the user. It produces an 805 nm laser beam at a threshold current greater than 33 mA, and produces 5 mW of output power from an operating voltage of 1.8 volts.

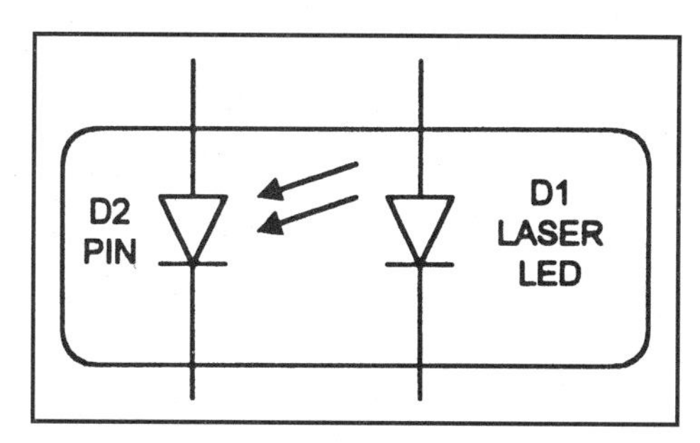

Figure 14-8. Typical laser diode package and circuit symbol (D1).

A stud-mounted laser diode package is shown in **Figure 14-9a**. This diode uses a threaded stud for mounting and making the negative electrical connection. The positive electrical connection (anode) is made to an electrical terminal protruding from the rear of the mounting stud. Because laser diodes tend to be low-efficiency devices, producing large amounts of waste heat, it is often necessary to mount them on a metallic heat sink (**Figure 14-9b**)to carry away excess heat that could destroy the diode. A variant shown in **Figure 14-10** is a stud-mounted laser diode in which the output is coupled to a fiber optic pigtail (Chapter 15 discusses fiber optics).

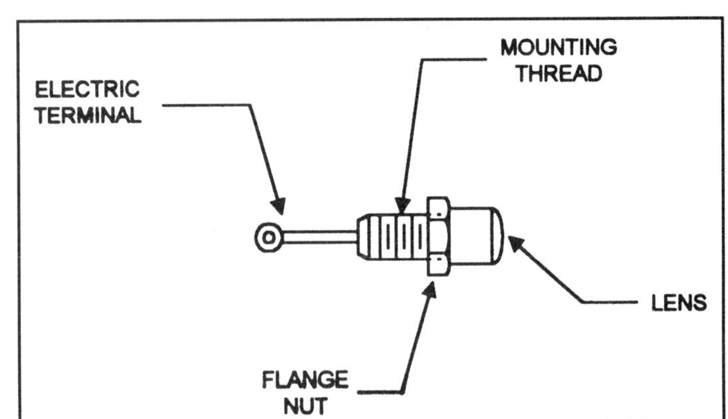

Figure 14-9a. A typical stud-mounted laser diode.

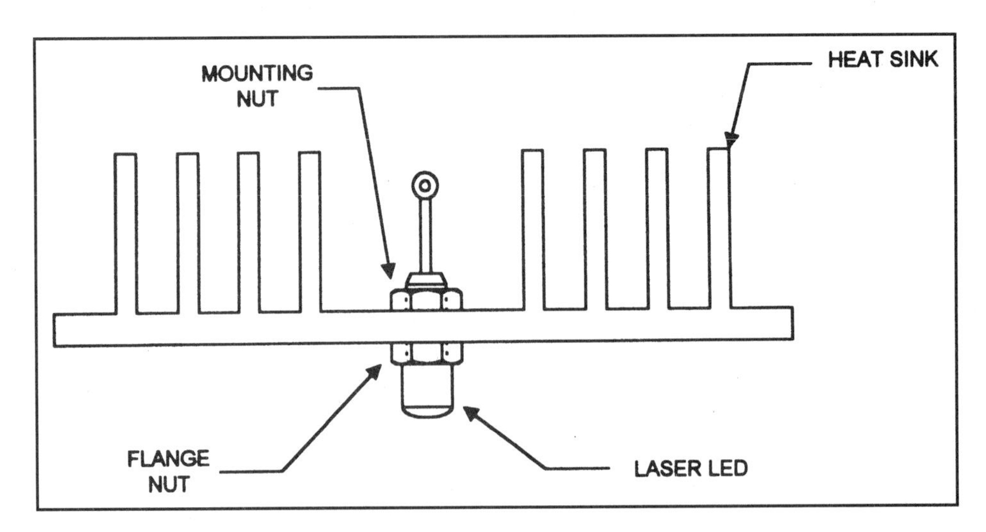

Figure 14-9b. Stud-mounted laser diode mounted on heat sink to improve heat dissipation.

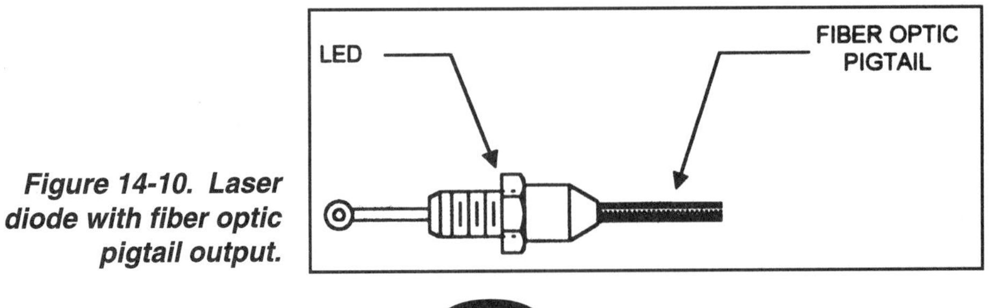

Figure 14-10. Laser diode with fiber optic pigtail output.

Driver Circuit for Solid-State Laser Diodes

The circuits used to drive laser diodes are similar to those used to drive ordinary LEDs, as shown in **Figure 14-11**. In this circuit, the laser diode is connected between V+ and ground, with a resistor network in series to limit current. In standard practice, resistor R1 can be adjusted to vary laser diode current, hence laser output, while R2 is used to set the maximum current (for safety) that can flow in the diode. A problem with this configuration is that laser diodes tend to draw large amounts of current, so the resistors must be 5 to 20-watt types, depending on the diode and associated voltage and current levels. As a result, it is more common for laser diodes to use a high power transistor driver circuit to control the current.

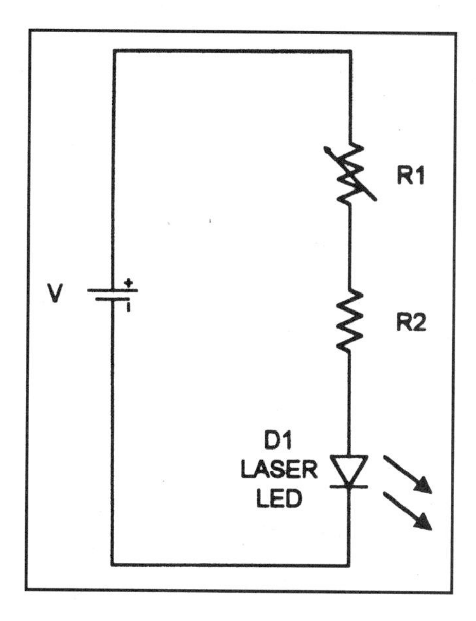

Figure 14-11. Simple laser diode driver circuit.

Figure 14-12a shows a two-stage transistor direct-coupled amplifier used as a laser diode driver. The output transistor Q1 is a large power transistor such as the 2N3055. It can dissipate 110 watts of collector power, and has a collector current rating of 15 amperes. The beta gain of the 2N3055 is about 45. The driver transistor Q2 is a smaller NPN power transistor in a plastic package. It will dissipate over 5 watts when connected to a heat sink, and 1.33 watts in free air, at a collector current of 1 ampere; the beta gain exceeds 120.

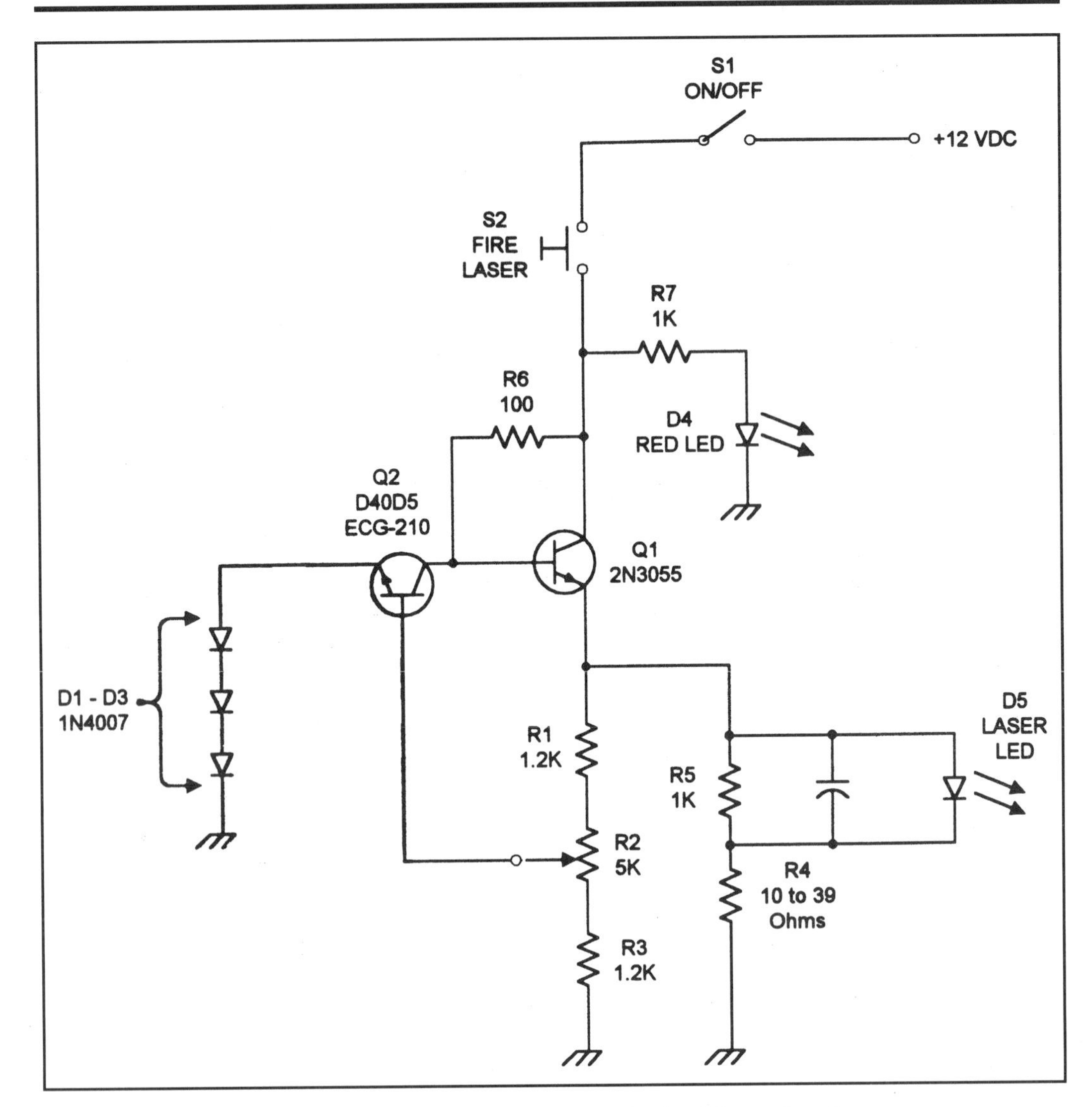

Figure 14-12a. All-transistor version of a laser diode driver circuit.

The emitter of Q2 is kept at a potential close to +2 volts by three silicon 1-ampere rectifier diodes in series (although 1N4007 are shown, any of the 1N400x series from 1N4001 to 1N4007 can be used). Each of these diodes has a forward voltage drop of 0.60 to 0.70 volt, resulting in a total of about 2 volts at the emitter of Q2. The base voltage for Q2 is derived from the emitter of Q1, which is at a potential near 5 volts. The voltage divider consisting of R1, R2, and R3 produces a potential near 2.6 volts for the base of the driver transistor Q2, and since this voltage is partially dependent on the voltage across the laser diode D5 and the series resistor R4, it provides a small amount of negative feedback.

The driver transistor excites the output transistor by varying the voltage applied to the base of Q1. When Q2 draws a heavy current, the base voltage of Q1 drops, so the collector current applied to the laser diode also drops. When Q2 draws less current, the collector current rises and provides more current to the laser diode.

Some laser diodes emit infrared light which cannot be seen by the human eye, and thus present a vision hazard. The light emitting diode (D4) provides a red light to warn that collector power is applied to the laser diode driver circuit.

Resistor R4 limits the current available to the laser diode in the event that Q1 shorts from collector to emitter, Q2 opens from base to emitter or from collector to base, or the adjustment of R2 is too high. This resistor should have a value between 10 and 39 ohms, depending on the laser diode's maximum current, and is generally a 5- or 10-watt power resistor.

Two series-connected switches are in the collector power circuit. The main power switch S1 is used to turn the circuit off entirely. Some users prefer to use a key operated model in order to limit access to the laser by those who are authorized. (Remember, lasers are dangerous devices, and are NOT toys.) Switch S2 is a pushbutton switch used to turn on the laser when a burst of light is needed.

A variation of the circuit is shown in **Figure 14-12b**. This circuit uses the same type of power transistor output stage as the previous circuit, but the driver is replaced by an operational amplifier. The op-amp is connected in the unity-gain noninverting follower configuration. The noninverting input (+IN) is connected to a potentiometer that sets the positive voltage applied to the +IN terminal. The op-amp output terminal will be at a potential that reflects the setting of R1, and in turn drives the bias network (R2 and R3) for the output transistor Q1. Good initial values for R2 and R3 are 33 kΩ for R2 and 3.3 kΩ for R3. (The correct values can be found empirically, or calculated if the beta gain of the transistor and the current range needed by the laser diode are known.)

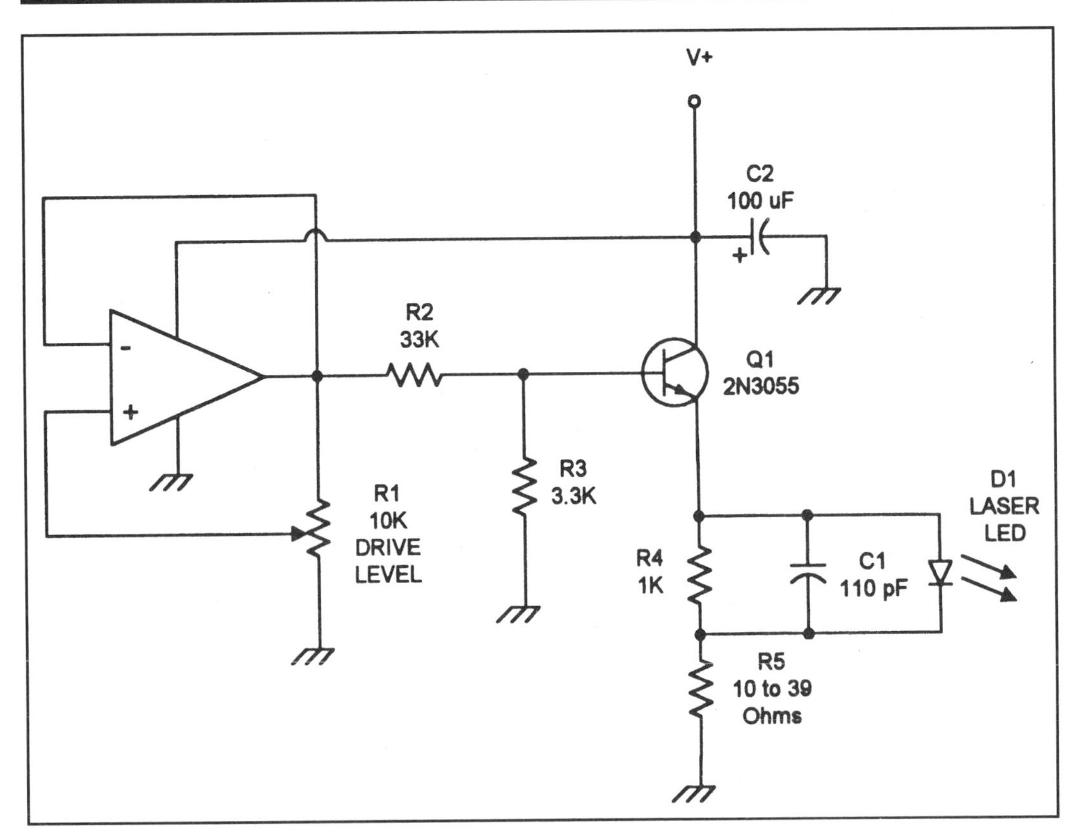

Figure 14-12b. Op-amp/transistor version of a laser diode driver circuit.

Summary

Before leaving the subject of lasers, it is prudent to reiterate the warnings about their hazards. The coherent light from a laser, even at very low power levels, is dangerous to eyes and skin. Avoid viewing laser light either directly or through either specular or diffuse reflections. Read, understand and follow all safety instructions provided by the manufacturer of any laser device that you put into operation. Before beginning to use lasers, read and understand the laser safety guidelines in the ANSI standard.

Notes

[1]This laser is sold mail-order by *Digi-Key*, P.O. Box 677, Thief River Falls, MN 56701-0677; 1-800-344-4539. For other sources, consult advertisements in magazines such as *Radio-Electronics, Popular Electronics,* or *Nuts 'n' Volts*, or mail order catalogs such as the *Edmund Scientific Company*.

Chapter 15

Fiber Optic Technology

Fiber optics is the latest in communications and instrumentation technology. Briefly stated, fiber optics is the passing of light through a plastic or glass fiber so that it can be directed to a specific location. If the light is encoded (*modulated*) with an information signal, then that signal is transmitted over the optical path. There are many advantages to the optical fiber communications or data link, including:

Very high bandwidth (accommodates video signals, multiple voice channels, or high data rate computer communications).

Very low weight and small size.

Low loss compared with other media.

No electromagnetic interference (EMI).

High degree of electrical isolation.

Explosion proof.

Good data security.

Improved "fail-safe" capability.

The utility of the high-bandwidth capability of the fiber optical data link is that it can handle a tremendous amount of electronically transmitted information simultaneously. For example, it can handle more than one video signal, which typically requires 500 kHz to 6 MHz of bandwidth, depending on resolution. Alternatively, it can handle a tremendous number of voice communication telephone channels. A high-speed computer data communications capability is also possible. Either a few channels can be operated at extremely high speeds, or a larger number of low-speed parallel data channels are available. Fiber optics is so significant that one can expect to see it proliferate in the communications industry for years to come.

The light weight and small size, coupled with relatively low loss, gives the fiber optic communications link a great economic advantage when large numbers of channels are contemplated. To obtain the same number of channels using coaxial cables or "paired wires" the system would require a considerably larger, and heavier, infrastructure.

Electromagnetic interference (EMI) has been an annoying factor in electronics since Marconi and DeForest interfered with each other in radio trials for the Newport Yacht Races prior to the turn of the twentieth-century. Today, EMI can be more than merely annoying; it can cause tragic accidents. For example, airliners are operated more and more from digital computers. Indeed, one airline copilot recently quipped (about modern aircraft) that one doesn't need to know how to fly anymore, but one does need to be able to type on a computer keyboard at 80 words per minute. This points out just how dependent aircraft have become on modern digital computers and intercommunication between digital devices. If a radio transmitter, radar or electrical motor is near one of the intercommunications lines, it would be possible to either introduce false data or corrupt existing data unless the proper precautions were taken in the design. Because the EMI is caused by electrical or magnetic fields coupling between electrical cables, optical fibers, being free of such fields, produce dramatic freedom from EMI.

Electrical isolation is required in many instrumentation systems either for user safety, or to avoid electronic circuit damage. In some industrial processes, high electrical voltages are used, but the electronic instruments used to monitor the process are both low-voltage and ground referenced. As a result, the high voltage can damage the monitor instruments. In fiber optical systems, it is possible to use an electrically floating sensor, and then transmit the data over a fiber link to an electrically grounded, low-voltage computer, instrument or control system.

Optical fibers can transmit light beams generated in non-contacting electronic circuits, making fiber optic systems ideal for use around flammable gases or fumes, nuclear power generators and other hazardous environments. Regular mechanical switches or relays arc on contact or release, and those sparks may ignite flammable gases or fumes. Occasional operating room explosions in hospitals occurred prior to 1980, and electrical arcs in switches have caused gasoline station explosions.

System security is enhanced because optical fibers are difficult to tap. An actual physical connection must be made to the system, resulting in a detectable power loss. In wire systems, capacitive or inductive pickups can acquire signals with less than total physical connection. Similarly, a system is more secure in another sense of the word because the fiber optic transmitters and receivers can be designed "fail-safe" so that one fault does not take down the system. I recall a hospital coronary-care unit data system that used parallel wire connections between the data output ports on bedside monitors and the central monitoring computer at the nurses' station. A single short circuit would reduce the system to chaos! That is less likely to happen in a fiber optic system.

Fiber optic principles were first noted in the early 1870s when John Tyndall introduced members of Britain's Royal Society to his experimental apparatus (**Figure 15-1**). An early, but not very practical, color television system patented by J.L. Baird used glass rods to carry the color information. By 1966 C. Hockham and C. Kao demonstrated a system in which light beams carried data communications via glass fibers. The significant fact that made the Hockham/Kao system work was the reduction of loss in the glass dielectric material to a reasonable level. By 1970, practical fiber optic communications was theoretically possible.

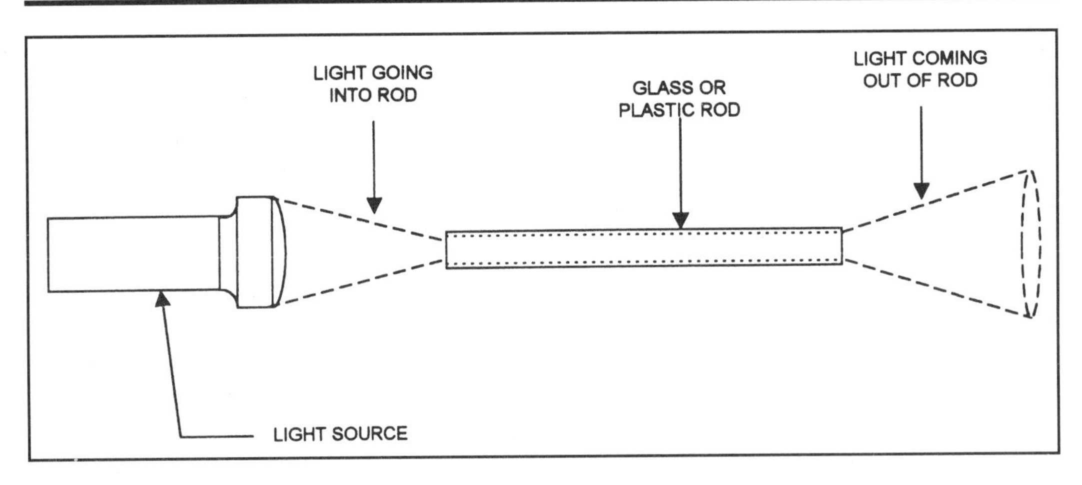

Figure 15-1. Light transmission through a glass or plastic rod.

Medicine has made use of fiber optics for more than two decades. Fiber optic *endoscopes* can be passed into various orifices of the body, either natural or surgically made, to inspect the interior of a patient's body. Typically there are two bundles, one for viewing and one for passing a light from a so-called "cold" - or physically harmless - light source into the body. For example, gynecologists can inspect and operate on certain internal organs in females using a *laparoscope* introduced through a small "band-aid" incision in the abdomen. Knee surgeons can use a fiber optic arthroscope to perform nearly miraculous operations on the human knee with far less trauma than previous procedures. Other physicians use fiber optic endoscopes to inspect the stomach and gastric tract. A probe is passed through the mouth or nose, down the esophagus and into the stomach so that tumors or ulcers can be inspected without resorting to surgery. In more recent times, miniature TV cameras using CCD (charge-coupled device) arrays have been used to view internal areas of the body, with optical fibers carrying the light and the sensors inside.

Fiber optic inspection is used in other areas as well. For example, I recall an advertisement for a septic tank service company that used fiber optics and television to inspect the tank. Plumbers use it for similar operations. Other industrial and residential services also use fiber optics to inspect areas that are either inaccessible or too dangerous for direct viewing.

Before examining fiber optic technology, it might be useful to review some of the basics of optical systems as applied to fiber optics.

Review of the Basics

The *index of refraction*, or *refractive index*, is the ratio of the speed of light in a vacuum to the speed of light in the medium of interest (glass, plastic, water); for practical purposes, the speed of light in air is close enough to its speed in a vacuum to be considered the same. Mathematically, the index of refraction, n, is:

$$n = \frac{c}{v_m} \qquad \text{eq. (15-1)}$$

Where:

c is the speed of light in a vacuum (3×10^8 m/s).

v_m is the speed of light in the medium.

Refraction is the change in direction of a light ray as it passes across the boundary surface, or "interface," between two media of differing indices of refraction. In **Figure 15-2**, two materials, N1 and N2, have indices of refraction n_1 and n_2, respectively. In this illustration, N1 is optically less dense than N2. Consider incident light ray A, approaching the interface through the less dense medium (N1 $\rightarrow$ N2). As it crosses the interface it changes direction toward a line normal (at a right angle) to the surface. Ray B approaches the interface through the denser medium (N2 $\rightarrow$ N1). In this case, the light ray is similarly refracted from its original path, but the direction of refraction is *away* from the normal line.

In refractive systems the angle of refraction is a function of the ratio of the two indices of refraction, i.e., it obeys *Snell's law*:

$$n_1 \sin \theta_i = n_2 \sin \theta_r \qquad \text{eq. (15-2)}$$

or,

$$\frac{n_1}{n_2} = \frac{\sin \theta_r}{\sin \theta_i} \qquad \text{eq. (15-3)}$$

Where:

θ_i is the angle of incidence.

θ_r is the angle of refraction.

Of particular concern in fiber optics is the situation of a light ray passing from one medium to a less dense medium. This can be illustrated with either a water-to-air system, or a system in which two different glasses, having different indices of refraction, are interfaced, see **Figure 15-2** Ray B.

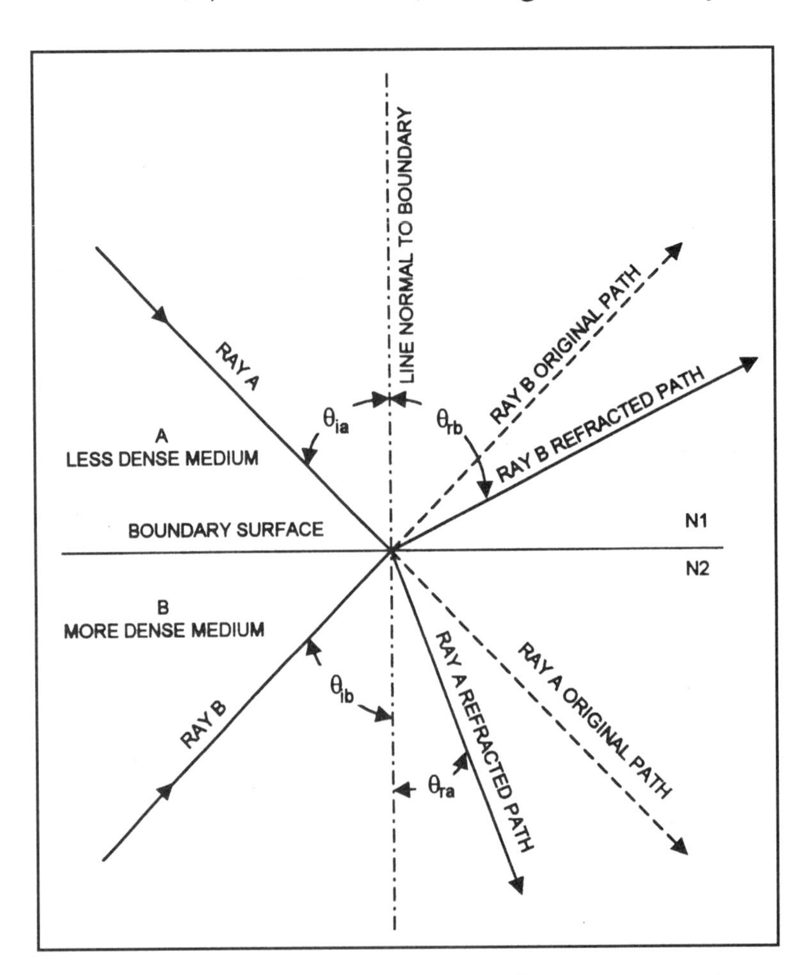

Figure 15-2. Refraction from a less dense to more dense medium (Ray A) and from more dense to less dense medium (Ray B).

Figure 15-3 shows a similar system with three different light rays approaching the same point on the interface from three different angles. (θ_{ia}, θ_{ib} and θ_{ic}, respectively). Ray A approaches at a *subcritical angle*, θ_{ia}, so it will split into two portions (A′ and A″). The reflected portion, A″, contains a relatively small amount of the original light energy, and may indeed be nearly indiscernible. The major portion, A′, of the light energy is transmitted across the boundary, and refracts at an angle θ_{ra}' in the usual manner.

Light Ray B, on the other hand, approaches the interface at a *critical angle*, θ_{ib}, and is refracted along a line that is orthogonal to the normal line; that is, it travels along the interface boundary surface. This angle is labelled θ_c in optics textbooks.

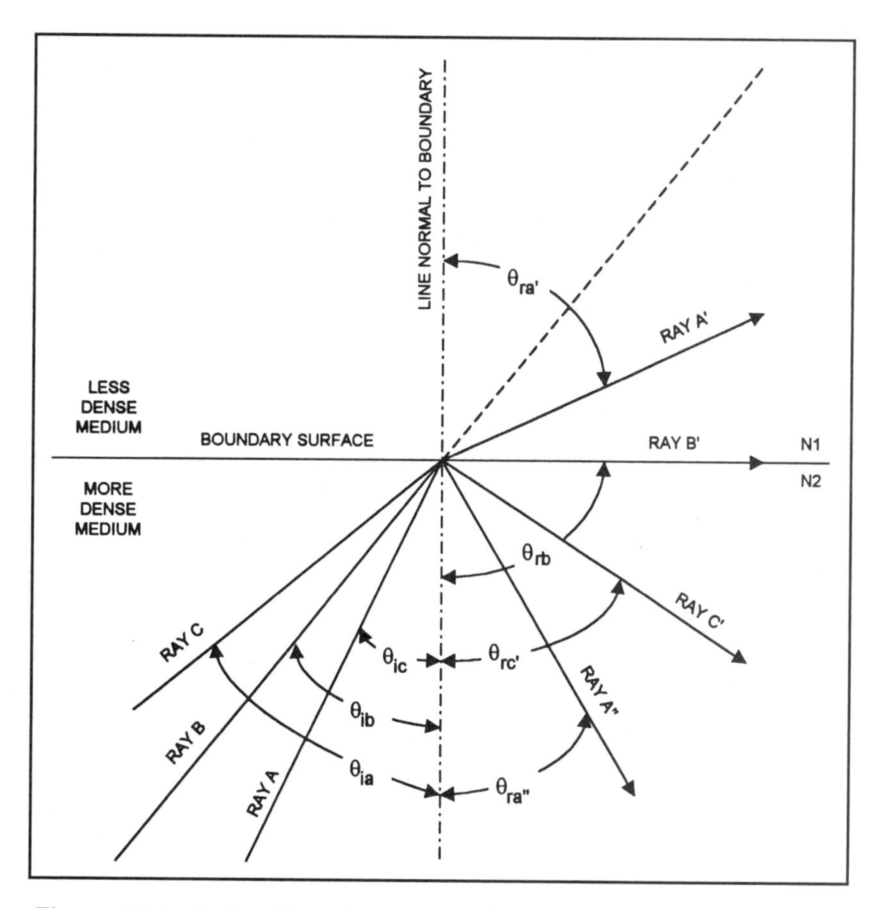

Figure 15-3. Refraction phenomena from more dense to less dense material showing subcritical (Ray-A), critical (Ray-B) and supercritical (Ray C) refraction.

Finally, Ray C approaches the interface at an angle greater than the critical angle, known as the *supercritical angle*. None of this ray is transmitted, but rather it is turned back into the original medium. This phenomenon is called *total internal reflection* (TIR)[1]. It is this phenomenon that allows fiber optics to work.

Fiber Optics

An optical fiber is similar to a microwave *waveguide*, and an understanding of waveguide action is useful in understanding fiber optics. A schematic model of an optical fiber is shown in **Figure 15-4**. A slab of material (N1) is sandwiched between two slabs of a less dense material (N2). Light rays that approach from a supercritical angle are totally internally reflected from the two interfaces (N2 → N1 and N1 → N2). Although only one "bounce" is shown in the illustration, the ray will be subjected to successive TIR reflections as it propagates through the N1 material. The proportion of light energy that is reflected through the TIR mechanism is around 99.9 percent, which compares favorably with the 85 to 96 percent typically found with plane mirrors.

Fiber optic lines are cylindrical, as shown in **Figure 15-5**. The illustrated components are called *clad optical fibers* because the inner core is surrounded by a less dense layer called *cladding*. Shown in **Figure 15-5** are two rays, each of which is propagated into the system at supercritical angles. These rays will propagate through the cylindrical optical fiber with little energy loss.

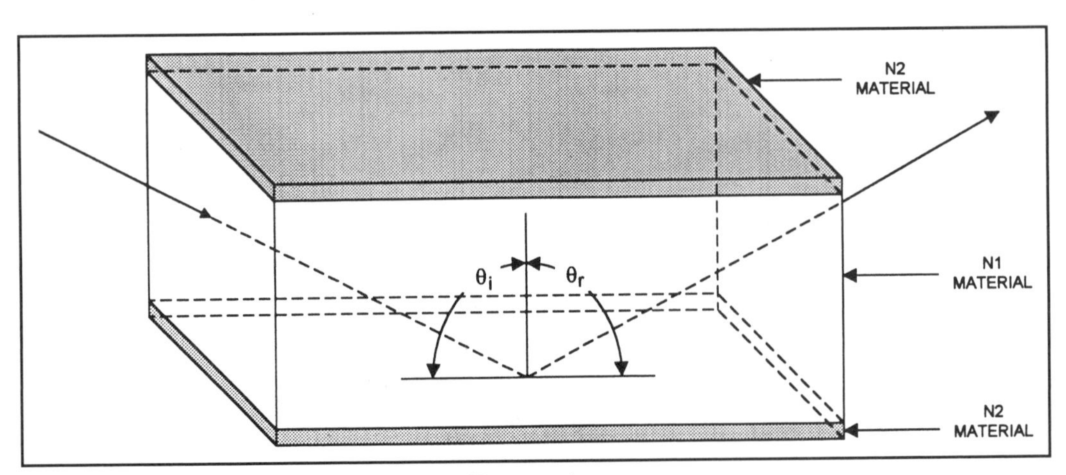

Figure 15-4. Waveguide analogy for fiber optics.

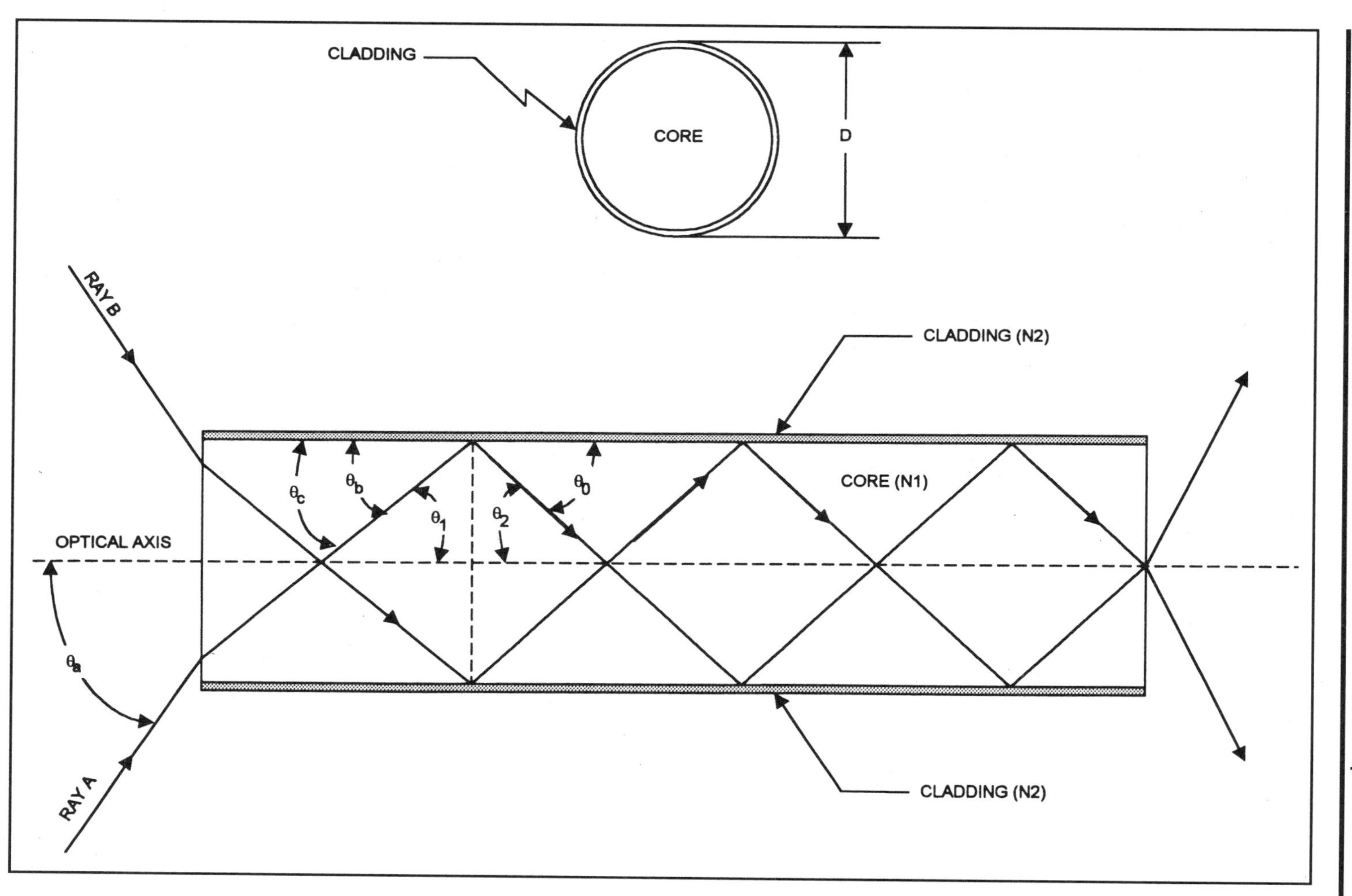

Figure 15-5. Total internal reflection forms basis for propagation in cylindrical optical fibers.

There are actually two forms of propagation. The minority form, called *meridional rays*, are easier to understand and mathematically modelled in textbooks because all rays lie in a plane with the optical axis (**Figure 15-6a**). The more numerous *skew rays* follow a helical path, so are somewhat more difficult to discuss (**Figure 15-6b**).

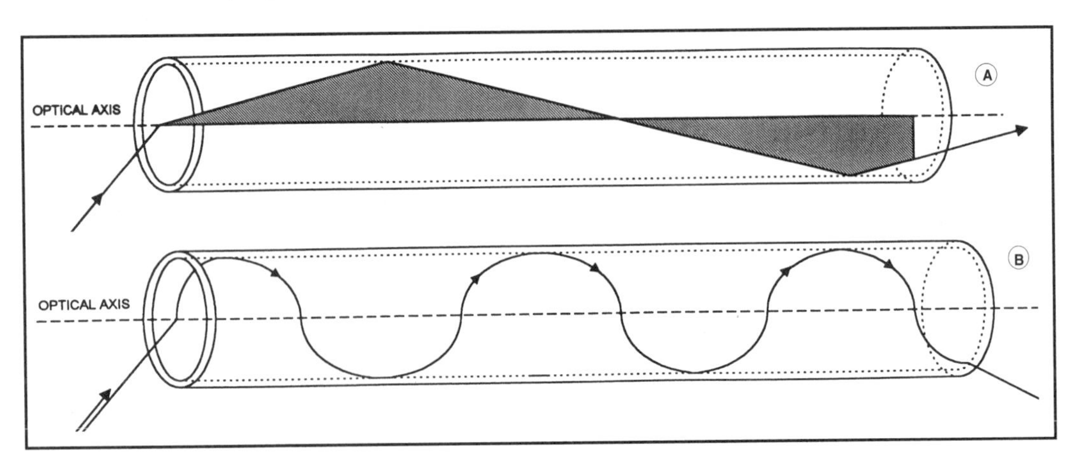

Figure 15-6. Two types of fiber optic light propagation: a) meridional propagation; b) skew propagation.

The *cone of acceptance* of the optical fiber is a conical region centered on the optical axis (**Figure 15-7**). The *acceptance angle* θ_a is the critical angle for the transition from air ($n = n_a$) to the core material ($n = n_1$). The ability to collect light is directly related to the size of the acceptance cone, and is expressed in terms of the *numerical aperture* (NA), which is:

$$NA = \sin \theta_a \qquad \text{eq. (15-4)}$$

The refraction angle of the rays internally, across the air-n_1 interface, is given by Snell's law:

$$\theta_{b1} = \arcsin\left(\frac{n_a \sin \theta_a}{n_1}\right) \qquad \text{eq. (15-5)}$$

It can be shown that:

- $$\theta_{a1} = \theta_{a2}$$ eq. (15-6)

- $$\theta_{b1} = \theta_{b2}$$ eq. (15-7)

- $$\theta_{a1} = \frac{\theta_a}{n_1}$$ eq. (15-8)

In terms of the indices of refraction of the ambient environment outside the fiber, the core of the fiber and the cladding material, the numerical aperture is given by:

- $$NA = \sin\theta_a = \frac{1}{n_a}\sqrt{(n_1)^2 - (n_2)^2}$$ eq. (15-9)

If the ambient material is air, then the numerical aperture equation reduces to:

- $$NA = \sqrt{(n_1)^2 - (n_2)^2}$$ eq. (15-10)

Internally, the angles of reflection q_{a1} and q_{a2}, at the critical angle, are determined by the relationship between the indices of refraction, n_1 and n_2, of the two materials:

- $$\theta_{a1} = \frac{\arcsin\sqrt{(n_1)^2 - (n_2)^2}}{n_1}$$ eq. (15-11)

Typical optical fiber materials have numerical apertures of 0.1 to 0.5; typical multimode fibers have diameters D of 25 μm to 650 μm. The ability of the device to collect light is proportional to the square of the numerical aperture multiplied by the diameter:

$$\zeta \propto (NA \times D)^2 \qquad \text{eq. (15-12)}$$

Where:

- ζ is the relative light collection ability.
- NA is the numerical aperture.
- D is the fiber diameter.

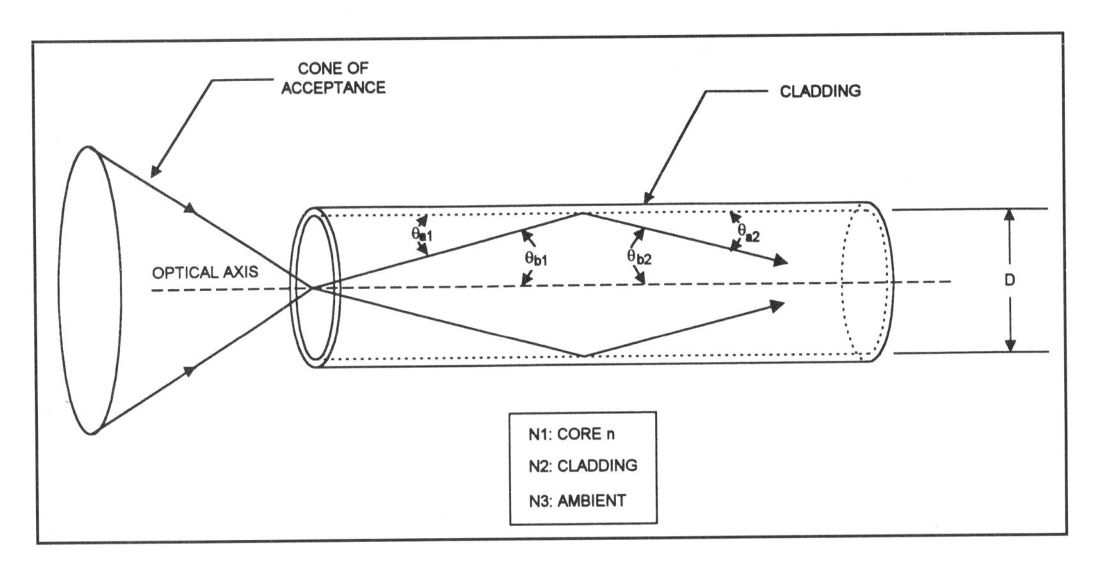

Figure 15-7. Cone of acceptance of an optical fiber.

Intermodal Dispersion

When a light ray is launched in an optical fiber it can take any of a number of different paths, depending in part on its launch angle (**Figure 15-8**). These paths are known as *transmission modes*, and vary from very-low order modes parallel to the optical axis of the fiber (Ray A), to the highest-order mode close to the critical angle (Ray C); in addition, there are many paths between these two limits. An important feature of the different modes is that the respective path lengths vary tremendously, being shortest with the low-order modes and longest with high-order modes. If an optical fiber has only a single core and single layer of cladding, it is called a *step index* fiber because the index of refraction changes abruptly from the core to the cladding. The number of modes, N, that can be supported is given by:

$$N = \frac{1}{2}\left(\frac{\pi D\,[NA]}{\lambda}\right)^2 \qquad \text{eq. (15-13)}$$

Any fiber with a core diameter D greater than about ten wavelengths (10λ) will support a very large number of modes, and is typically called a *multimode fiber*. A typical light beam launched into such a step index fiber will simultaneously find a large number of modes available to it. This may or may not affect analog signals, but it has a deleterious effect on digital signals called *intermodal dispersion*.

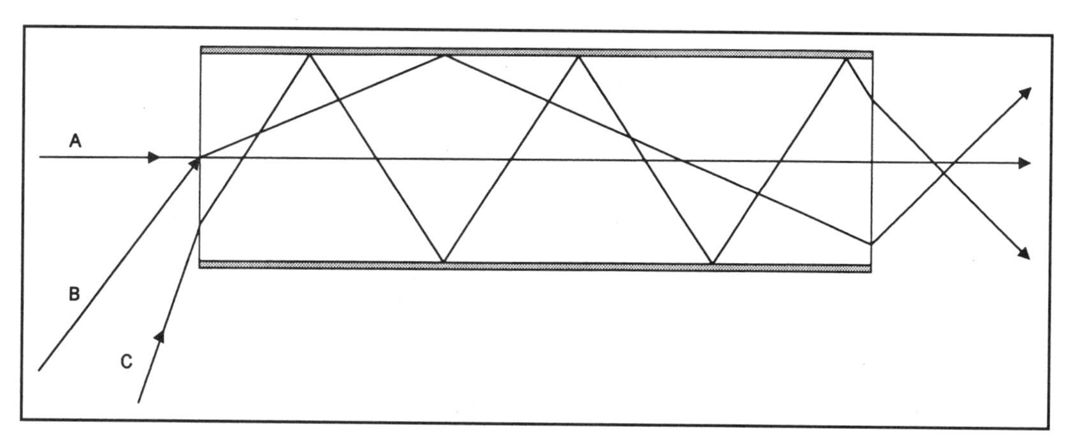

Figure 15-8. Multimode light propagation.

Figure 15-9 illustrates the effect of intermodal dispersion on a digital signal. When a short-duration light pulse (**Figure 15-9a**) is applied to an optical fiber that exhibits a high degree of intermodal dispersion, the received signal (**Figure 15-9b**) is dispersed, or "smeared," over a wider area. At slow data rates this effect may prove negligible because the dispersed signal can die out

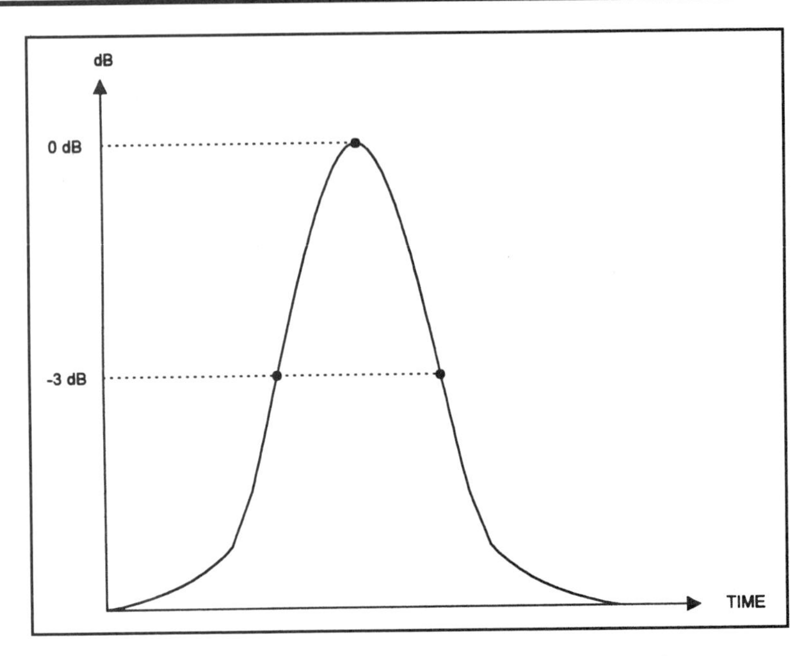

Figure 15-9a. Transmodal dispersion of an input pulse.

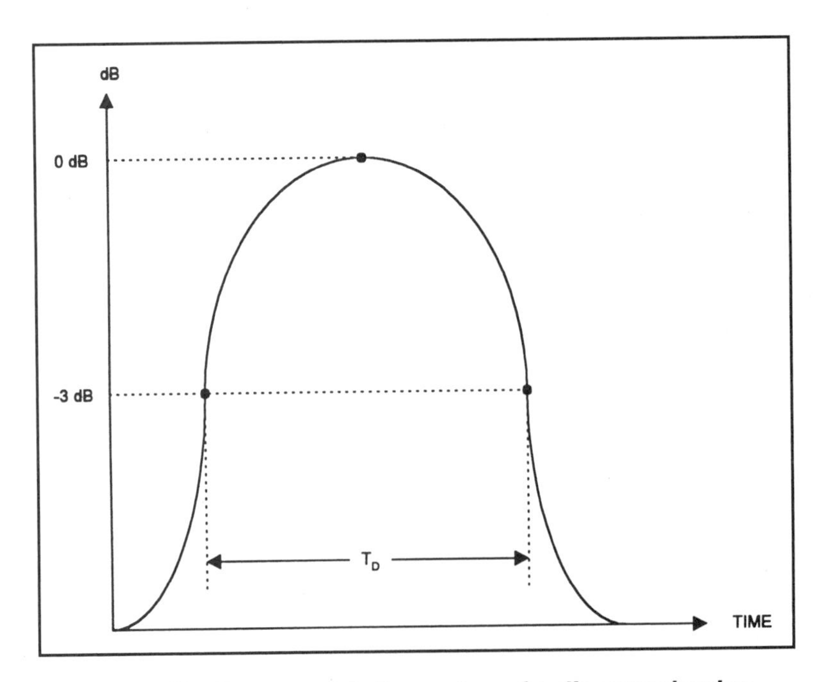

Figure 15-9b. Transmodal dispersion of a dispersed pulse.

before the next pulse arrives. But at high speeds, the pulses may overrun each other (**Figure 15-10**), producing an ambiguous signal that exhibits a high data error rate.

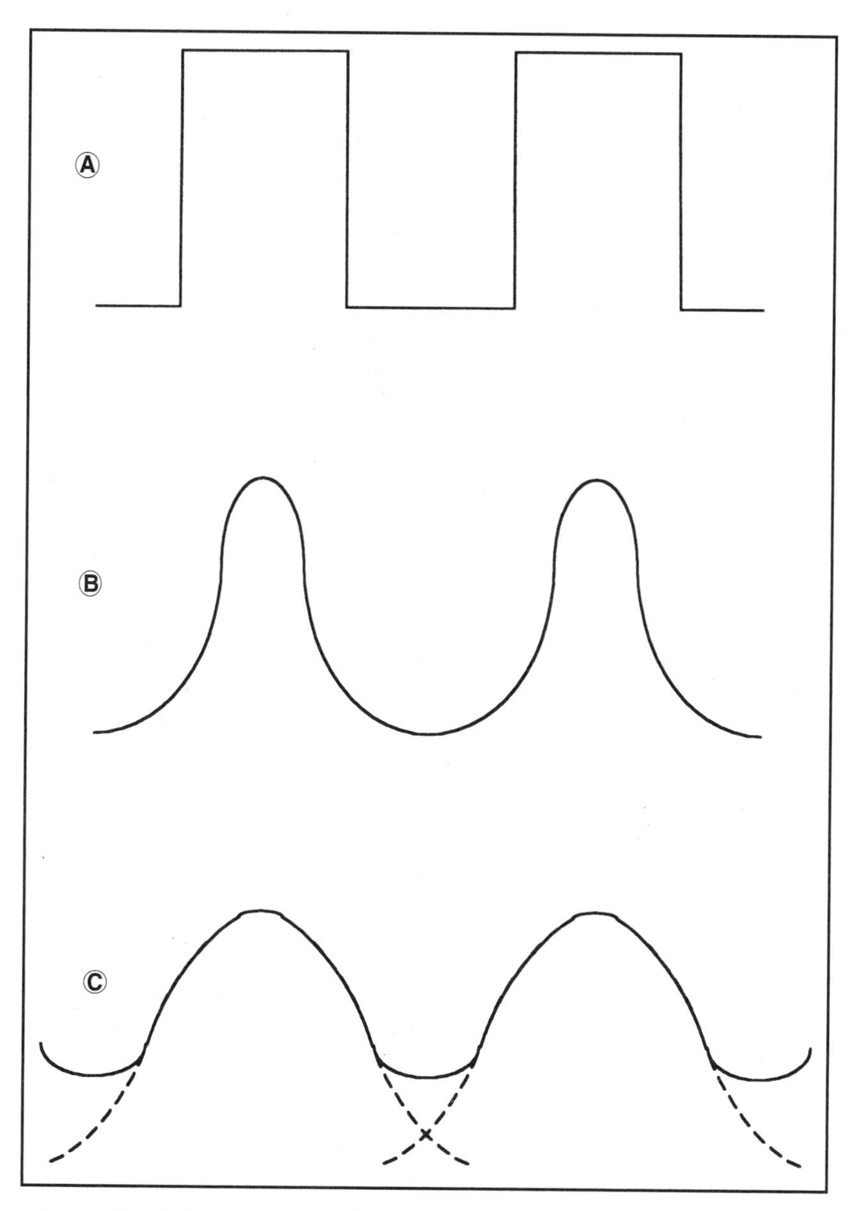

Figure 15-10. Transmodal dispersion creates problems in data fiber optic data systems: a) input data square wave; b) resultant light pulse signal; c) dispersed signal causes overriding of pulses.

Intermodal dispersion is usually measured relative to the widths of the pulses at the -3 dB (half-power) points. In **Figure 15-9**, the time between -3 dB points on the incident pulse transmitted into the optical fiber is T, while in the received pulse the time between -3 dB points is T_d. The dispersion is expressed as the difference, or:

- $$Dispersion = T - T_d$$ eq. (15-14)

A means for measuring the dispersion for any given fiber optic element is to measure the dispersion of a Gaussian (normal distribution) pulse at those -3 dB points. The cable is then rated in nanoseconds of dispersion per kilometer of fiber (ns/km).

The bandwidth of the fiber, an expression of data rate in megahertz per kilometer (MHz/km), can be specified from knowledge of the dispersion, using the expression:

- $$BW\,(MHz/km) = \frac{310}{Disp.(ns/km)}$$ eq. (15-15)

Graded Index Fibers

A solution to the dispersion problem is to build an optical fiber with a continuously varying index of refraction that decreases with increasing distance from the optical axis. While such smoothly varying fibers are not easy to build, it is possible to produce an optical fiber with layers of differing indices of refraction (**Figure 15-11**). Such a fiber is known as a graded index fiber.

The overall index of refraction determines the numerical aperture, and is taken as the average of the different layers.

With graded index fibers, the velocity of propagation of the light ray in the material is faster in the layers away from the optical axis than in the lower layers. As a result, a higher-order mode wave will travel faster than a wave in a lower order. The number of modes available in the graded index fiber is:

$$N = \frac{1}{4}\left(\frac{\pi D [NA]^2}{\lambda}\right) \qquad \text{eq. (15-16)}$$

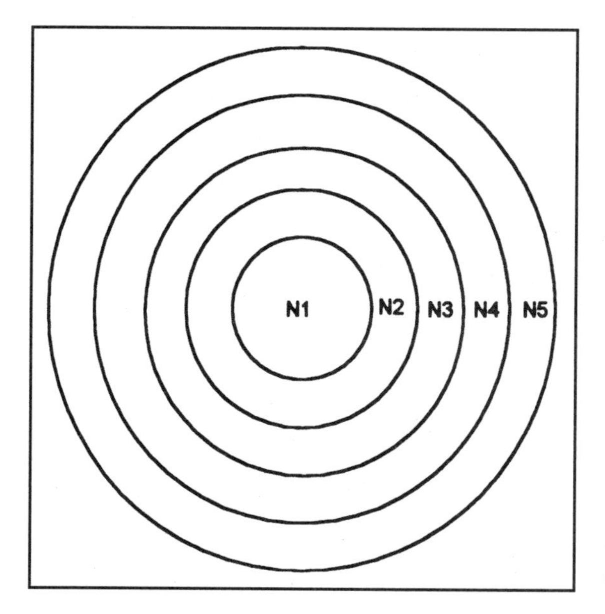

Figure 15-11. Graded index optical fiber.

Some cables operate in a *critical mode*, designated HE_{11} (to mimic microwave terminology), in which the cable is very thin compared with multimodal cables. As the diameter of the core decreases, so does the number of available modes, and eventually the cable becomes *single-mode* (sometimes, especially in the United Kingdom, called *monomodal*). If the core is as small as 3 to 5 mm, then only the HE_{11} mode becomes available. The critical diameter required for single-mode operation is:

$$D_{crit} = \frac{2.4\,\lambda}{\pi [NA]} \qquad \text{eq. (15-17)}$$

Because a single-mode cable reduces the number of available modes, it also reduces intermodal dispersion. Thus, the monomode fiber is capable of extremely high data rates or analog bandwidths.

Losses in Fiber Optic Systems

Understanding and controlling losses in fiber optic systems is integral to making the system work properly. Before examining the sources of such losses, we will briefly examine the vocabulary used to describe losses in the system, as well as the gains of the electronics systems used to process the signals applied to, or derived from, the fiber optic system. This vocabulary uses the *decibel* as the system of measurement.

The subject of "decibels" almost always confuses the newcomer to electronics, and even many an old-timer seem to have occasional memory lapses regarding the subject. For the benefit of both groups, and because the subject is so vitally important to understanding electronics systems, we will examine the decibel.

The decibel measurement originated with the telephone industry, and was named after telephone inventor Alexander Graham Bell. The original unit was the *bel.* The prefix *deci* means "1/10," so the *decibel* is one-tenth of a bel. The bel is too large for most common applications, so it is rarely, if ever, used. Thus, we will concentrate only the decibel (abbreviated "dB").

The decibel is nothing more than a means of expressing a ratio between two signal levels, for example the "output-over-input" ratio of an amplifier. Because the decibel is a ratio, it is also dimensionless. Consider the voltage amplifier as an example of dimensionless gain; its gain is expressed as the output voltage over the input voltage (V_o/V_{in}).

Example 15-1

A voltage amplifier outputs 6 volts when the input signal has a potential of 0.5 volt. Find the gain (A_v).

$$A_v = V_o/V_{in}$$

$A_v = (6 \text{ volts})/(0.5 \text{ volt})$

$A_v = 12$

Note in the above example that the "volts" units appeared in both numerator and denominator, and "canceled out" leaving only a dimensionless "12" behind.

In order to analyze systems using simple addition and subtraction, rather than multiplication and division, a little math trick is used on the ratio. We take the base-10 logarithm of the ratio, and then multiply it by a scaling factor (either 10 or 20). For voltage systems, such as our voltage amplifier, the expression becomes:

$$dB = 20 \log\left(\frac{V_o}{V_{in}}\right) \qquad \text{eq. (15-18)}$$

Example 15-2

Example 15-1 shows a voltage amplifier with a gain of 12 because 0.5-volt input produces a 6-volt output. How is this same gain (V_o/V_{in} ratio) expressed in decibels?

$$dB = 20 \log (V_o/V_{in})$$

$$dB = 20 \log (6 \text{ volts}/0.5 \text{ volt})$$

$$dB = 20 \log (12)$$

$$dB = +21.6 \text{ decibels}$$

The fact that the quantity represented is a gain is indicated by the plus sign. If the quantity represented a loss ($V_o < V_{in}$), then the sign of the result would be negative. Working the problem above for the ratio 0.5/6 results in a loss of -21.6 dB. The numerical result for a loss using the same voltages is the same as for a gain, but the sign is reversed.

Despite the fact that the ratio has been converted to a logarithm, the decibel is nonetheless nothing more than a means for expressing a ratio. Thus, a voltage gain of 12 can also be expressed as a gain of 21.6 dB. A similar expression can be used for current amplifiers, where the gain ratio is I_o/I_{in}:

$$dB = 20\log\left(\frac{I_o}{I_{in}}\right) \quad \textit{eq. (15-19)}$$

For power measurements, which are what is important in light and fiber optic systems, a modified expression is needed in order to account for the fact that power is proportional to the square of the voltage or current:

$$dB = 10\log\left(\frac{P_o}{P_{in}}\right) \quad \textit{eq. (15-20)}$$

We now have three basic equations for calculating decibels, one each for current ratios, voltage ratios, and power ratios. The usefulness of decibel notation is that it can make nonlinear power and gain equations into linear additions and subtractions.

Converting Between dB Notation and Gain Notation

Sometimes gain is expressed in dB, and it is necessary to calculate the gain in terms of the output/input ratio. For example, suppose we have a +20 dB amplifier with a 1 millivolt input signal. To find the expected output voltage, rearrange the expression (dB = 20 log [V_o/V_{in}]) to solve for output voltage V_o, giving the new expression:

$$V_o = V_{in}10^{(dB/20)} \quad \textit{eq. (15-21)}$$

For the reader's convenience, **Table 15-1** shows common voltage and power gains and losses expressed as ratios and in decibels.

Ratio (Out/In)	**Voltage/Current Gain (dB)**	**Power Gain (dB)**
1/1000	-60	-30
1/100	-40	-20
1/10	-20	-10
1/2	-6.02*	-3.01†
1/1	0	0
2/1	+6.02*	+3.01†
5/1	+14	+7
10/1	+20	+10
100/1	+40	+20
1,000/1	+60	+30
10,000/1	+80	+40
100,000/1	+100	+50
1,000,000/1	+120	+60

Table 15-1. Decibel conversion for certain ratios. *Usually rounded to 6 dB. † Usually rounded to 3 dB.

Special dB Scales

Various groups have defined special dB-based scales that meet their own needs. A special scale is made by defining a certain signal level as 0 dB, and referencing all other signal levels to the defined 0 dB point. Several such scales commonly used in electronics are:

dBm. Used in RF measurements, this scale defines 0 dBm as one milliwatt of power dissipated in a 50-ohm resistive load.

Volume Units (VU). The VU scale is used in audio work, and defines 0 VU as 1 mW of 1000-Hz audio signal dissipated in a 600-ohm resistive load.

dBmV. Used in television antenna coaxial cable systems with a 75-ohm impedance, the dBmV scale uses 1 mV across a 75-ohm resistive load as the 0 dBmV reference point.

dBkm. This scale can be used for fiber optics, and refers to the gain or loss relative to the attenuation in a standard optical fiber one kilometer long. Alternatively, either dBmi (dB loss relative to attenuation over 1 mile) or dBl (a normalized unit length) can be used.

The light power P_o at the output end of an optical fiber is reduced from the input light power P_{in} because of internal losses. As in many natural systems, light loss in the fiber material tends to be exponentially decaying (**Figure 15-12a**), and obeys an equation of the form:

$$P_o = P_{in} e^{(-\Lambda/L)} \qquad \text{(eq. 15-22)}$$

Where:

Λ is the length of the optical fiber being considered.

L is the length for which $e^{-\Lambda/L} = e^{-1}$.

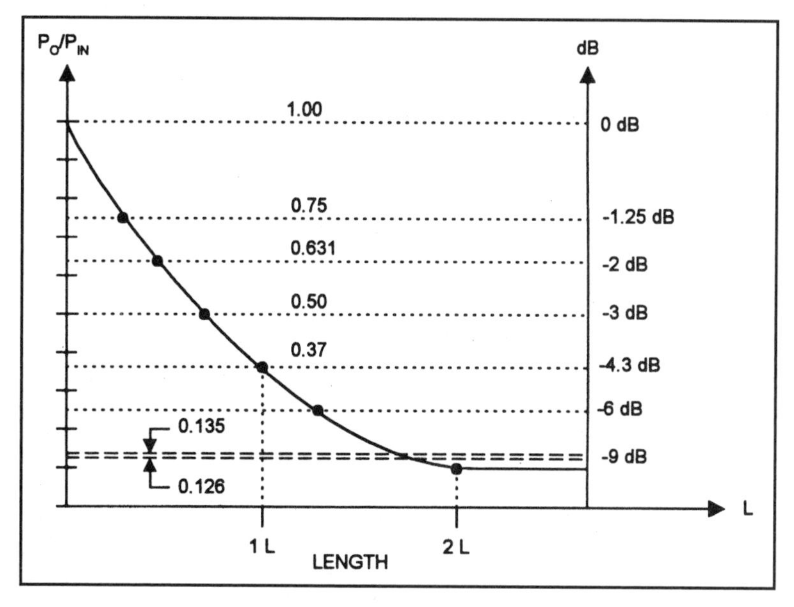

Figure 15-12a. Log-scale graph of signal attenuation.

From **Equation 15-22** above, and by comparing **Figure 15-12a** and **Figure 15-12b**, the reader can see why decibels are preferred. The decibel notation eliminates the exponential notation, allowing losses in any given system to be calculated by simply adding or subtracting dB.

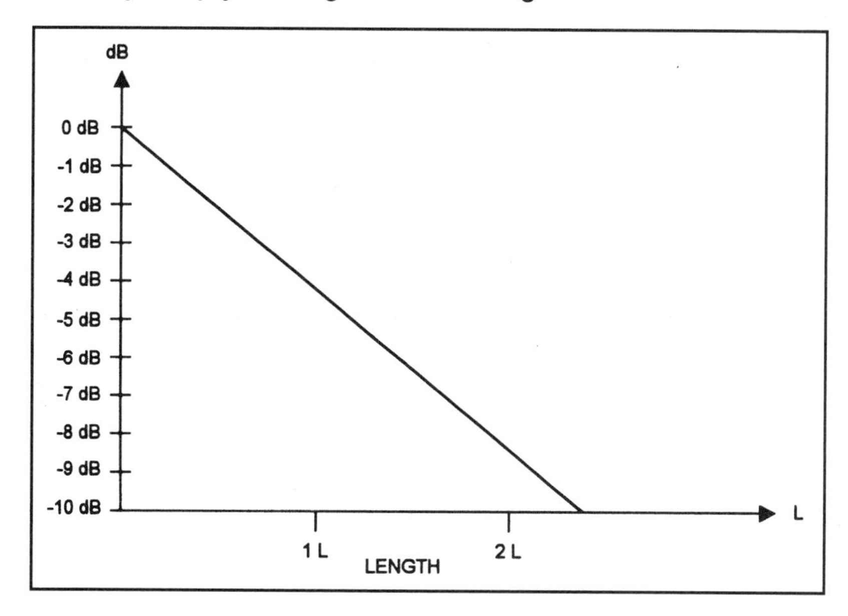

Figure 15-12b. Decibel scale of attenuation.

There are several mechanisms for loss in fiber optics systems. Some of these are inherent in any optical system, while others are a function of the design of the specific system being considered.

Defect Losses. Figure 15-13 shows several possible sources of loss due to defects in the fiber itself. In unclad fibers, surface defects (nicks or scratches) that breach the integrity of the surface will allow light to escape. Also in unclad fibers, grease, oil or other contaminants on the surface of the fiber may form an area with an index of refraction different from what is expected, causing the light direction to change. If the contaminant has an index of refraction similar to that of glass, then it may act as if it were glass and cause loss of light to the outside world. Finally, there is always the possibility of *inclusions:* objects, specks or voids in the material making up the optical fiber. Inclusions can affect both clad and unclad fibers. When light hits an inclusion, it tends to scatter in all directions, causing a loss. Some of the light rays scattered from the inclusion may recombine either destructively or constructively with the main ray, but most do not.

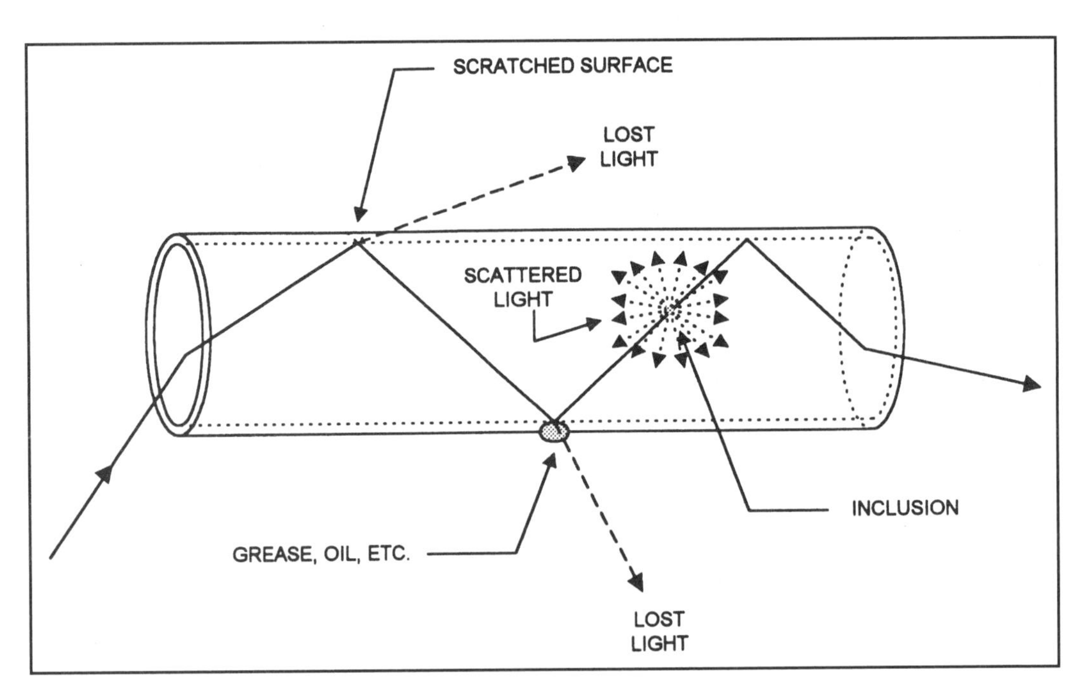

Figure 15-13. Causes of loss in fiber optics include scratches, contaminants, and inclusions.

Inverse Square Law Losses. In all optical systems, there is the possibility of losses due to spreading of the beam. Light power per unit area is inversely proportional to the square of the distance ($1/D^2$) from the source. If you shine a flashlight at a wall from a distance of one meter, and measure the power per unit area at the wall, and then move the light back to twice the distance and measure again, you will find that the power has dropped to one-fourth of its original level.

Transmission Losses. These losses are due to light that is caught in the cladding material of clad optical fibers. This light is either lost to the outside, or is trapped in the cladding layer and is thus not available to be propagated in the core.

Absorption Losses. This form of loss is due to the nature of the core material, and is inversely proportional to the transparency of the material. In some materials, absorption losses may not be uniform across the entire light spectrum, but may instead be wavelength sensitive.

Coupling Losses. Another form of loss is due to coupling systems. All couplings have an associated loss. Several different losses of this sort are identified.

Mismatched Fiber Diameters. This form of loss is due to coupling a large-diameter fiber (D_L) to a small-diameter fiber (D_S) so that the larger fiber transmits to the smaller one. In decibel form, this loss is expressed by:

- $$dB = -10\log\left(\frac{D_S}{D_L}\right)$$ eq. (15-24)

Numerical Aperture Coupling Losses. Another form of coupling loss occurs when the numerical apertures of the two fibers are mismatched. If NA_r is the numerical aperture of the receiving fiber, and NA_t is the numerical aperture of the transmitting fiber, then the loss is expressed as:

- $$dB = -10\log\left(\frac{NA_r}{NA_t}\right)$$ eq. (15-24)

Fresnel Reflection Losses. These losses occur at the interface of an optical fiber with air (**Figure 15-14a**), and are due to the large change in index of refraction between glass and air. There are actually two losses to consider: the loss caused by internal reflection from the inner surface of the interface, and that caused by reflection from the opposite surface across the air gap in the coupling. Typically the internal reflection loss is on the order of 4 percent, while the external reflection loss is about 8 percent.

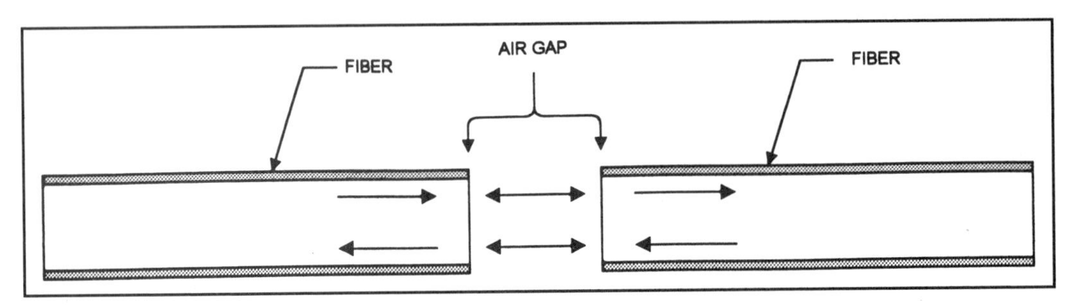

Figure 15-14a. In air, reflections may occur.

Any form of reflection in a transmission system may be modeled similarly to the modeling of reflections in a radio transmission line. Studying *standing waves* and related subjects in books on RF systems can yield some understanding of these problems. The amount of reflection in coupled optical systems uses similar arithmetic:

$$\Gamma = \left(\frac{n_1 - n_2}{n_1 + n_2}\right)^2 \qquad \text{eq. (15-25)}$$

Where:

Γ is the coefficient of reflection.

n_1 is the index of refraction for the receiving material.

n_2 is the index of refraction for the transmitting material.

The mismatching of refractive indices is analogous to the mismatch of impedances problem seen in transmission line systems, and the cure is also analogous. Where a transmission line uses an impedance matching coupling device, an optical fiber will use a coupler that matches the "optical impedances," the indices of refraction. **Figure 15-14b** shows a coupling between the ends of two fibers (lenses may or may not be used, depending on the system) made with a liquid or gel having an index of refraction similar to that of the fibers. The reflection losses are reduced or even eliminated.

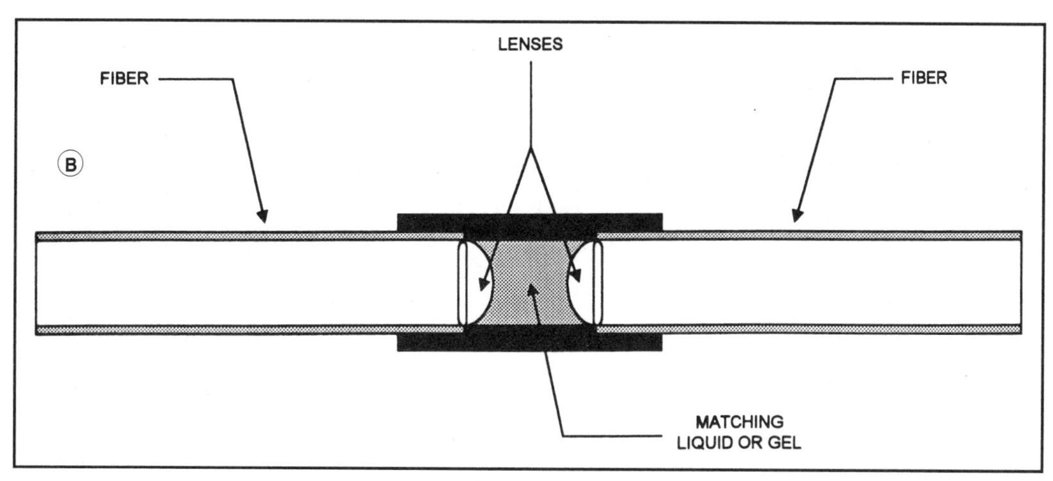

Figure 15-14b. "Impedance matched" coupler.

Fiber Optic Communications Systems

A communications system requires an information signal source (such as voice, music, digital data or an analog voltage representing a physical parameter), a transmitter, a propagation medium (in this case optical fibers), a receiver and an output. In addition, the transmitter may include any of several different forms of *encoder* or *modulator*, and the receiver may contain a *decoder* or *demodulator*.

Figure 15-15 shows two main forms of communications link. The *simplex* system is shown in **Figure 15-15a**. In this system, a single transmitter sends information over the path in only one direction to a receiver set at the other end. The receiver cannot reply or otherwise send data back the other way. The simplex system requires only a single transmitter and a single receiver per channel.

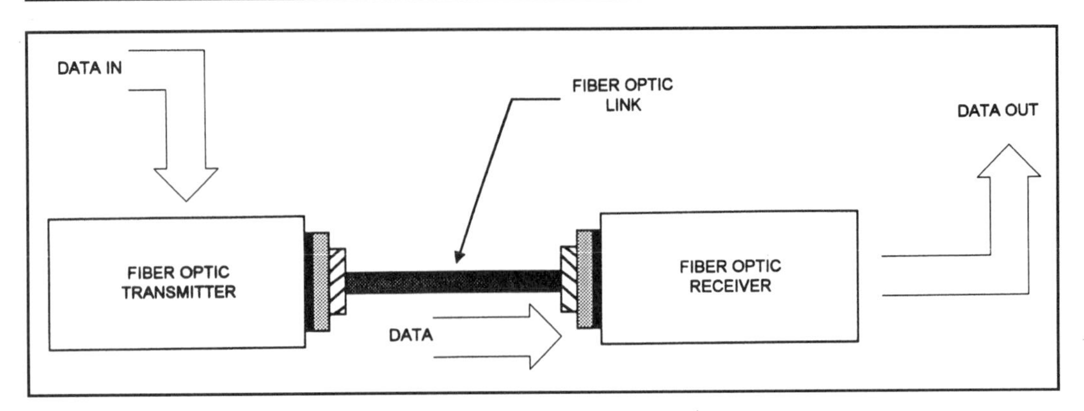

Figure 15-15a. Simplex system provides one way communication.

A *duplex* system (**Figure 15-15b**) is able to simultaneously send data in both directions, allowing both send and receive capability at each end. The duplex system requires a receiver, a transmitter and a two-way beam splitting Y-coupler at each end.

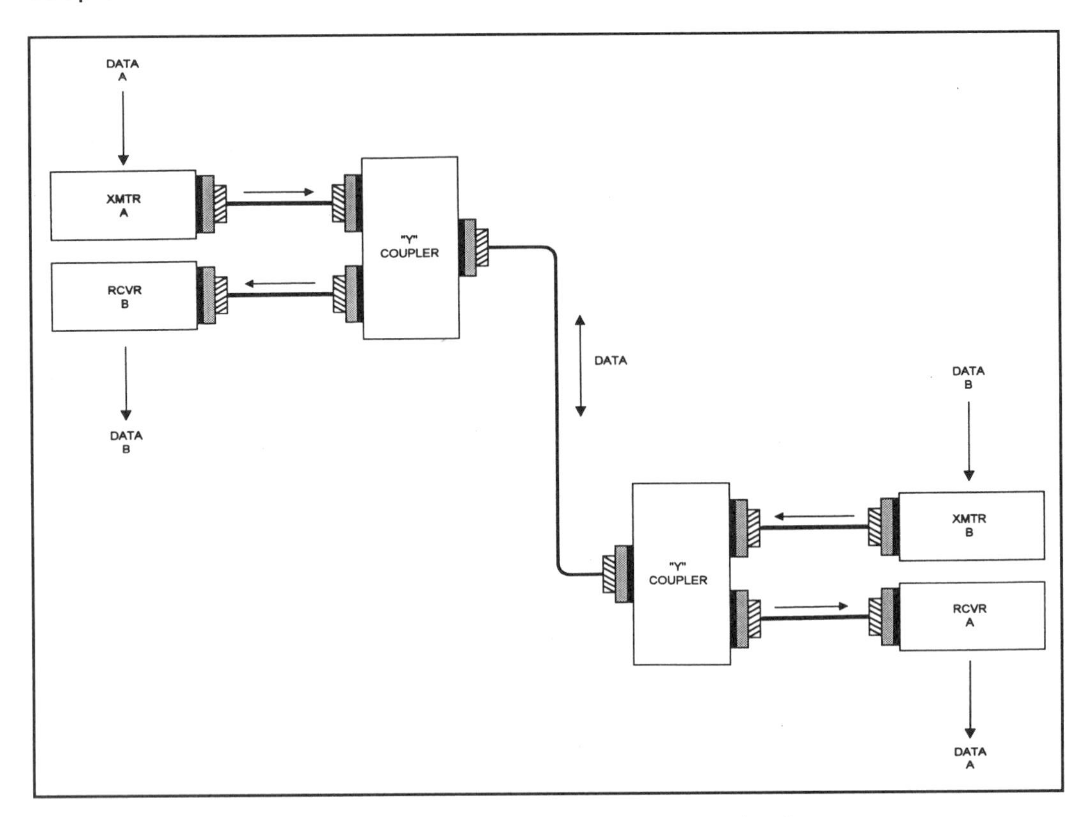

Figure 15-15b. Duplex system provides two-way communications.

Receiver Amplifier and Transmitter Driver Circuits

Before an optical fiber system is useful for communications, a means must be provided to convert electrical (analog or digital) signals into light beams. Also necessary is a means for converting the light beams back into electrical signals. These jobs are done by *driver* and *receiver preamplifier* circuits, respectively.

Figure 15-16 shows two possible driver circuits. Both circuits use LEDs as the light source. The circuit in **Figure 15-16a** is useful for digital data communications. These signals are characterized by on/off (HIGH/LOW or 1/0) states in which the light emitting diode is either ON or OFF, indicating which of the two possible binary digits is required at the moment.

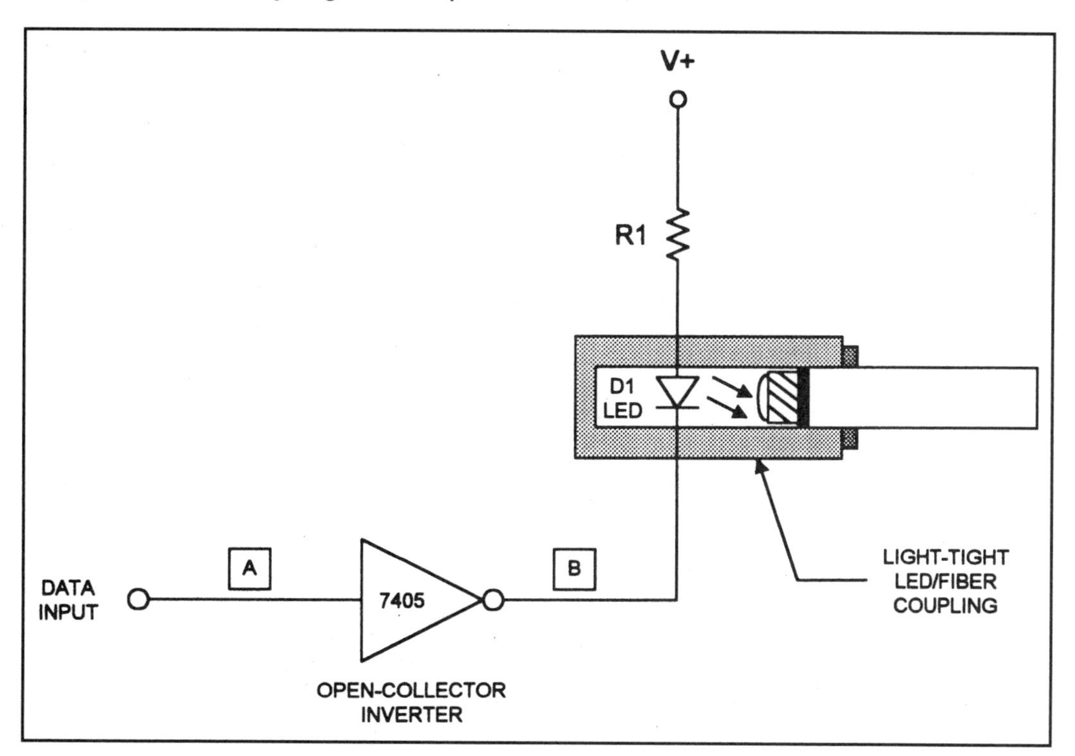

Figure 15-16a. Digital fiber optic driver circuit.

The driver circuit consists of an open-collector digital inverter device in a light-tight container. These devices obey a very simple rule: if the input A is HIGH, then the output B is LOW, and vice versa. Thus, when the input data signal is HIGH, the cathode of the LED is grounded, and the LED turns on and sends a light beam along the optical fiber. When the input data line is LOW, the LED is ungrounded (and therefore turned off), so no light enters the fiber. The resistor R1 is used to limit the current flowing in the LED to a safe value. Its resistance is found from Ohm's law and the maximum allowable LED current:

$$R1 = \frac{(V+) - 0.7}{I_{max}} \quad \text{eq. (15-26)}$$

An analog driver circuit suitable for voice and instrumentation signals is shown in **Figure 15-16b.** This circuit is based on the operational amplifier (see Chapter 10). There are two aspects to this circuit: the *signal path* and the *dc offset bias.* The offset bias is needed to place the output voltage at a point where the LED is lighted at about one-half of its maximum brilliance when the input voltage V_{in} is zero. That way, negative polarity signals will reduce the LED brightness, but won't turn it off (see **Figure 15-16c**). In other words, biasing avoids clipping off the negative peaks. If the expected signals are monopolar, then V1 should be set to barely turn on the LED when the input signal is zero.

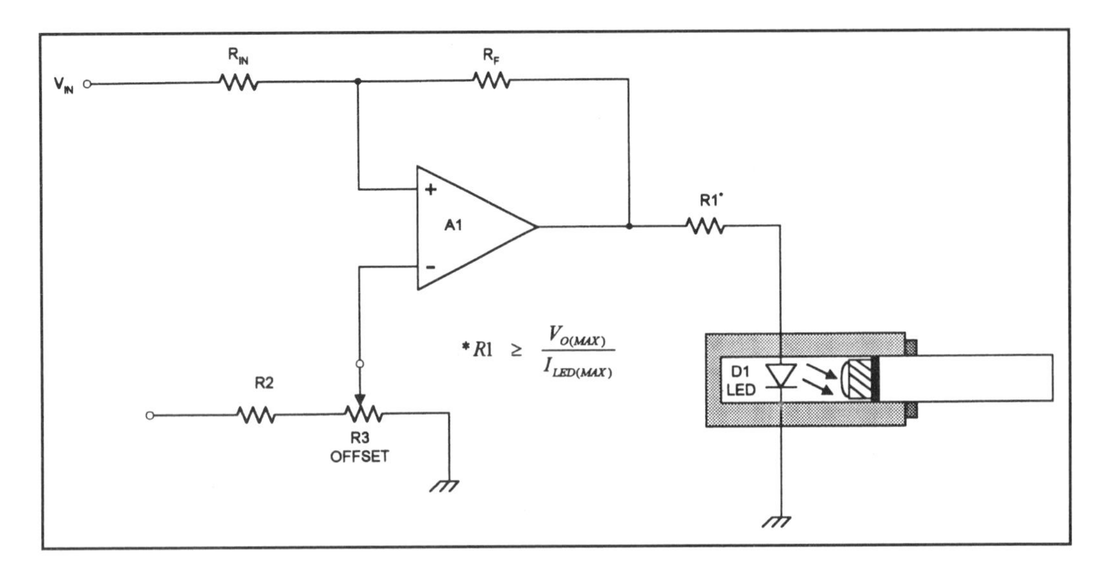

Figure 15-16b. Analog fiber optic driver circuit.

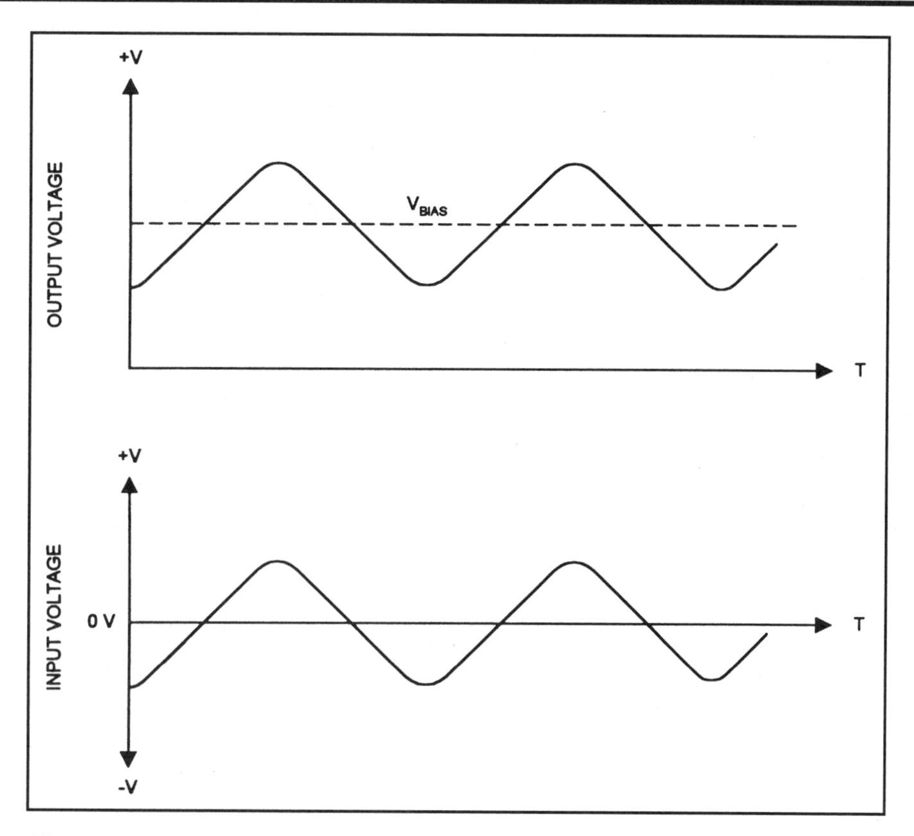

Figure 15-16c. Effect of dc offset in preventing signal distortion due to clipping in driver circuit.

The signal V_{in} sees an inverting follower with a gain of $-R_f/R_{in}$, so the total output voltage (accounting for the dc bias) is:

$$V_o = \left(\frac{-V_{in} R_f}{R_{in}}\right) + V1\left(\frac{R_f}{R_{in}} + 1\right) \qquad \text{eq. (15-27)}$$

Because the network R2/R3 is a resistor voltage divider, the value of V1 will vary from 0 volts to a maximum of:

$$V1 = \frac{(V+)R3}{R2 + R3} \qquad \text{eq. (15-28)}$$

Therefore, we may conclude that $V_{o(max)}$ is:

$$V_{o(\max)} = \left(\frac{-V_{in} R_f}{R_{in}}\right) + \left(\frac{(V+) R3}{R2 + R3}\right)\left(\frac{R_f}{R_{in}} + 1\right) \qquad \text{eq. (15-29)}$$

Three different receiver preamplifier circuits are shown in **Figure 15-17**: analog versions are shown in **Figure 15-17a** and **Figure 15-17b**, while a digital version is shown in **Figure 15-17c**. The analog versions of the receiver preamplifiers are based on operational amplifiers. Both analog receiver preamplifiers use a photodiode as the sensor. These PN or PIN junction diodes (see Chapters 7 and 8) produce an output current I_o that is proportional to the light shining on the diode junction.

The version shown in **Figure 15-17a** is based on the inverting follower circuit. The diode is connected with its noninverting input grounded, thereby set to zero volts potential, and the diode current is applied to the op-amp's noninverting input. The feedback current I_f exactly balances the diode current, so the output voltage will be:

$$V_o = -I_o R_f \qquad \text{eq. (15-30)}$$

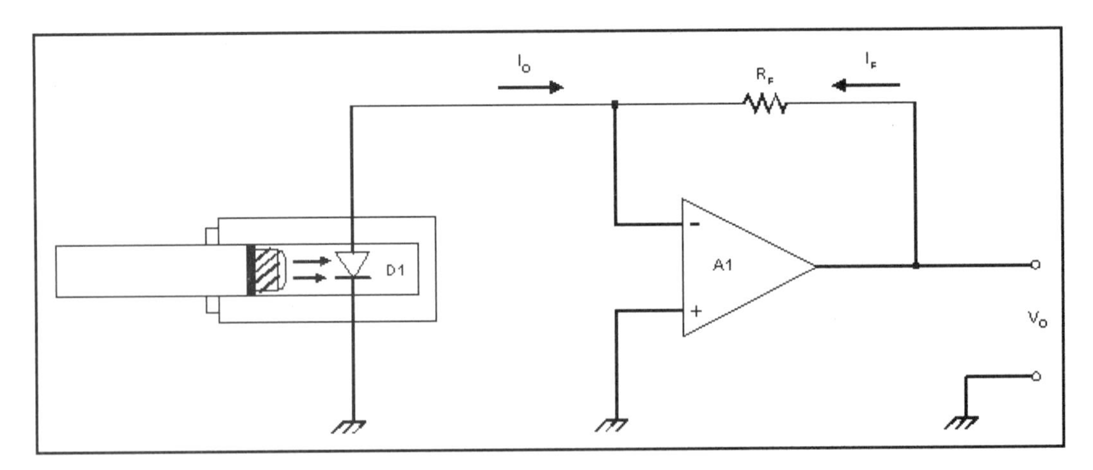

Figure 15-17a. Inverting fiber optic receiver circuits.

The noninverting follower version shown in **Figure 15-17b** uses the diode current to produce a proportional voltage drop (V1) across a load resistance R_L. The output voltage for this circuit is:

$$V_o = I_o R_L \left(\frac{R_f}{R_{in}} + 1 \right) \qquad \text{eq. (15-31)}$$

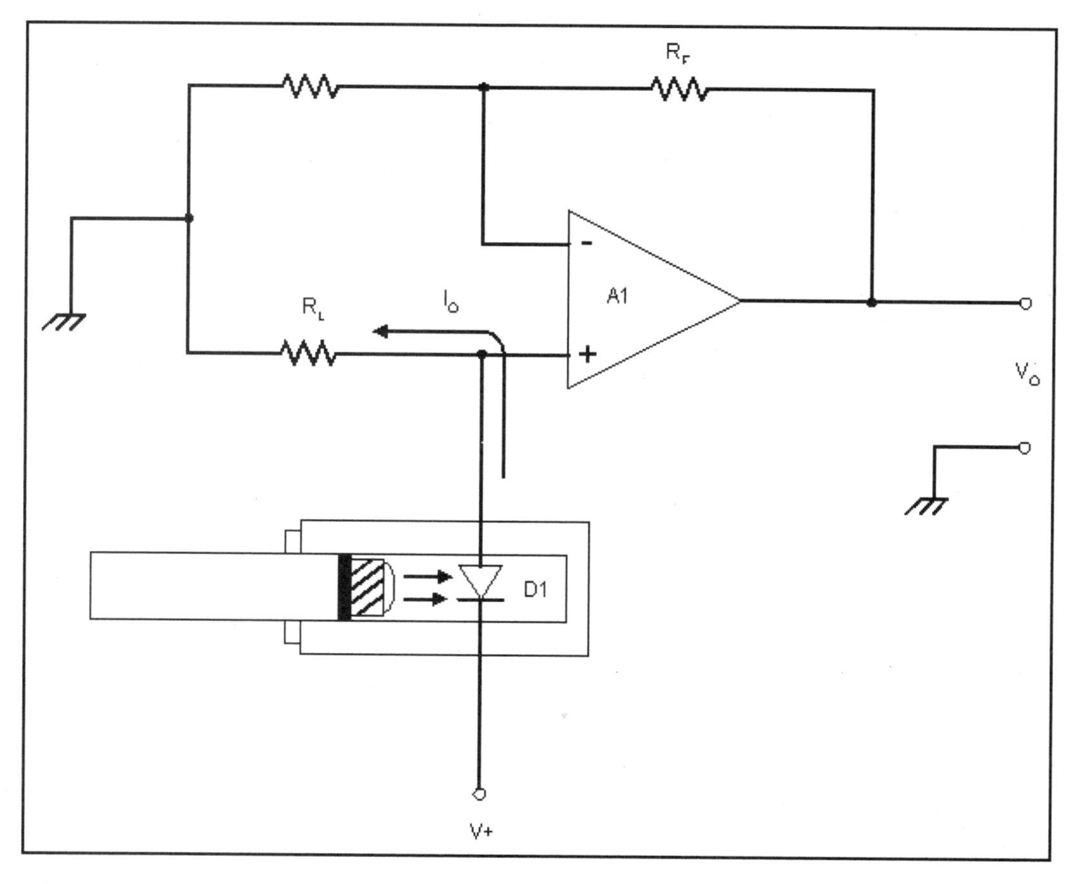

Figure 15-17b. Noninverting fiber optic receiver circuit.

Both analog circuits will respond to digital signals, but they are not optimum for that type of signal. Digital signals will have to be reconstructed because of uncertainties caused by dispersion. A better circuit is that of **Figure 15-17c**, in which the sensor is a phototransistor connected in the common emitter configuration. When light shines on the base region, the transistor conducts, causing its collector to be at a potential only a few tenths of a volt above ground. Conversely, when there is no light shining on the base, the collector of the transistor is at a potential close to V+, the power supply potential.

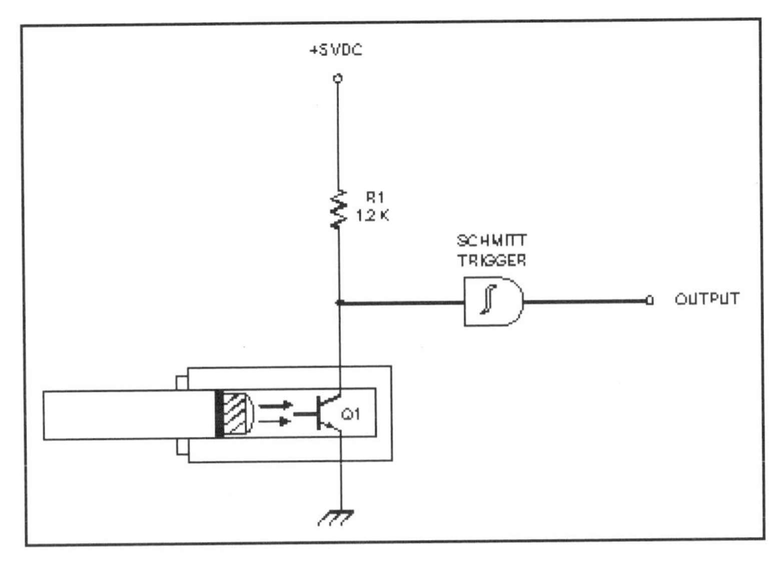

Figure 15-17c. Digital receiver.

Clean-up action occurs in the following stage, a digital Schmitt trigger. The output of such a device will snap HIGH when the input voltage exceeds a certain minimum threshold, and remain HIGH until the input voltage drops below another threshold (snap-HIGH and snap-LOW thresholds are *not* equal). Thus, the output of the Schmitt trigger is a clean digital signal, even though the sensed signal may be blurred.

Notes

[1]TIR is called total internal *reflection*, but it is actually a refraction phenomenon.

Chapter 16

Television Sensor Tubes

Television is the art of sending light images over telecommunications channels such as cable or radio waves. Everyone is familiar with the entertainment uses of television. Television also has military, industrial, scientific, medical and trade uses. The plumber or septic tank servicer can send a television camera on cables through pipes or into septic tanks to inspect their condition. The medical community can use miniature TV cameras to inspect the interior of the human body to look for cancerous tumors, ulcers and other threats to physical health. This chapter will briefly examine the process of converting a light image into the electrical impulses that produce a television picture.

The Basic Television System

The television system used in the United States is based on a standard called the *National Television Standards Committee* (NTSC) system. Although now a color standard, the original television (TV) standard was black and white (B&W) only. Some "humorists," noting that some European television systems have superior color, especially when transmitted over long cables, are known to call the NTSC system "Never Twice Same Color." Be that as it may, the basic system is the same for both B&W and color because of an original Federal Communications Commission (FCC) requirement that the color TV system be compatible with B&W receivers. The policy of the FCC is that a new feature added to an old service must be designed so that existing equipment is not rendered obsolete when the new signal is transmitted.

Figure 16-1a shows the basic block diagram of a TV camera and image generation system. The sensor tube is a photosensor similar to those discussed earlier in this book. The sensor is aimed at successive points on the object by a *deflection* section that sweeps the sensor tube both vertically and horizontally (see **Figure 16-1b**).

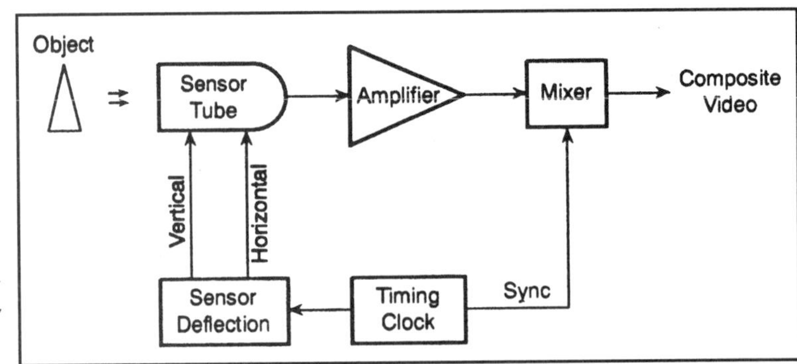

Figure 16-1a. Block diagram of basic TV camera system.

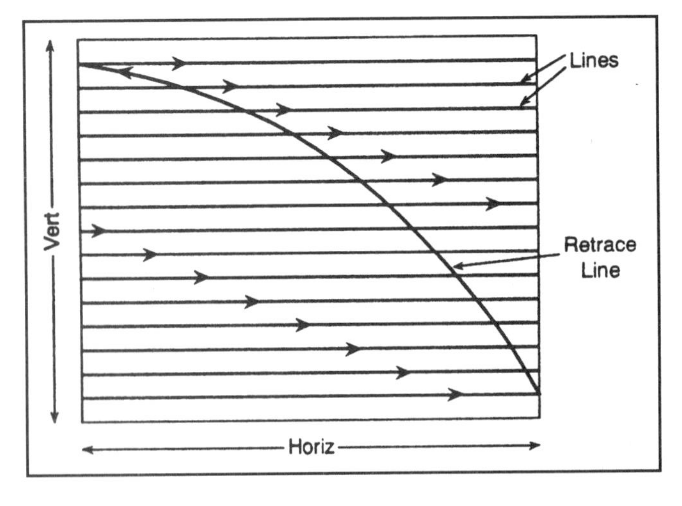

Figure 16-1b. Scan pattern across face of tube.

The horizontal sweep drags the sensor from left to right to create a *line*. The line produces a continuously varying video signal in which the amplitude of the signal is a function of the light level arriving at that point. **Figure 16-1c** shows one line of video as it scans across a pattern of black, gray and white bands. The signal amplitude varies with the amount of light arriving from the televised scene. The video signal varies between two extremes: full black and full white. The "blacker than black" region is used for timing signals.

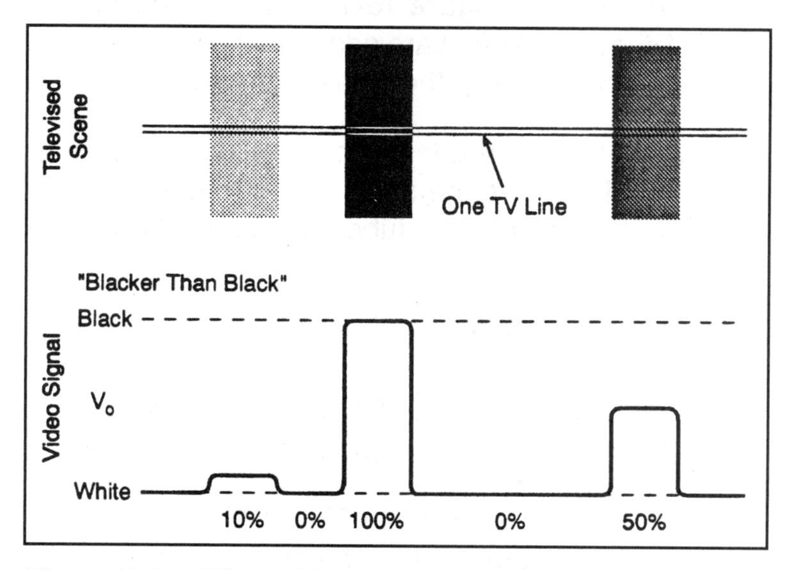

Figure 16-1c. Effect of line brightness on video signal.

At the end of each line, the vertical deflection system drops down one space so that the next line is swept out slightly offset from the previous line. When the last line is swept out, and arrives at the lower right corner, the vertical sweep signal resets the beam to the upper left corner. This process continues over and over on a continuous basis.

When they are all combined, these lines form the picture. In the U.S. system, there are 525 horizontal lines (usually 520 are visible). These lines can be seen by observing a large-screen B&W TV set close up, or by using even a weak magnifying glass on a small B&W set (one cannot see them as easily on a color TV set because a matrix of blue, green and red primary color dots is used to produce the color). The horizontal sweep rate is 15,750 Hz on B&W, and 15,734 Hz on color TV sets, while the vertical rates are 60 Hz and 59.94 Hz, respectively.

The *resolution* of the TV system depends on the number of lines used to scan the object to form an electronic image. The greater the number of lines, the better the resolution. The U.S. 525-line broadcast TV system has less resolution than the 600 and 800-line systems used in some countries and in industrial and medical TV systems.

The vertical and horizontal deflection signals are controlled by a timing system that sets the frequencies (**Figure 16-1a**). This stage also produces vertical and horizontal sync pulses that are added to the video signal to produce a *composite video signal*, which is the signal transmitted to the receiver.

A TV receiver receives the composite video signal, separates the video from the sync pulses, and then uses the sync pulses to control its own internal deflection system for the cathode ray tube used to display the video image.

TV Camera Sensor Tubes

This section examines several different B&W sensor tubes: *iconoscopes*, *image orthicons*, *vidicons* and *charge coupled arrays*.

Iconoscope. An iconoscope is shown in **Figure 16-2**. This tube was one of the earliest forms of TV camera sensor. The basic elements of this tube are the *electron gun*, *magnetic deflection coils* and a *photosensor* consisting of a collector ring and a target. The electron beam generated in the electron gun is swept vertically and horizontally by a pair of magnetic deflection coils.

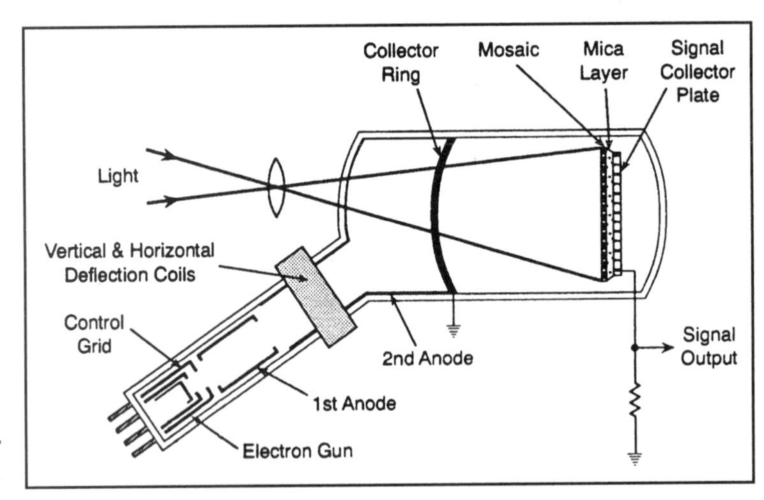

Figure 16-2. Iconoscope sensor tube.

The iconoscope target consists of a photosensitive mosaic separated from a signal collector plate by a thin sheet of mica insulator. The mosaic is formed by depositing light-sensitive cesiated silver globules, each about 0.001 inch in diameter, onto the surface of the mica. These globules become charged when light strikes them, causing them to interact with electrons from the scanning system. Electrons are transmitted to the output according to the level of charge on the globules, forming the signal current.

The silver globules in the iconoscope are smaller than the diameter of the electron beam. The beam hits an area about 0.007 inch in diameter, forming the smallest possible area of image. These areas are called *picture elements* or *pixels* for short (like those used in modern graphics displays). An iconoscope contains about 200,000 pixels.

A defect of the iconoscope design is the so-called *dark spot signal.* The process of collecting electrons released in the mosaic is only 25 to 30 percent efficient, that is, only that percentage of electrons reaches the collector ring and the external circuitry. The remaining 70 to 75 percent of the released electrons are redistributed somewhat randomly over the mosaic surface. When this layer of secondary electrons reaches a certain level, the signal output decreases, creating the dark spot signal. External *shading generator* circuits are designed to counteract this effect.

Image Orthicon. An image orthicon camera tube is shown in **Figure 16-3a.** This tube is 100 to 1000 times more sensitive to lower light levels than the iconoscope, but does not have quite the same level of resolution. It is consid-

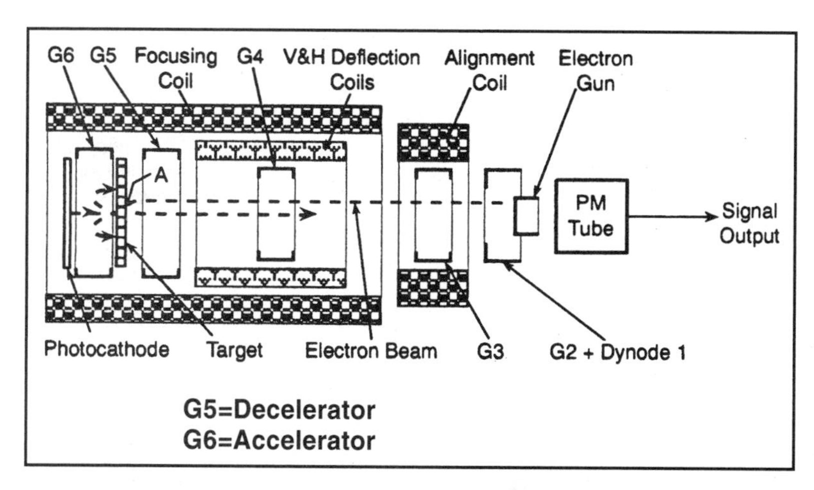

Figure 16-3a. Image orthicon sensor tube structure.

ered superior to the iconoscope for most practical purposes. The elements of the image orthicon are a photocathode, a target, a deflection system, a focusing system and an electron gun with associated positively charged accelerator elements.

Light from the scene being imaged is coupled through a lens to a photocathode immediately inside the tube structure. This cathode is a photoemissive surface, and releases electrons in an amount proportional to the level of light striking it. These electrons are accelerated toward a target by a positive potential on grid G6, and kept on a parallel path by the focusing coils. A cloud of secondary electrons is emitted by the target when the primary electrons from the photocathode give up their kinetic energy. This action creates areas of varying positive (or less negative) charges on the surface of the target. The target (**Figure 16-3b**) consists of two meshes separated by a thin membrane of optical glass or magnesium oxide (MgO).

Electrons from an electron gun in the rear of the tube are accelerated toward the back of the target, but are decelerated by grid G5, stopping at a point just short of the target (point A in **Figure 16-3a**). When the electron beam strikes the target, some of the electrons are removed to neutralize the areas of positive charge on the target surface. The remaining electrons are turned back toward the positive potential applied to grid G4, and are accelerated to the back of the tube where they are collected by the dynode electrodes. These dynodes (only one is shown) form a photomultiplier tube (PMT) section, which accounts in part for the sensitivity of the image orthicon. The output of the photomultiplier is the video signal.

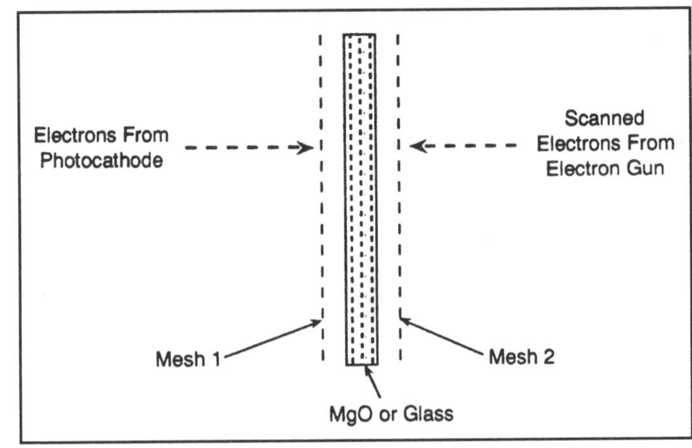

Figure 16-3b. Image orthicon target.

Vidicons. The most common camera tube today is the *vidicon*, invented by RCA. This tube is smaller than the iconoscope and image orthicon, and can be purchased with target electrodes that are sensitive to infrared, visible, visible and infrared or ultraviolet wavelengths. **Figure 16-4a** shows the basic structure of the vidicon. In some respects, its operation is very much like the orthicon: a scanned electron beam is aimed at a target, decelerated to a slow speed to permit it to interact with the positively charged target, and then returned to the rear of the tube for collection. The video signal is generated by modulation of the electron beam by the charge on the target.

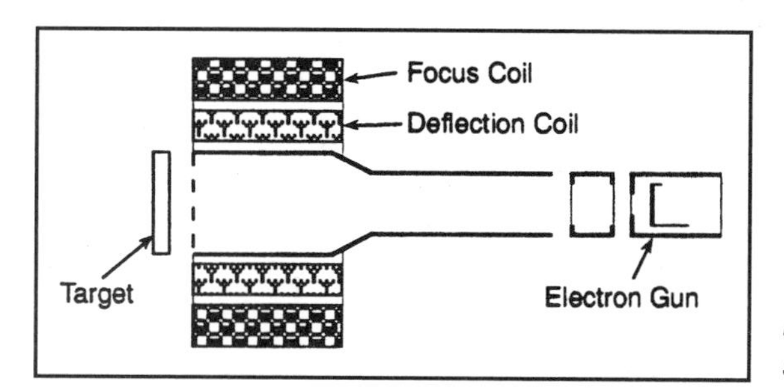

Figure 16-4a. Vidicon sensor tube structure.

A standard vidicon target is shown in **Figure 16-4b**, along with an equivalent circuit. The target consists of a thin faceplate of optical glass, followed by a semiconductive layer of silicon dioxide (SiO_2) glass, and a photoconductive layer of tin dioxide (SnO_2). The semiconductive glass and photoconductive layer are separated from each other. The equivalent circuit of each picture element consists of a capacitance shunted by a light-variable resistance.

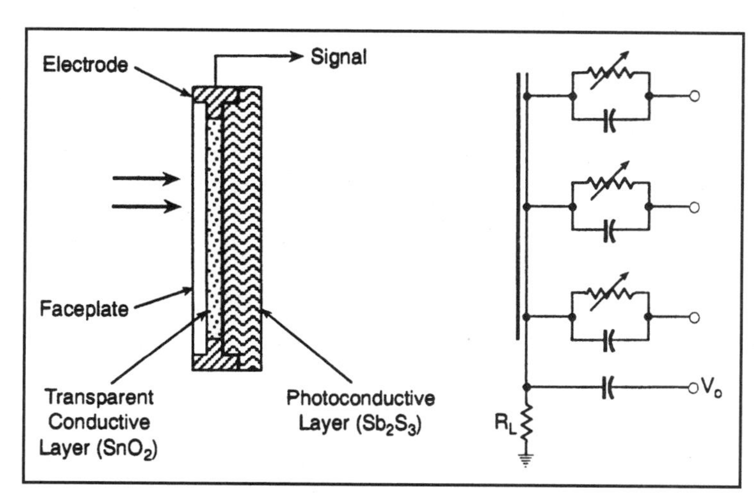

Figure 16-4b. Vidicon target.

When light strikes the photoconductive layer, it reduces the element resistance, thereby reducing the charge on the "capacitor." The scanning electron beam essentially "reads" the charge on each of these capacitances.

A variation on the theme is the *plumbicon target* shown in **Figure 16-4c**. This target forms a PIN diode of lead dioxide (PbO_2), each section of which is appropriately doped to make it an N-type, P-type, or intrinsic (I-type) semiconductor. In the plumbicon vidicon, the elements are modeled as light-sensitive photo PIN diodes. A silicon dioxide target is available that also contains elements that are photodiodes.

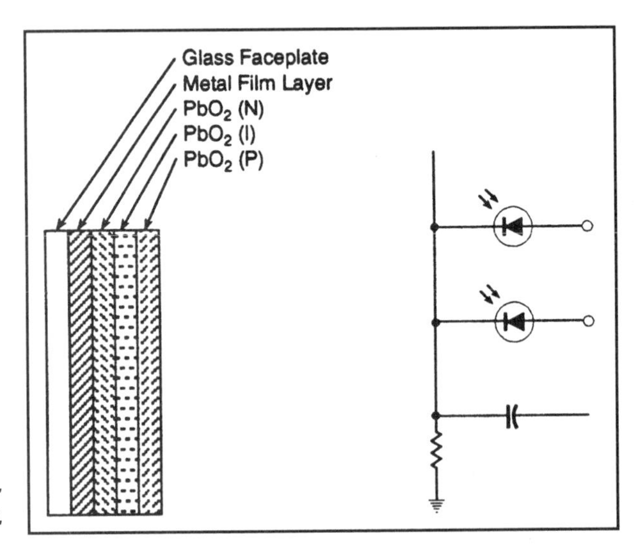

Figure 16-4c. Plumbicon target.

CCD Array. The charge coupled device sensor is an array of closely spaced metal oxide semiconductor diodes. The basic structure (**Figure 16-5**) consists of a region of N-type silicon separated from metallic electrodes by a layer of silicon dioxide. It serves to move charges from one well to another "bucket brigade" style under the influence of clock pulses, and to collect them at the output end of the chain. The CCD is a small, flat sensor that is the basis of the modern video cameras made for consumers, as well as modern industrial, scientific and medical TV cameras. Tiny CCD cameras have been placed on the tips of probes for exploring the human stomach. Light from an adjacent optical fiber cable illuminates the scene, allowing the physician to explore for tumors or ulcers.

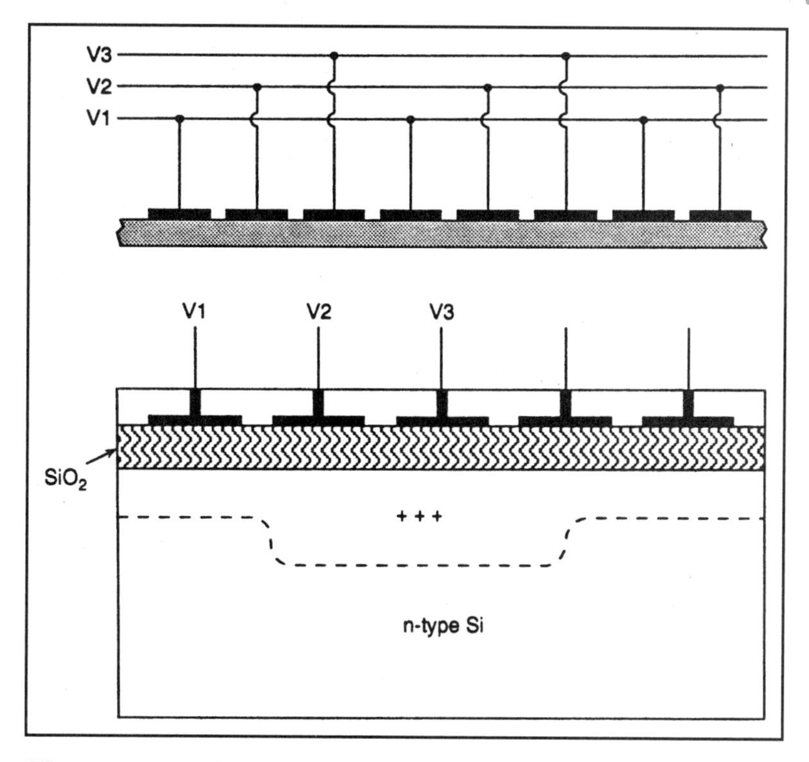

Figure 16-5. Charge Coupled Device (CCD) sensor basic sructure.

Chapter 17

More Optical Instruments

This chapter will discuss various optical instruments, some of which are directly pertinent to electro-optical sensors, and others which are intended mostly for visual observation. It is not unusual to find electro-optical sensors replacing the human eye in some cases. The designs of these instruments are presented in sufficient detail for you to be able to produce versions of your own with a bit of effort.

Flame Detectors

Furnaces that burn fuel-air mixtures can be dangerous, especially if the fuel is force-fed either by gravity or by a motor-driven pump or injector. In such furnaces, it is common to have either a pilot flame or electrical arc mechanism to ignite the fuel at the proper time. For example, in an oil-burning furnace, fuel is "atomized" in a nozzle and sprayed into the combustion chamber, where an arc device ignites the mixture. Some furnaces, especially those used in industrial boiler systems, include a flame sensor to prevent spraying excess fuel into the combustion chamber unless a flame is present. Otherwise, a dangerous amount of fuel will build up in the bottom of the chamber, and ignite improperly when the igniter actually operates.

In many gas furnaces, a pilot light is used to ignite the gas when the thermostat opens the gas valve. If the valve opens, but the pilot light is not working, then the chamber will fill with gas, which can spill over into the room. If the pilot light turns on (or you ignite it!), or an electrical arc occurs in the room, then the excess gas will ignite explosively.

In many of these systems some sort of flame sensor is used, either for the main burner flame or a pilot lamp. In some cases, the sensor is a *thermocouple* embedded in the flame. Thermocouples are devices that produce an electrical potential that is proportional to the junction temperature. These devices are used in many residential natural or LP gas furnaces. In other cases, the sensor is an electro-optical device.

Figure 17-1 shows an E-O sensor used for flame detection. The sensor shown here is a phototransistor (Q1), but other types such as photovoltaic cells and photoconductors are also used. The photosensor is located outside the combustion chamber because of the excessive heat inside the chamber. Typically, a window of quartz or other transparent material is used to separate the sensor from the flame, to permit light to enter but restrain heat. In modern systems, optical fibers are undoubtedly used to couple light from the furnace to a remotely located sensor.

The sensor will not do any good whatsoever unless it is connected to an external alarm circuit. The alarm circuit uses a current-mode comparator (A1) based on an operational amplifier. In an op-amp used as a comparator, the output voltage V_o will be one of three levels depending on the voltage applied to the inputs (-IN and +IN). If $V_{-IN} = V_{+IN}$, then the output is zero; if $V_{-IN} > V_{+IN}$, then the op-amp sees it is a positive input, so the output is saturated at a large negative value $-V_{sat}$; if $V_{-IN} < V_{+IN}$, then the output is $+V_{sat}$.

In the *current mode* of comparator operation, one input is held at ground potential (zero volts). In this case the noninverting input +IN is grounded. At the other input, resistors R1 and R2 are connected at the summing junction. The operation is similar to the voltage mode. In a typical alarm system, the

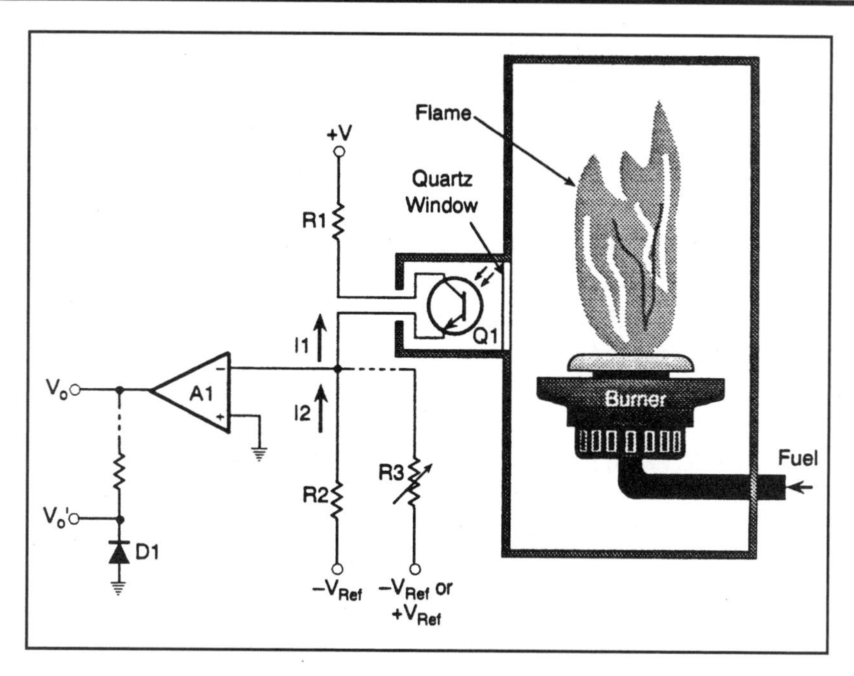

Figure 17-1. Optical furnace monitor system.

output will deliver either zero or one polarity in the safe condition, and the opposite polarity in the unsafe condition. For example, when I1 = I2, the output will be zero. When I2 < I1, the output is $-V_{sat}$, but when I2 < I1, the output snaps HIGH to $+V_{sat}$. Thus, when V_o is either zero or negative, the circuit indicates the safe condition, but when V_o is positive, a dangerous condition exists.

The output circuit of the alarm can be altered by using a series current limiting resistor and diode D1 so that the output potential is not allowed to become negative, but remains close to zero in the safe condition because D1 clips the negative excursion to approximately 0.6 volt. Another alternative is to use comparator integrated circuits that are inherently unable to output more than two states. For example, the LM-311 device uses an open-collector NPN transistor output. The output terminal must be connected to V+ through an external pull-up resistor for it to work. Because of that arrangement, the output can be only zero or V+.

By making R2 a variable resistor, it is possible to make the circuit sensitivity adjustable. An alternative method is to use variable resistor R3 in parallel with R2 to $-V_{ref}$, or to a positive reference source (depending on the situation). These configurations produce an offset current that must be overcome before the comparator operates. One reason for such an arrangement is to prevent fluctuation around a zero condition from tripping the alarm and causing false output indications. The offset is set to a value that is above the random noise fluctuations, but below the point that would be immune to tripping in the event the flame went out.

Photodensitometers

A *densitometer* is an instrument that measures the optical density of a sample material that is placed in a light path between a source and a sensor (**Figure 17-2a**). Uses of this instrument are numerous. A common use is for the measurement of the optical density of photographic negatives. Both custom photographic labs and mass-production "snapshot" labs use variants of this instrument in their production of photographic prints. The output of the densitometer is used to measure the relative darkness or lightness of the negative.

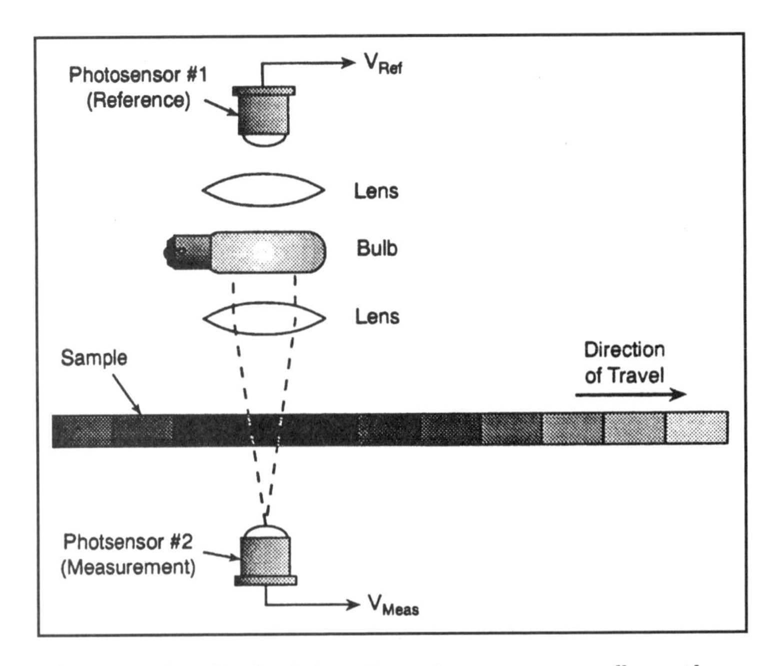

Figure 17-2a. Optical densitometer sensor configuration.

The processing of radiation exposure badges is a related application. Radiation workers, X-ray technicians, radiologists, research scientists and others who work around various forms of ionizing radiation wear badges to gauge their exposure. The badges consist of a piece of photographic film that becomes fogged by exposure to X-rays or other radiation. The badges are collected periodically, developed in the usual manner, and examined under a densitometer that detects the amount of exposure that the film received.

Chemists also use densitometry in order to measure the absorption of light by materials. It is common to use a filter between the sample and the sensor in order to restrict the examination to specific wavelengths of light. In most of these cases, however, the colorimeter is used (see Chapter 11). Densitometers are used for situations in which the optical density varies with position on the sample. There may be little difference between a colorimeter and a densitometer, other than moving the sample and the sensor with respect to each other.

In **Figure 17-2a**, the sample has a varying optical density. The sample is pulled beneath a fixed set of sensors, producing a voltage output (V_{meas}) that is a function of both the optical density and the point at which the measurement is taken. In some systems, the motion is two-dimensional, so the sample stage can be positioned underneath the sensor in two directions. When viewed from above, the motion would map to an X-Y cartesian coordinate system. Both automated positioning and "micrometer dial" manual positioning systems are used.

The basic sensor shown in **Figure 17-2b** is a "generic" schematic. Any sensor can be used, provided it produces a voltage output (which may require circuitry along with the sensor), and is sensitive to the wavelengths of the light source.

The light absorption of the sample must be compared to the light source. After all, one cannot guarantee that the light source will remain constant over its lifetime. We can also not be certain that a lamp used as a replacement for the original lamp will be so similar in output level that current readings of density can be compared with previous readings. As a result, it is common to provide a second *reference photosensor* to monitor the light output while the measurement sensor is measuring the light level transmitted through the sample. The two sensor output signals, V_{meas} and V_{ref}, are compared in a ratiometric circuit that takes their quotient (see end of this chapter for a detailed discussion of ratiometric measurements).

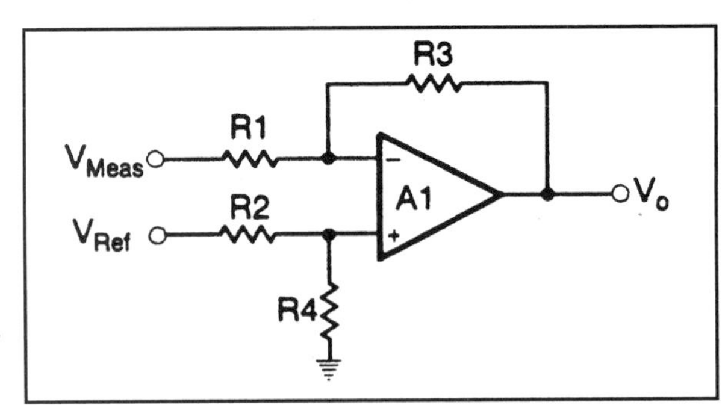

Figure 17-2b. Optical densitometer sensor amplifier.

Dynamic range may also pose a problem. When the optical density of the sample is low, the light will fall almost unimpeded on the sensor. Yet when the optical density is very high, then the sensor will receive very little light. A difference amplifier circuit will help overcome this problem, especially since one or both voltages can be scaled before being applied to the difference amplifier.

A difference amplifier is nothing more than a dc differential amplifier found in operational amplifier technology (**Figure 17-2b**). It produces an output voltage V_o that is proportional to the voltage gain of the circuit and the difference between the input voltage. If R1 = R2, and R3 = R4, then the gain of the difference amplifier is found by making the calculation:

$$A_v = \frac{R3}{R1} = \frac{R4}{R2} \qquad \text{eq. (17-1)}$$

Where:

A_v is the voltage gain of the circuit.

R1, R2, R3, and R4 are the values of the resistors.

Example 17-1

Find the dc voltage gain of a difference amplifier in which the following resistor values are found: R1 = 100 kΩ, R2 = 100 kΩ, R3 = 4.7 MΩ, R4 = 4.7 MΩ.

Solution:

- $$A_v = \frac{R3}{R1}$$

- $$A_v = \frac{4.7\ M\Omega}{100\ k\Omega\ x \left(\frac{1\ M\Omega}{10\ k\Omega} \right)}$$

- $$A_v = \frac{4.7\ M\Omega}{0.1\ M\Omega} = 47$$

When the differential input voltage, V_{meas} - V_{ref}, is added to the equation, we have the transfer equation for the output voltage:

- $$V_o = (V_{meas} - V_{ref})\ x \left(\frac{R3}{R1} \right) \quad \textit{eq. (17-2)}$$

In the densitometer system shown in **Figure 17-2**, the output voltage from the difference amplifier at any given point on the sample will be constant, even when the lamp source varies. The only variation in the output voltage occurs when the sample optical density changes.

In some systems, it is common to use a transducer to produce a signal that is a function of the position on the sample (**Figure 17-3**). In this example, the sample transport stage is coupled to a small, low-speed motor that drags the sample beneath the sensor pair. The motor is also coupled to a potentiometer that is connected between ground and a reference voltage, V+.

The display and recording device is an X-Y recorder. These instruments plot two voltages on ordinary graph paper, which is mounted onto the recorder's paper platform. An arm carrying an ink pen is arranged over the paper. The position of the pen arm is a function of the horizontal (H) input voltage, while the position of the pen on the arm is a function of the vertical (V) input voltage. Thus, in the configuration shown in **Figure 17-3a**, the pen charts a line that represents the optical density as a function of position on the sample. A typical result for a sample similar to this is shown in **Figure 17-3b**.

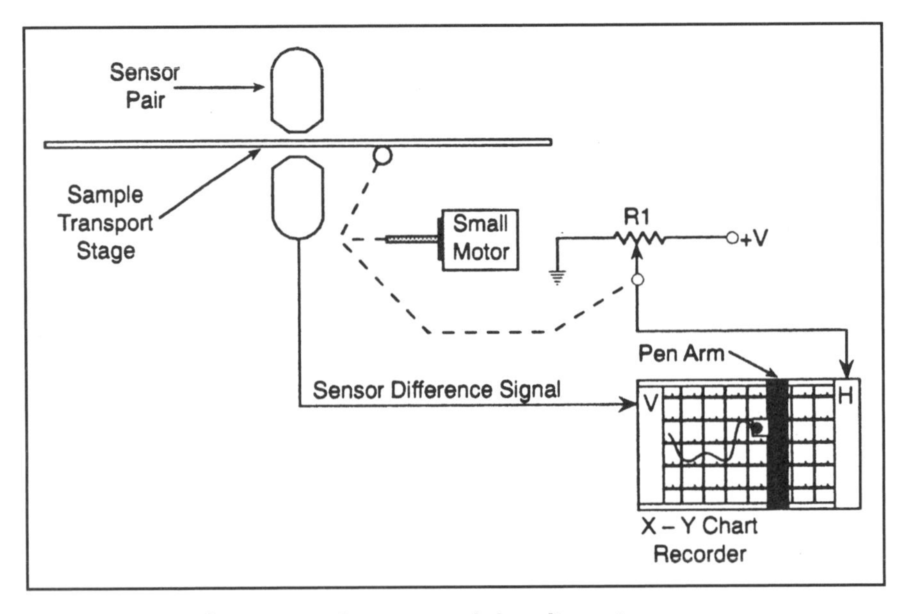

Figure 17-3a. Structure of a scanned densitometer.

Telescopes

A *telescope* is an instrument that allows the user to see objects at a distance by magnifying the image of them. Three basic forms are found: *refractor*, *Newtonian reflector*, and *Cassegrainian reflector*.

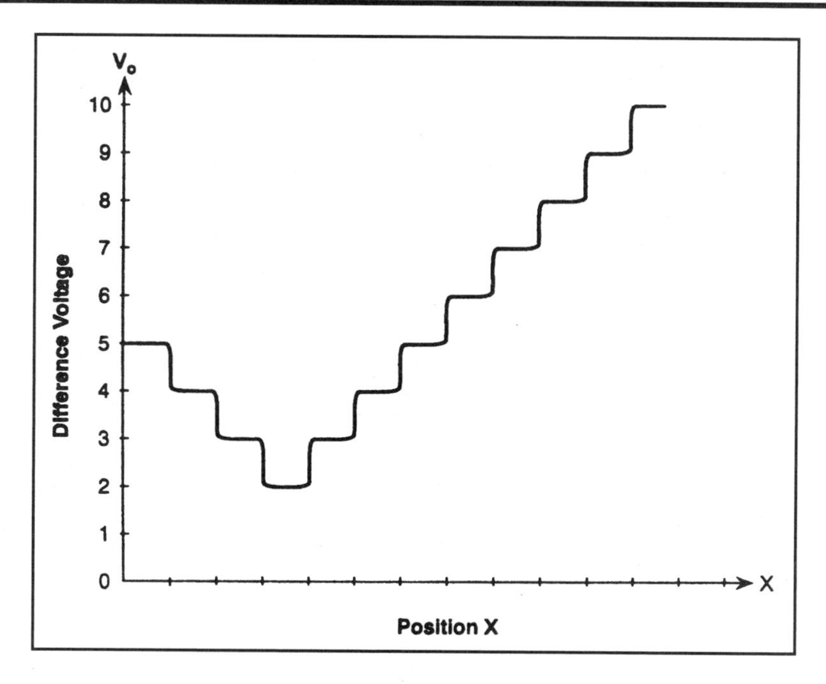

Figure 17-3b. Output as a function of position for a sample such as used in Figure 17-2a.

The refractor telescope is the oldest form, and is found in both astronomical and "spyglass" forms (**Figure 17-4**). It consists of at least two lenses: a main *objective lens* and a smaller *ocular* or *eyepiece lens*. The objective lens is a large converging lens with a very long focal length. When trained on a distant object, QE, it will form an inverted real image PE′ in the focal plane of the ocular lens. The ocular lens then produces a magnified virtual image RE″ that the observer sees.

The *angular magnification* of the object is determined by the apparent angle the object subtends in the observer's eye, and is a function of the focal lengths of the two lenses:

$$Magnification = \frac{FL_{objective}}{FL_{ocular}} \qquad \text{eq. (17-3)}$$

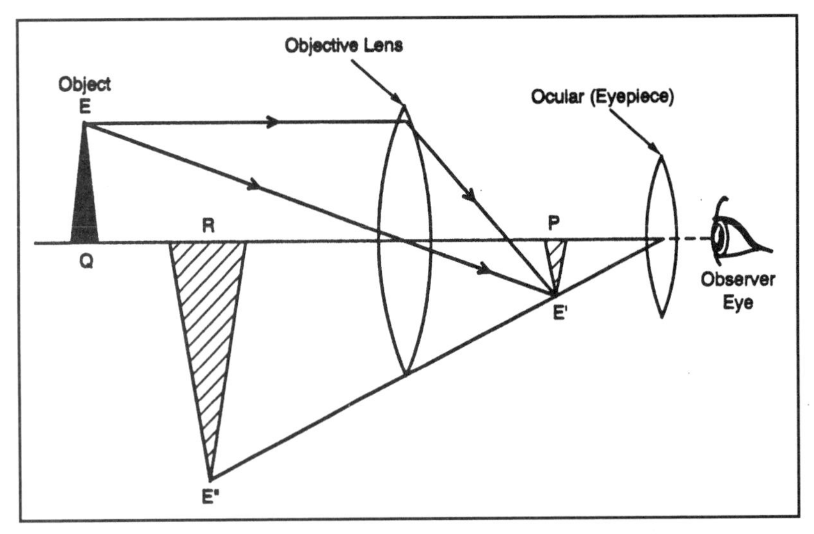

Figure 17-4. Refractor telescope.

Example 17-2

A refractor telescope has an objective lens with a 200 cm focal length, and an ocular lens with a 25-mm focal length. Find the “power” (angular magnification) of this telescope.

- $$Magnification = \frac{FL_{objective}}{FL_{ocular}}$$

- $$Magnification = \frac{200\ cm}{25\ mm \_ \left(\frac{1\ cm}{10\ mm}\right)}$$

- $$Magnification = \frac{200\ cm}{2.5\ cm} = 80$$

Such a telescope would be called an “80×” or “80 power” telescope.

A Newtonian reflector telescope (**Figure 17-5**) is constructed differently, and in many applications is superior to the refractor telescope. This telescope uses a finely ground, long- focal-length parabolic mirror to reflect the essentially parallel light rays entering from a distant object to its focal point, F. The mirror may be as small as 3.5 inches, or as large as the 200-inch reflector at the Mount Palomar Observatory.

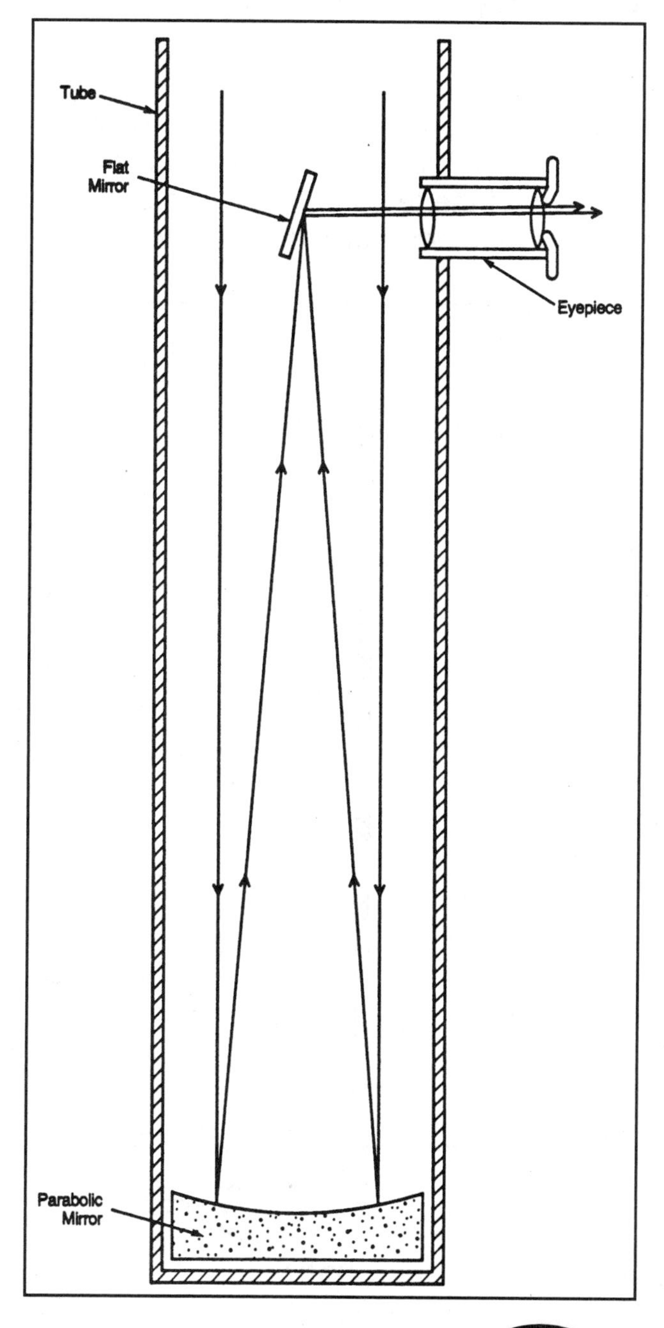

A small secondary flat mirror is placed at the focal point (F) of the parabolic mirror. The secondary mirror reflects light into the eyepiece used by the observer.

The entire assembly is placed inside a protective tube that is painted flat black inside in order to reduce stray reflections from the walls of the tube. The tube can be made from metal, plastic, cardboard, fiberglass or any other stable material.

A disadvantage of the Newtonian reflector is that the eyepiece is at the top end of the assembly. In astronomical telescopes of large size, this arrangement causes the astronomer to stand on a ladder or in a basket on a crane at the ob-

Figure 17-5. Newtonian reflector telescope.

server level. A Cassegrainian telescope provides a solution to the problem that preserves the advantages of the reflector while also allowing the astronomer to observe from ground level.

A Cassegrainian telescope is shown in **Figure 17-6**. This telescope uses a main parabolic reflector that has a small hole in its optical center to accommodate the eyepiece. Parallel light rays entering the aperture are reflected from the main mirror to a secondary mirror called a *subreflector*, which reflects the light into the eyepiece in the main mirror. This arrangement not only allows the observer to be at the bottom of the telescope, but also shortens the overall length of the instrument. This system is also the basis for "reflector"-style photographic telephoto lenses.

Simple Photographic Camera

Simple photographic cameras are familiar to nearly all readers. Indeed, George Eastman's invention is ubiquitous, from the disposable models that are thrown away once the film is developed, to very costly professional and scientific models. A simple camera is shown in **Figure 17-7**. The "sensor" in this instrument is a piece of film that is coated with a light- sensitive emulsion. When the emulsion is exposed to light, it is changed chemically. The process of developing washes away a part of the emulsion as a function of how much light fell on each spot, leaving an image of the scene photographed. The

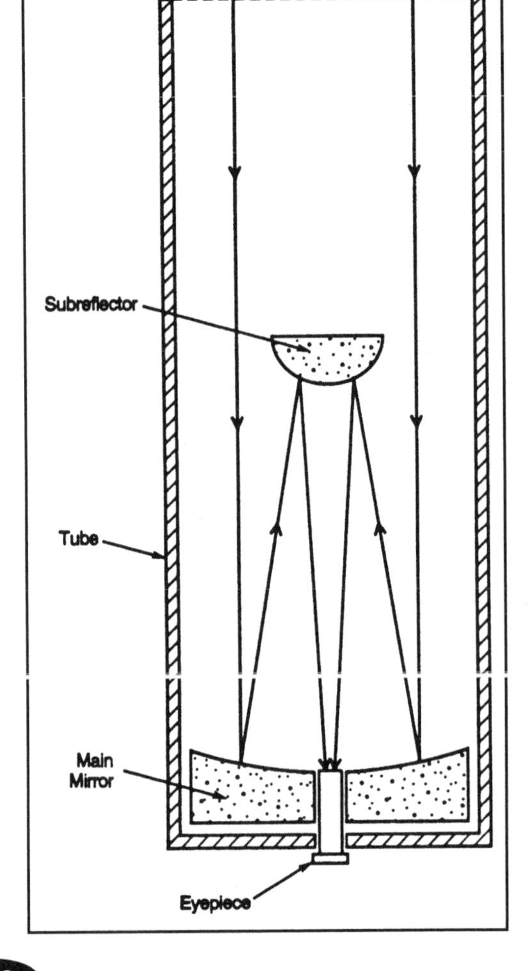

Figure 17-6. Cassegrainian reflector telescope/telescopic lens.

film in the camera is positioned in the focal plane of a compound lens. The position of the lens can be adjusted in order to focus the image.

Light enters the camera through an aperture called the *iris*, which has a diameter *d*. An optical flat or filter may be positioned in front of the iris in order to alter the entering light, or to protect the lens. A *shutter* between the lens and the film opens and closes to admit light to the film. A parameter called the *f-number* (or *f-stop*) is significant to the operation of cameras and their lenses. The f-stop is the ratio of the focal length of the lens to the iris' aperture opening. The larger the f-number, the longer the film must be exposed.

The four variables in photography are the film speed (how fast it will react to a given light level), which is specified in ASA or DIN number, the shutter opening/closing speed, the iris diameter and the amount of light entering from the outside. The technical art of photography involves manipulating these variables.

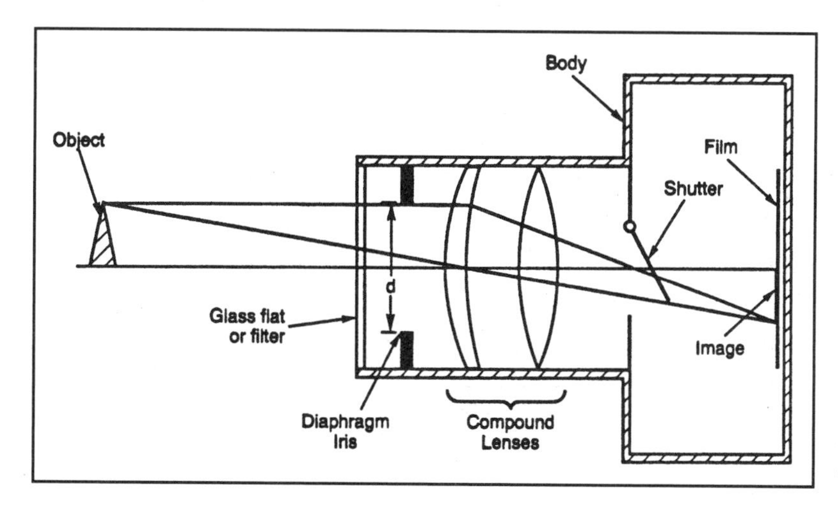

Figure 17-7. Photographic camera.

Projectors

The *projector* or *projection lamp* is a device used to project an image of a nearby object to a distant viewing screen (**Figure 17-8**). We are familiar with these instruments from such appliances as slide projectors and opaque projectors used in schools. The projector consists of a high-intensity lamp backed with a reflector mirror, a pair of condenser lenses, a slide holder for the optical transparency being viewed and a projection lens. In order to form a real, enlarged, inverted image at a distance, the slide is placed just outside of the principal focus of the projection lens.

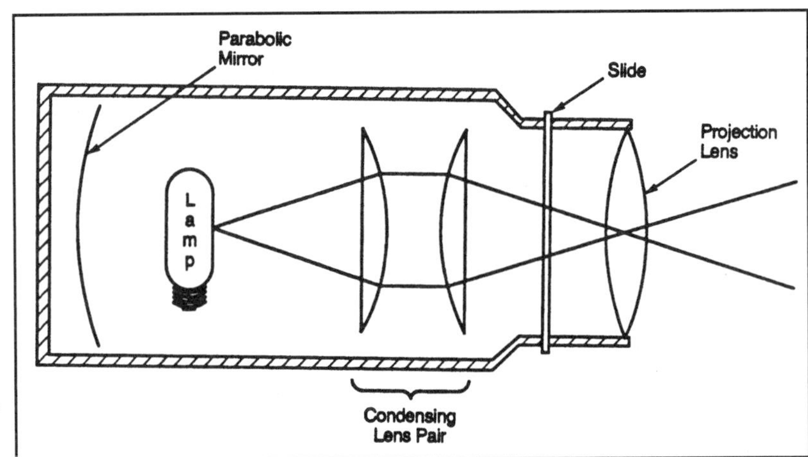

Figure 17-8. Transparency slide projector.

Compound Microscopes

The *compound microscope* is a collection of lenses arranged to magnify small objects close to the eye so that they can be viewed (**Figure 17-9**). The various lenses act together as if they were a single ideal converging lens with a very short focal length. The lens closest to the object (QE), is called the *objective lens*. It produces a real, inverted, magnified image PE′ of the object QE being viewed, provided that the object is just outside of its focal plane.

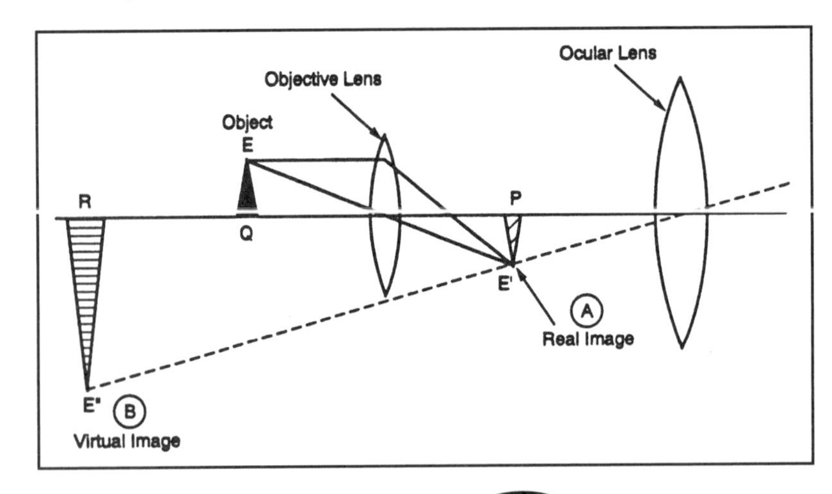

Figure 17-9. Compound microscope.

The image PE′ is viewed with an *ocular* or *eyepiece* lens, which produces an enlarged virtual image RE″ of the real image that was produced by the objective lens. The linear magnification is a function of the focal lengths of the two lenses.

Spectrophotometers

A *spectrophotometer* is an instrument that compares the relative absorptivity or reflectance of a sample at various wavelengths. By finding out which wavelengths of light are absorbed, and in what proportions, one can identify the chemical composition of the sample, especially certain liquids or gases. **Figure 17-10** shows the schematic for a simple *monochromator*, a subspecies of the class of instrumentation to which the spectrophotometer belongs. It uses a diffraction grating or prism to disperse collimated light. The light is generated by a lamp, passed through slit S1, and reflected from a mirror (the function of which is to reduce the instrument's physical size while retaining the length of the light path).

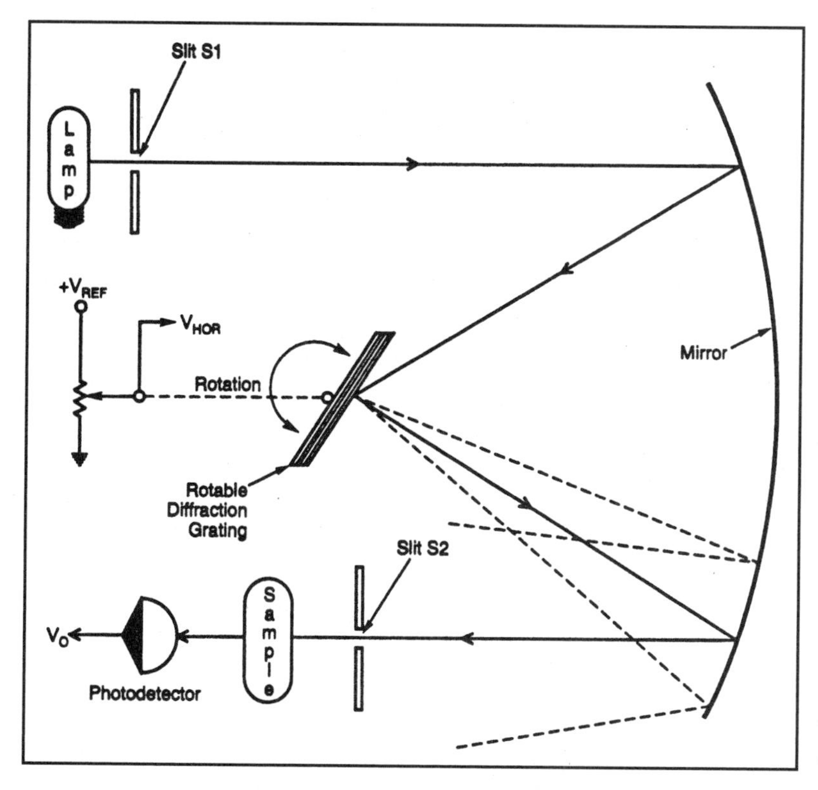

Figure 17-10. Monochromator.

Light from the mirror and S1 is reflected from the diffraction grating back to the mirror, and then is reflected through slit S2 to the sample. A detector distal to the sample measures the amount of light passing through the sample. The dimensions of the slit can be used as a filter to narrow the admissible spectrum, but the main discrimination comes from moving the diffraction grating to successively reflect different wavelengths into the second slit (recall that a diffraction grating breaks white light into its color spectrum).

The wavelength of the light passing through slit S2 to the sample is a function of the diffraction grating position. In other words, if the position of the diffraction grating is known, it can be used to identify the spectral components at various points. If a potentiometer or other sensor is used to keep track of the grating position (similar to what was done in **Figure 17-3a**) then one can electrically chart the absorption as a function of wavelength. Apply V_{hor} from the position transducer to the horizontal input of an X-Y chart recorder or oscilloscope. Similarly, apply V_o, the photosensor output voltage, to the vertical input of the chart recorder or oscilloscope.

A potential defect of the spectrophotometer as shown is that neither the response of the photosensor nor the emission spectrum of the light source are linear over the entire optical spectrum. As a result, zero and gain calibration must be done at each discrete wavelength. This job can be done by removing the sample between wavelengths and measuring the intensity of light at each wavelength either immediately before or immediately after the sample measurement. There are alternate path methods, i.e., those that use a partially silvered second mirror (or beam-splitting prism), and slit to keep track of the light-beam intensity as measurements are being made. These measurements are part of a general class of *ratiometric* measurements.

Ratiometric Instruments

Although it is easy to discuss electronic instruments (including E-O instruments) in ideal terms, those ideal conditions are rarely found in real measurement situations. The goal is to produce a signal voltage V_{meas} that is a function of only variations in the properties of the sample being measured. Unfortunately, there are also sources of measurement error.

A principal point of error is changes in the intensity of the light source used to make a measurement. The light source typically consists of an incandescent lamp (although fluorescent and gas-filled glow-lamps are also used), a small slit for collimation, and sometimes a wavelength-selective or wavelength blocking filter of some sort or another. If the voltage supply to the lamp varies, then the light intensity could vary also. In addition, light sources typically fade in intensity as they age.

A ratiometric instrument (**Figure 17-11**) will help overcome the light level variation. In this instrument, the light from the source is broken into two orthogonal (90 degree) paths by either a beam-splitting prism or a half-silvered mirror set at a 45 degree angle to the light path. In **Figure 17-11**, the beam-splitter produces two approximately equal beams, "A" and "B," at right angles to each other.

Beam A from the light source is passed through the sample to the measurement sensor, while beam B is passed directly to the reference sensor. In some cases, beam B may also pass through a filter that approximately matches the optical properties of the sample, so that only differences show up in the data. In some cases, the two light beams pass directly to the sensors, while in other cases, intermediate mirrors are used to steer the light.

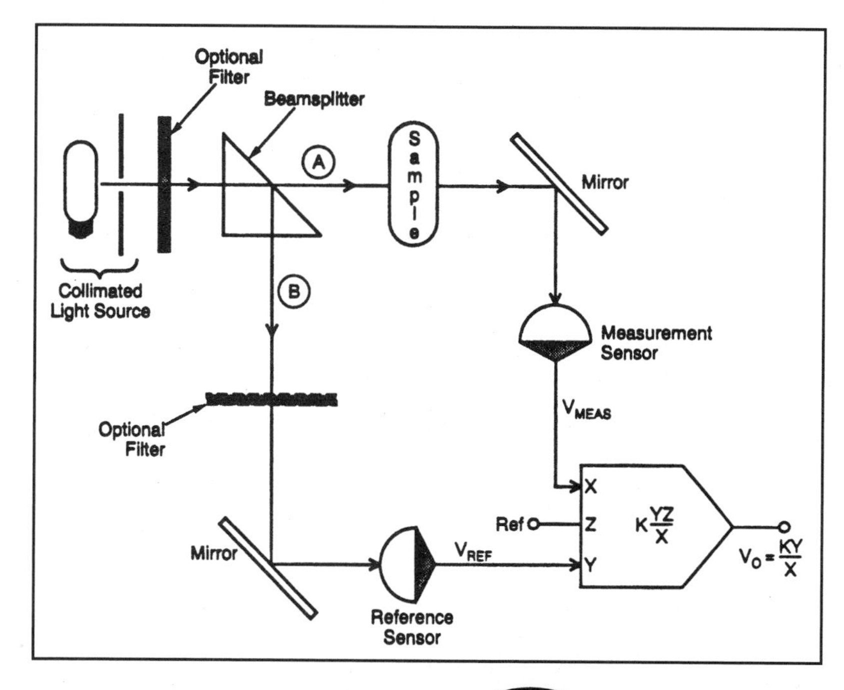

Figure 17-11. Ratiometric instrument.

The term "ratiometric" implies that a measurement is made by computing the *ratio* of two quantities. One could digitize the two voltages, V_{meas} and V_{ref}, and feed them to a computer to compute the ratio in software. Alternatively, one could also use a device called an *analog multiplier/divider* circuit. These devices come in a variety of monolithic integrated circuit and hybrid versions, and produce an output voltage that is a function of two or three input voltages according to the generic expression:

$$V_o = K\left(\frac{V_z V_y}{V_x}\right) \qquad \text{eq. (17-4)}$$

Where:

V_x, V_y, and V_z are the voltages applied to the X, Y, and Z inputs, respectively.

K is a constant (usually either 1/20 or 1/10).

In **Figure 17-11**, the Z-input is biased to a fixed reference value, so it becomes part of the value of K. In that case, we can reduce **Equation 17-4** to:

$$V_o = K\left(\frac{V_y}{V_x}\right) volts \qquad \text{eq. (17-5)}$$

If V_{meas} is applied to the X-input, and V_{ref} is applied to the Y-input, and K = 1/10, then the expression of **Equation 17-5** becomes:

$$V_o = \frac{1}{10} x \left(\frac{V_{ref}}{V_{meas}}\right) volts \qquad \text{eq. (17-6)}$$

Example 17-3

The idea behind this circuit is that an artifact variation in the light level will affect both voltages by an equal percentage, while variations in the sample will only affect V_{meas}. The sample and sensors are such that initially V_{meas} is +2.5 volts, and V_{ref} is +8 volts. From **Equation 17-6**, the output voltage V_o will be:

$$V_o = \frac{1}{10} \times \left(\frac{8.0}{2.5}\right) volts = 0.32\ volt$$

The light intensity now changes by increasing 10 percent. The two voltages V_{ref} and V_{meas} rise to +8.80 and +2.75 volts, respectively. Under these conditions the output voltage should remain the same because both paths are affected equally:

$$V_o = \frac{1}{10} \times \left(\frac{8.80}{2.75}\right) volts = 0.32\ volt$$

If the sample density decreases by 10 percent of the original level, then the measurement sensor voltage will increase by 10 percent to +2.75 volts; if the light level remains unchanged, the reference voltage will still be +8.0 volts. Under this new condition the output voltage will be:

$$V_o = \frac{1}{10} \times \left(\frac{8.0}{2.75}\right) volts = 0.29\ volt$$

In both cases, the measurement sensor output voltage changed, but only in the instance where the change was due to a change in the sample was the change reflected in the output signal.

A variation on this theme is shown in **Figure 17-12**. The multiplier is replaced with a pair of logarithmic amplifiers and a difference amplifier. By converting the two signal voltages V_{meas} and V_{ref} into their natural logarithms, and taking the difference between them, we effectively divide the reference voltage by the measurement voltage. An antilog amplifier converts the voltage back to the linear format of the original signal voltages.

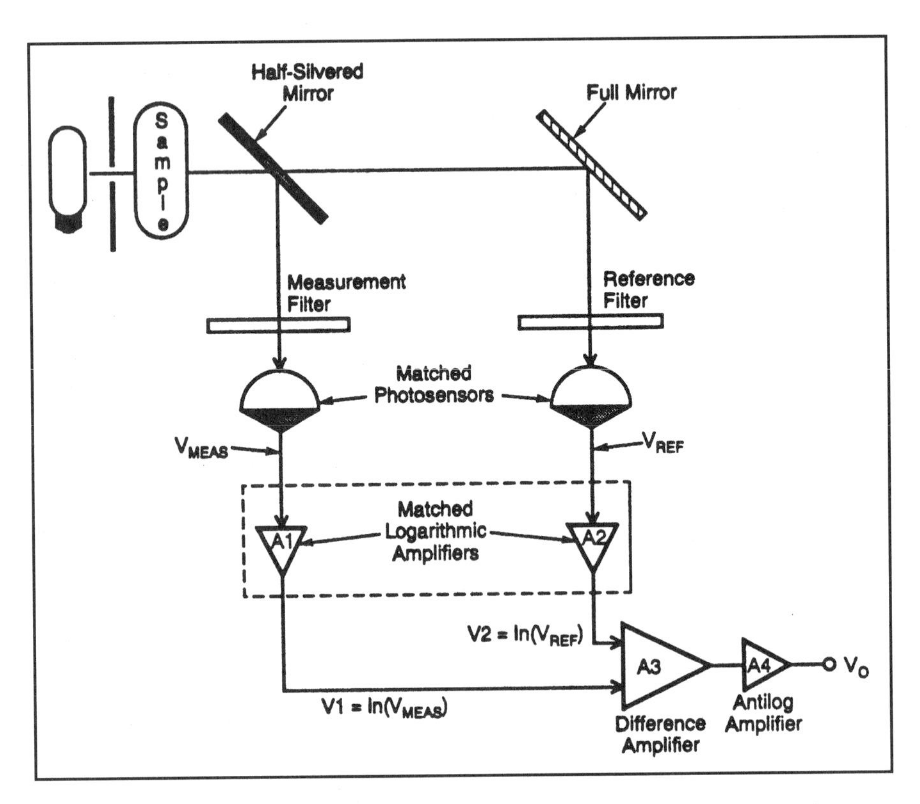

Figure 17-12. Alternate ratiometer.

Chapter 18

Optical Sensor Resolution Improvement Techniques

A primary attribute of any optical sensing system is the *resolution* (R) of the sensor. The definition of resolution (Webster 1979) is: "...the process of making distinguishable the individual parts of an object, closely adjacent optical images, or sources of light." High resolution implies that closely adjacent objects can be distinguished, while low resolution systems blur the objects together.

In sensor-based imaging systems, high resolution results in sharper images in which finer grain detail is visible. More information about the structure of the imaged object can be inferred from high resolution sensors than from low resolution sensors. In most instances, higher resolution is considered superior to low resolution, although that is not universally the case.

In airborne radar and infrared sensors (e.g., "forward looking infrared" or "FLIR") a human factors issue sometimes arises regarding resolution. The high-resolution sensor makes the image display too detailed for an overworked aircrew member. The best design is that which provides the user with resolution that is good enough, but not so much information as to impede rapid interpretation. In other cases, where there is time for interpretation, or where small details are of critical importance, higher resolution is desirable.

In addition to the ability to distinguish adjacent objects that are close together, there is another implication of sensor resolution: position ambiguity. When a sensor is used to precisely locate an object, the resolution of that sensor determines how accurately the position can be fixed.

Figure 18-1 shows both the position ambiguity problem as well as the source of the inability of low resolution sensors to separate two distinct objects within its field of regard. All sensors have a zone over which they detect targets, i.e., their field of regard (FOR). Targets outside of the FOR are not detected, while targets inside the FOR produce a sensor output. In the example of **Figure 18-1** a generic sensor traverses past the target along a line parallel to the X-axis, producing a voltage (V) output signal. As soon as the sensor's FOR reaches the target, the sensor output begins rising. The output peaks when the target is directly opposite the sensor, which is its most sensitive point. The sensor output as the sensor location changes, and the FOR moves away from the target.

In the example of **Figure 18-1** the sensor is moving, but that is not a necessary condition. If the target moves and the sensor is stationary, exactly the same output curve results. Similarly, if both the sensor and the target are in motion, the same thing is found. The critical issue is whether or not the sensor and target are moving with respect to each other.

The ability to resolve position of the target is related the ability to resolve differences in amplitude along the position curve. The slope of the sensor output is the discriminant in that case, i.e., when $\Delta V/\Delta X$ is positive ($\Delta V/\Delta X > 0$), then the sensor is moving towards the target; when $\Delta V/\Delta X$ is negative ($\Delta V/\Delta X < 0$), then the sensor is moving away from the target; when $\Delta V/\Delta X = 0$, then the sensor is immediately opposite the target and the position is precisely fixed.

Or so it works in theory. The problem arises due to the inability to distinguish amplitude differences on the position curve. Several sources contribute to resolution deterioration. In analog displays, whether an oscilloscope or paper chart recording, the ability to detect small changes is often impeded.

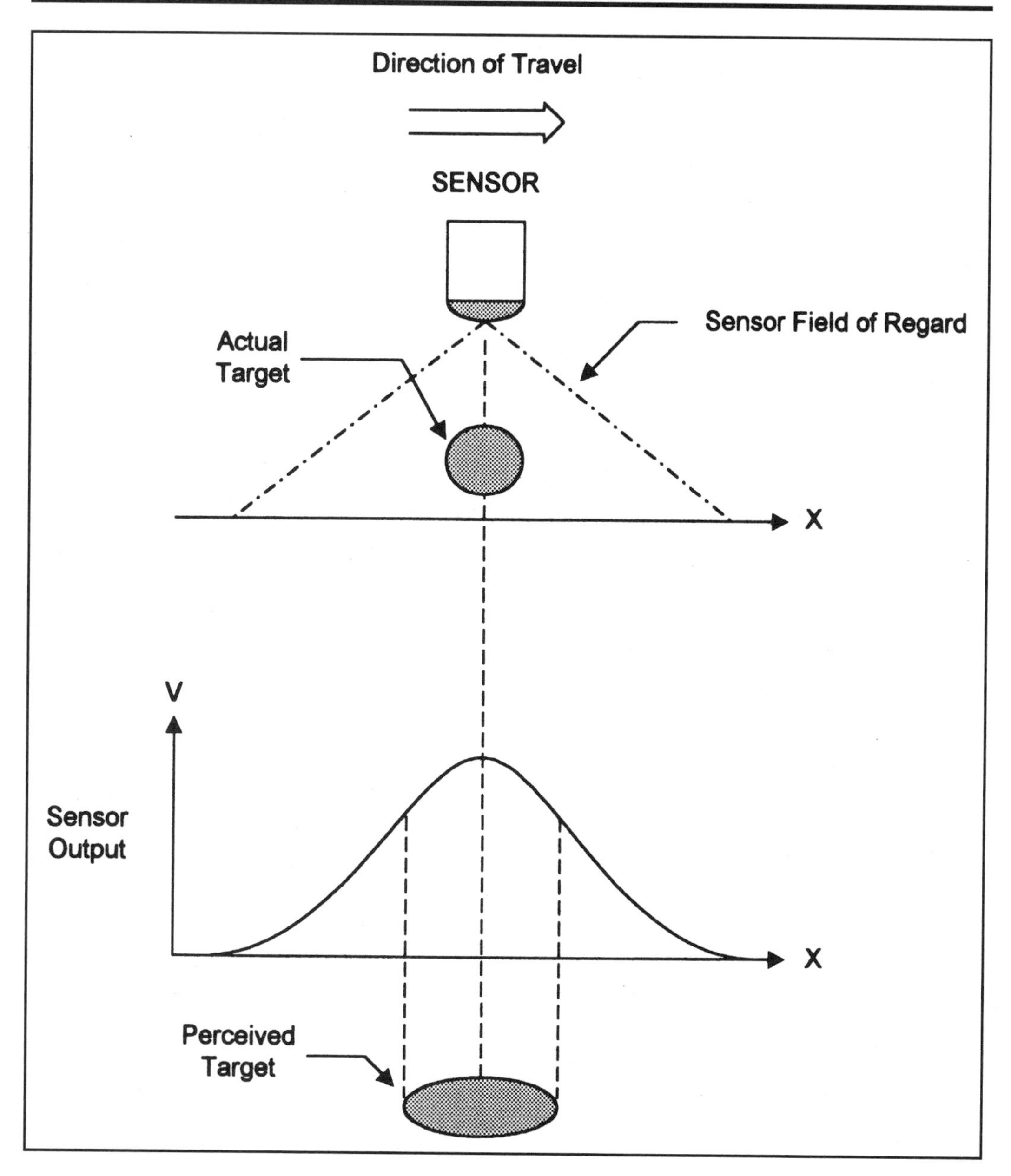

Figure 18-1. Sensor resolution depends on the sensor's field of regard.

There is also the situation where the peak of the signal is not so sharply defined as in **Figure 18-1**. In those cases, the flatness of the curve creates a very small, but non-zero value of $\Delta V/\Delta X$. In digital systems, the 1-LSB step of the analog-to-digital converter, the quantization error (Q), and the noise level can adversely affect the ability to recognize small values of $\Delta V/\Delta X$, or the exact point at which $\Delta V/\Delta X \rightarrow 0$. Although the effect is shown somewhat exaggerated in **Figure 18-1** for graphical reasons, the result of the problem is a smearing of the target, which distorts its shape and makes its position more difficult to fix.

The position ambiguity zone can sometimes be adjusted by a threshold arrangement similar to **Figure 18-2**. Threshold detector systems are sometimes seen in which an adjustment is permitted to remove false targets, noise components and other artifacts. The threshold is adjusted to be above the noise and error floor, but this does not affect the inability to detect the point at which $\Delta V/\Delta X$ goes to zero.

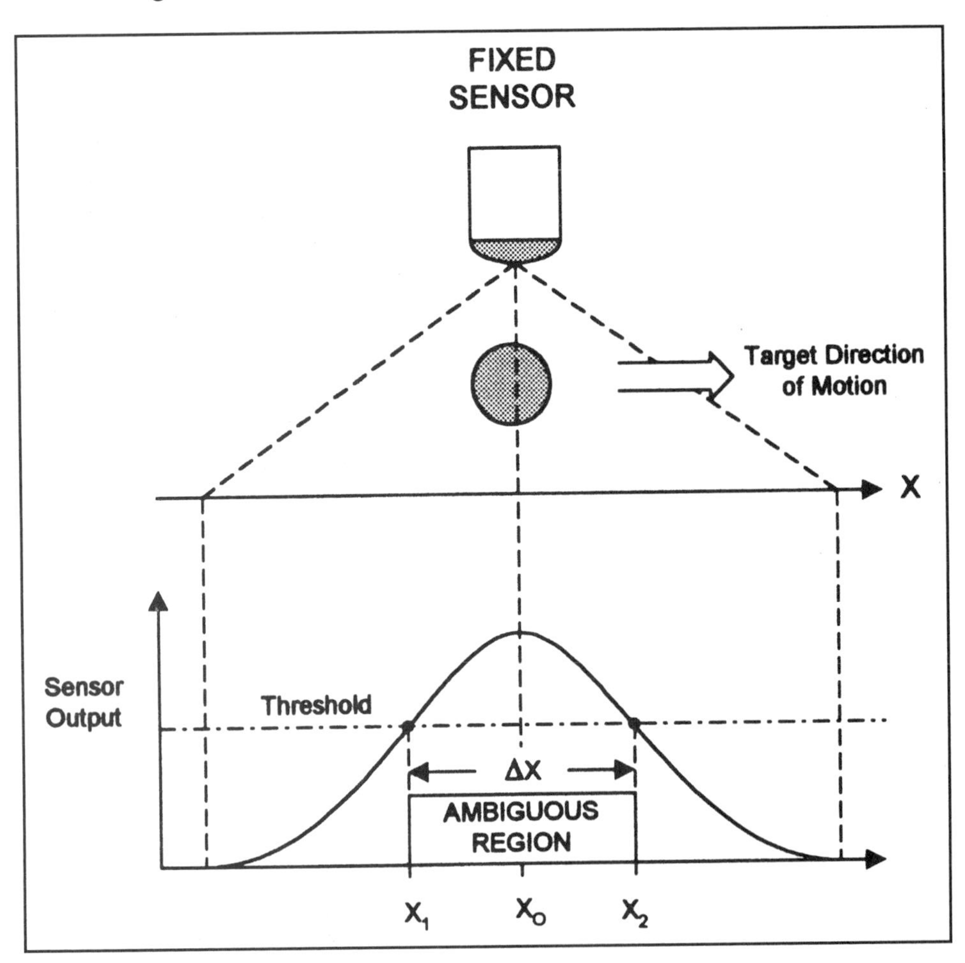

Figure 18-2. An ambiguous zone is created by sensor resolution.

A superior approach to reducing the ambiguity zone is to increase the resolution of the sensor, if it is possible with the specific sensor in question. This task is done by selecting a sensor with a smaller field of regard (**Figure 18-3**). The output voltage-vs-position (V-vs-X) is therefore narrower, and the maximum extent of the ambiguous region is correspondingly smaller. The distor-

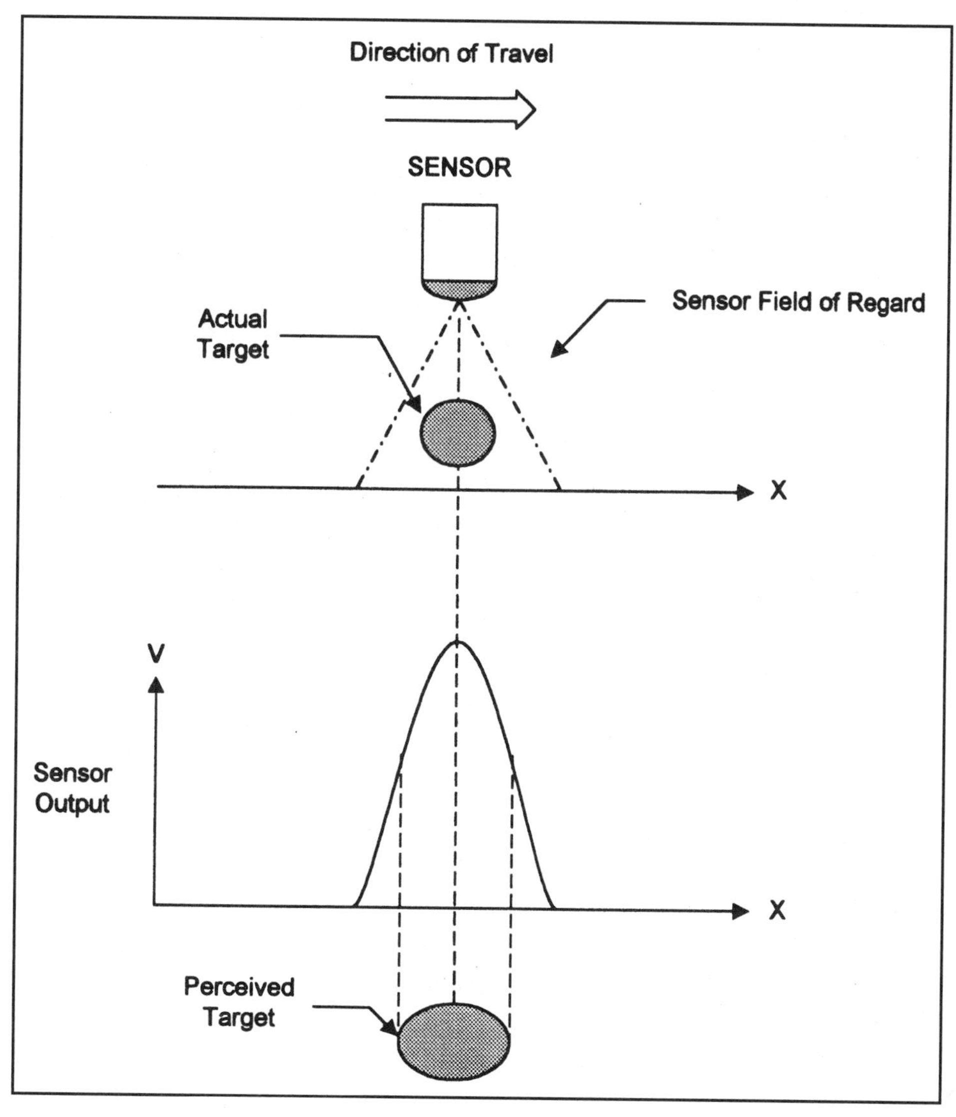

Figure 18-3. Narrowing the resolution reduces the ambiguity.

tion of the target shape, as well as the position ambiguity is improved. Again, this approach does not address the problem of detecting the $\Delta V/\Delta X = 0$ condition.

It must also be noted that reducing the FOR, as in **Figure 18-3**, may not be either possible or feasible in all cases. If the instrument already uses a sensor with the highest available resolution, but is nonetheless still insufficient, then the problem is not solved.

The target discrimination problem is shown in **Figure 18-4**. In this scenario, a pair of identical targets, T1 and T2, are spaced a distance (ΔX) apart, with T1 centered on X1 and T2 centered on X2 (**Figure 18-4a**). The mid-point between T1 and T2 is designated X_O. A sensor traveling past the targets, on a line parallel to the line of centers T1-T2, will begin producing an output as soon as the FOR encounters T1. This signal is shown as V1. The signal V1 will peak when the sensor is opposite the target, i.e., at X1.

When T2 enters the sensor's FOR, it begins to contribute to the sensor output. The contribution of the second target is shown as V2. The actual output of the sensor is a composite of V1 and V2, as shown in **Figure 18-4b**. The characteristic double-hump results from the difference in position between the peaks of V1 and V2. The graphical representation of the two targets is similar to **Figure 18-4c**. The peals are more or less distinguishable, but the two segments are connected together. The "dumbbell" shape makes it difficult to discern the boundaries between T1 and T2. When the resolution is poorer yet, then the dumbbell deteriorates into an oblong shape similar to **Figure 18-1**. In that instance, the dip seen in the signal curve of **Figure 18-4b** becomes more shallow, or disappears altogether. It is then impossible to distinguish the two targets.

If a higher resolution sensor is used to detect T1 and T2, as in **Figure 18-5**, then the dip on the composite waveform becomes far more pronounced, and the targets clearly break out on the display. The advantages of higher resolution are obvious. Unfortunately, it's not always possible to obtain sensors for any particular instrument with the resolution needed (of which, more later).

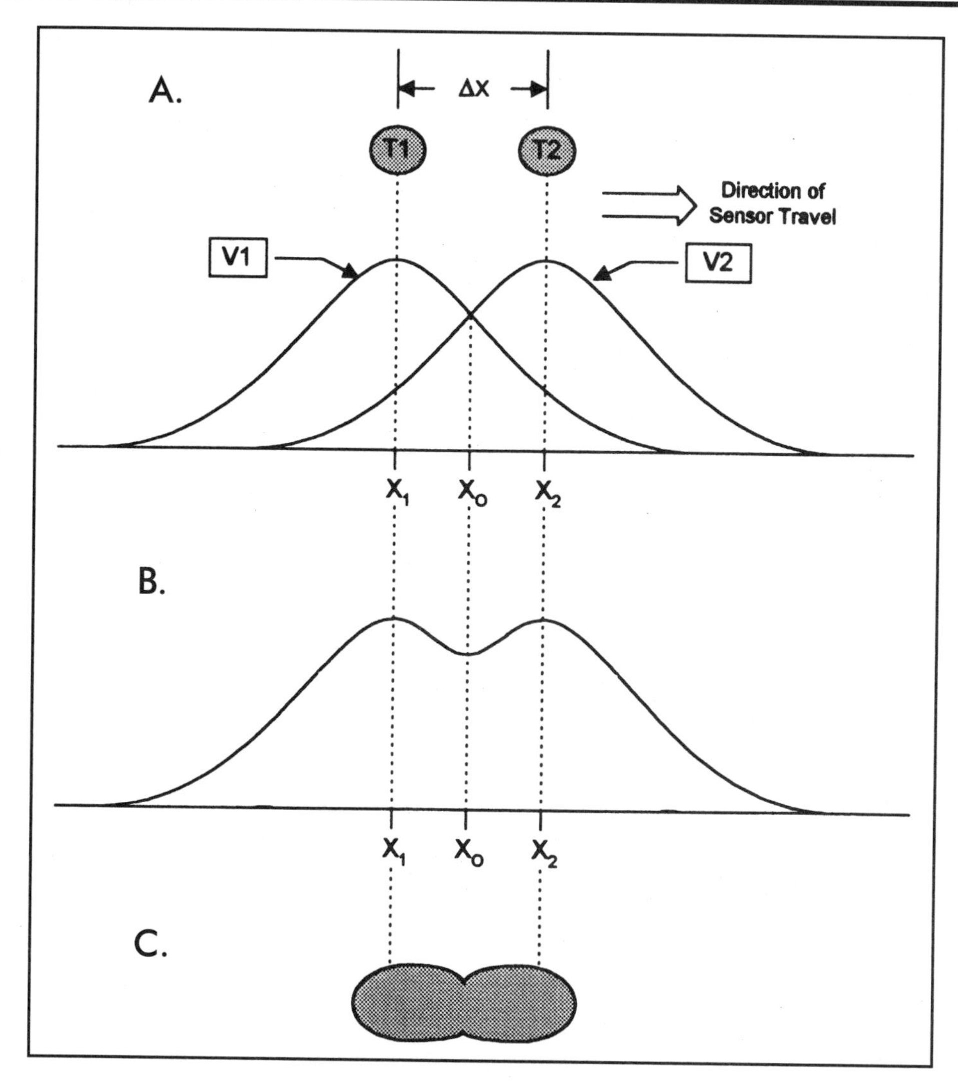

Figure 18-4. Poor resolution will obscure two adjacent targets.

Other Resolution Issues

The resolution of the sensor is not the only issue involved in overall system resolution. At least two other factors come into play: the *display resolution* and the *sampling rate* (digital systems).

Display resolution has always been a potential problem in sensor systems. Analog meters have inertia and pointer size ambiguities; digital meters have quantization error and "last digit bobble" problems; oscilloscopes have a minimum beam width that can be accommodated; paper recorders have these and other problems.

The modern sensor-based instrumentation system most likely uses a video display terminal similar to those found on computers. Indeed, the display may be identical to a computer terminal. The smallest unit that can be controlled on the display is the picture element, or pixel. A separation of a single pixel may not be discernible on the screen if the pixel size is too small. It may take a two or more pixel separation before the user's eye can discern the separation.

Another problem seen on digital or video displays is the light intensity or gray scale change that must occur before the operator can see the difference. In work on radar, it has been noted that situations arise in which the radar signal processor can distinguish two targets, and the data stream from the processor proves it, yet the operator cannot. In some cases, there was a separation of several pixels, but the problem turned out to be that the intensity of the light on the screen only reduced 1-LSB (which is less than the eye can discern). There may have to be a minimum specified change in signal intensity for the operator to see the difference, even though examination of the data stream would clearly show the difference.

The sampling rate problem is shown in **Figure 18-6**. The curves in **Figure 18-6a** are the responses of the sensor to the two targets. Their algebraic sum is the sensor output signal, as demonstrated earlier in **Figure 18-4** and **Figure 18-5**. In digital systems, an analog-to-digital converter (A/D) is used to convert the voltage signal from the converter to digital format for input to a computer (Carr 1992). Distinguishing the two targets depends on finding the minima in the value of data words from the A/D converter. As shown in the sampling signals of **Figure 18-6b** and **Figure 18-6c** (each of which has a different sampling rate), the depth and sharpness of the null is a function of the sampling rate. Because the process is asynchronous, there is no guarantee that the actual null will be found, but a high sampling rate improves the probability that a value close to the null, if not the null itself, is realized.

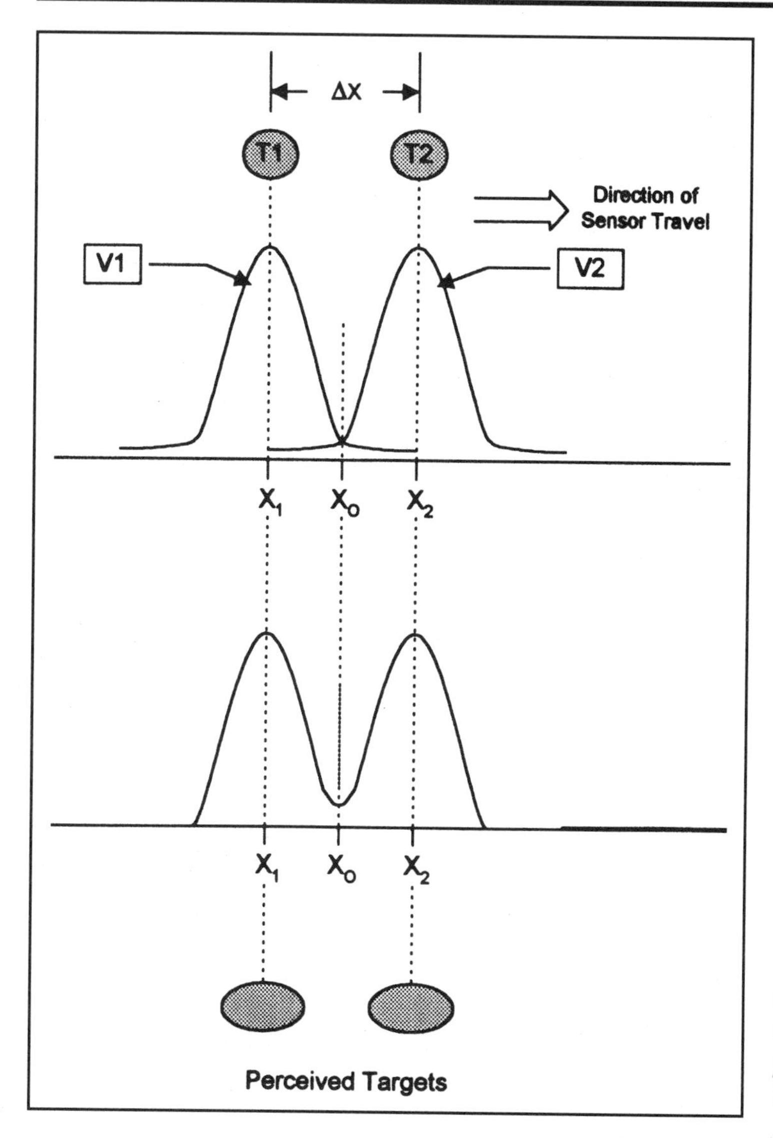

Figure 18-5. Improving resolution makes it possible to break-out both targets.

Sensor Resolution Improvement Techniques (SRI)

A technique from radar technology can be used with other forms of sensor to greatly increase the resolution of the sensor. Two forms exist. One uses two sensors, and the other uses a single sensor. The two-sensor method can be used in either digital or analog circuits, where the single-sensor method is used in digital circuits.

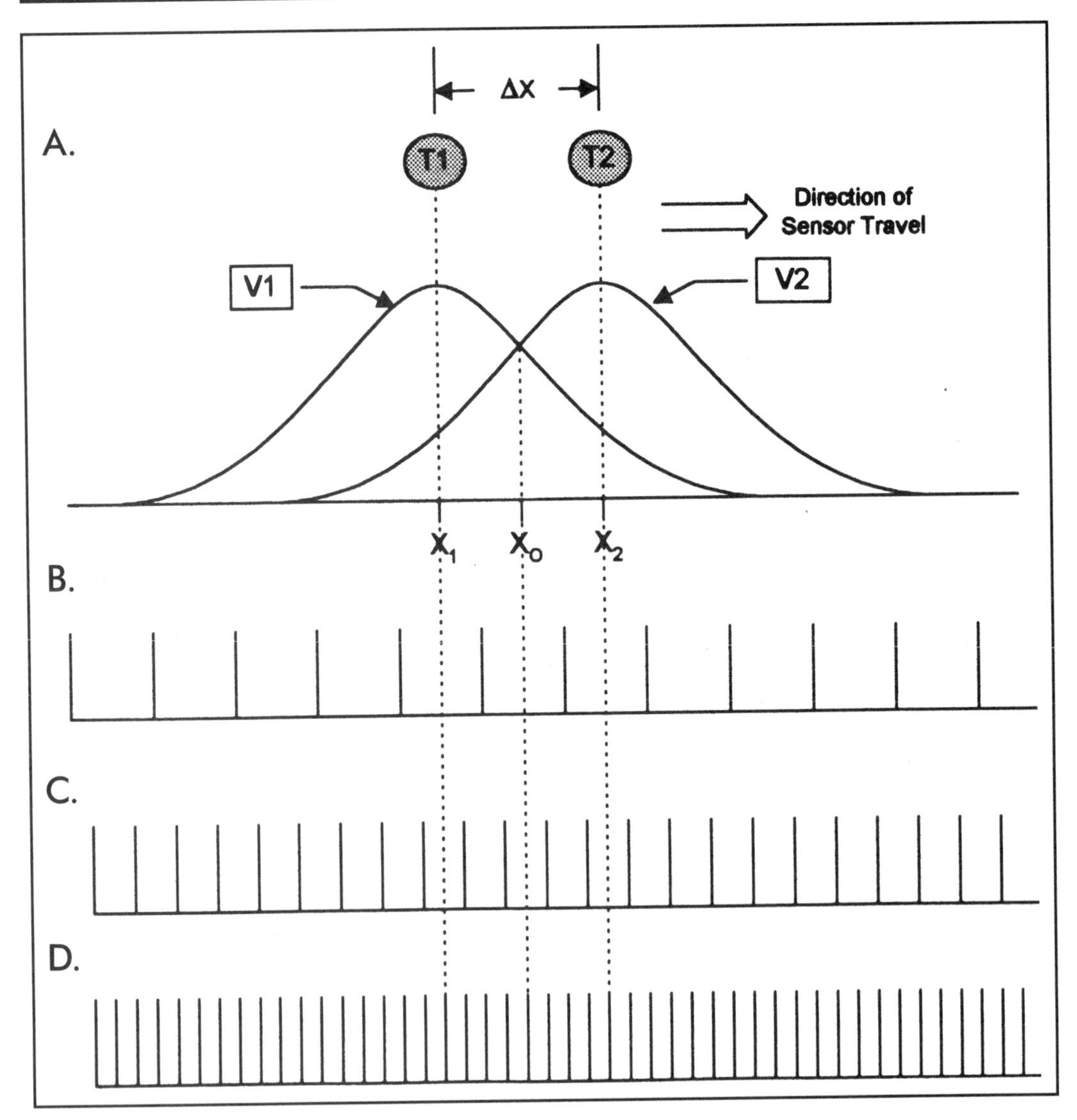

Figure 18-6. Sampling rate problems can also cause resolution problems.

Radar systems often use a technique called *monopulse resolution improvement*, or MRI (Skolnick 1980, Stimpson 1983). Two basic forms are used. Another method, called *lobing*, sequentially switches the main lobe (or "central maxima") of the beam from the radar antenna from one side of the target to the other. Comparisons can then be made to discern the true location of the target (Stimpson 1983). The monopulse methods accomplish the same effect by breaking the antenna into two sections, and simultaneously transmitting two different lobes towards the target (some systems use four lobes, two horizontal and two vertical, but the principle is the same). By comparing the return signals, the radar can locate the target. Two methods are used: *amplitude comparison* and *phase comparison.*

Amplitude comparison can be used for sensors other than radar. The block diagram for a two-sensor version is shown in **Figure 18-7**, and the associated signals are shown in **Figure 18-8**. The two sensors (S1 and S2) are aligned a distance ΔX apart such that their fields of regard overlap. In my experiments, the FORs were overlapped at approximately the -3 dB points (-6 dB voltage) of V1 and V2. As the target traverses along a line in front of the sensors, voltages V1 and V2 are produced (see **Figure 18-8a**). These signals have the characteric shapes seen earlier. The resolution of these individual signals are determined by the size and shape of the FOR for each sensor.

The processing for sensor resolution improvement (SRI) starts by combining the signals V1 and V2 in two different ways: the sum signal (V3 = V1 + V2) and the difference signal (V5 = V1 - V2).

Figure 18-8b shows the sum signal. Note that the width of this signal is considerably broader than that of either individual signal, which is expected because it is the algebraic sum of the two signals. If the sensors were separated more, then a dip would appear in the center of the waveform.

The difference signal is shown in **Figure 18-8c**. Note that this signal is bipolar. It is positive on one side of the center point, which represents X_O in the ΔX space between sensors, and negative on the other. The zero-crossing point represents a relatively good indication of the position, and is sometimes used for X_O discrimination.

The difference signal is further processed through an absolute value circuit (also called a full-wave rectifier or full-wave precision rectifier in the operational amplifier implementation) to produce signal V4 = ABS(V1 - V2). This waveform is shown in **Figure 18-8d**, and is nothing more than **Figure 18-8c** with the negative excursion flipped over to make both halves positive. Again, the zero point can be used for position discrimination, but does little for distinguishing two closely spaced targets.

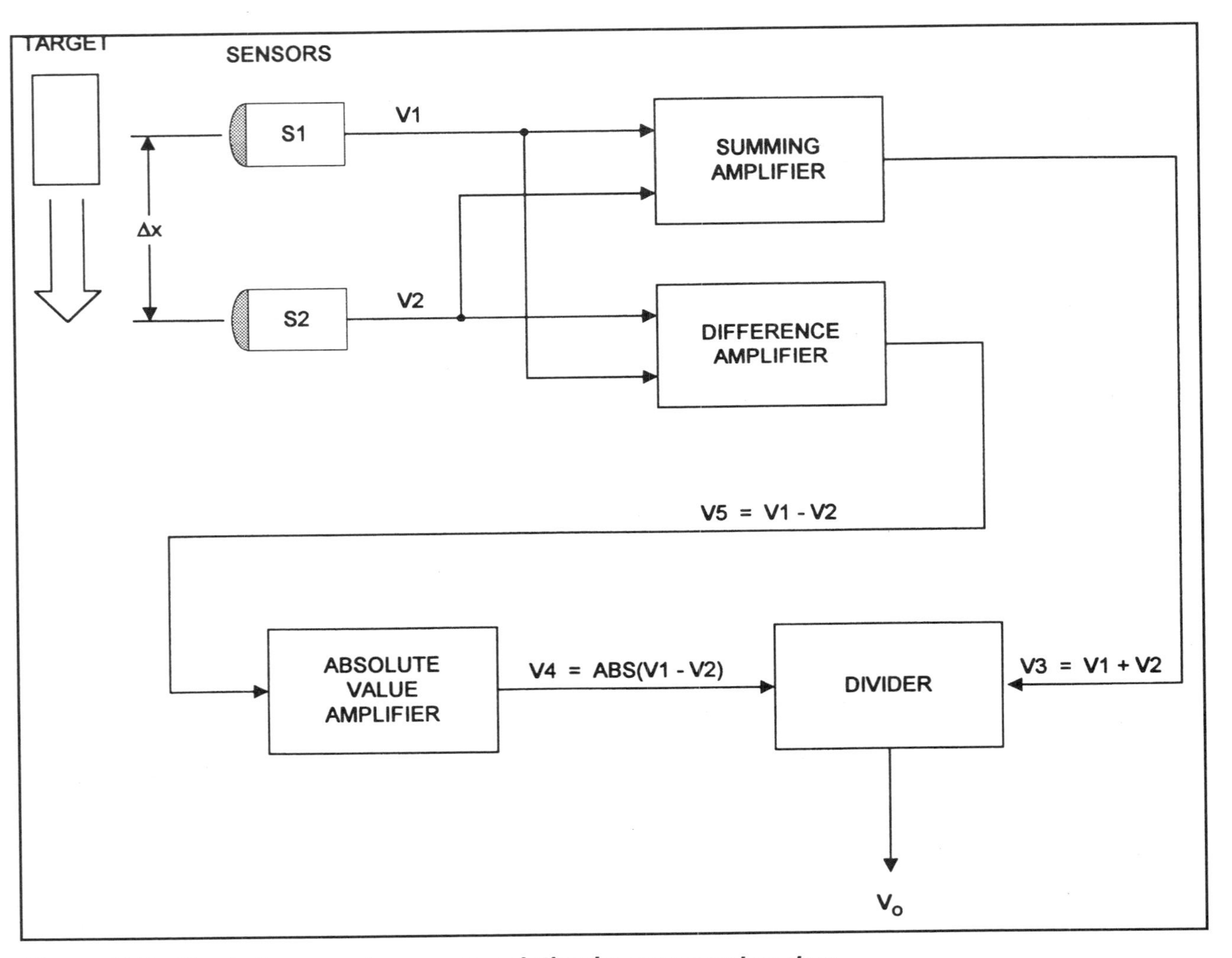

Figure 18-7. Block diagram of sensor resolution improvement system.

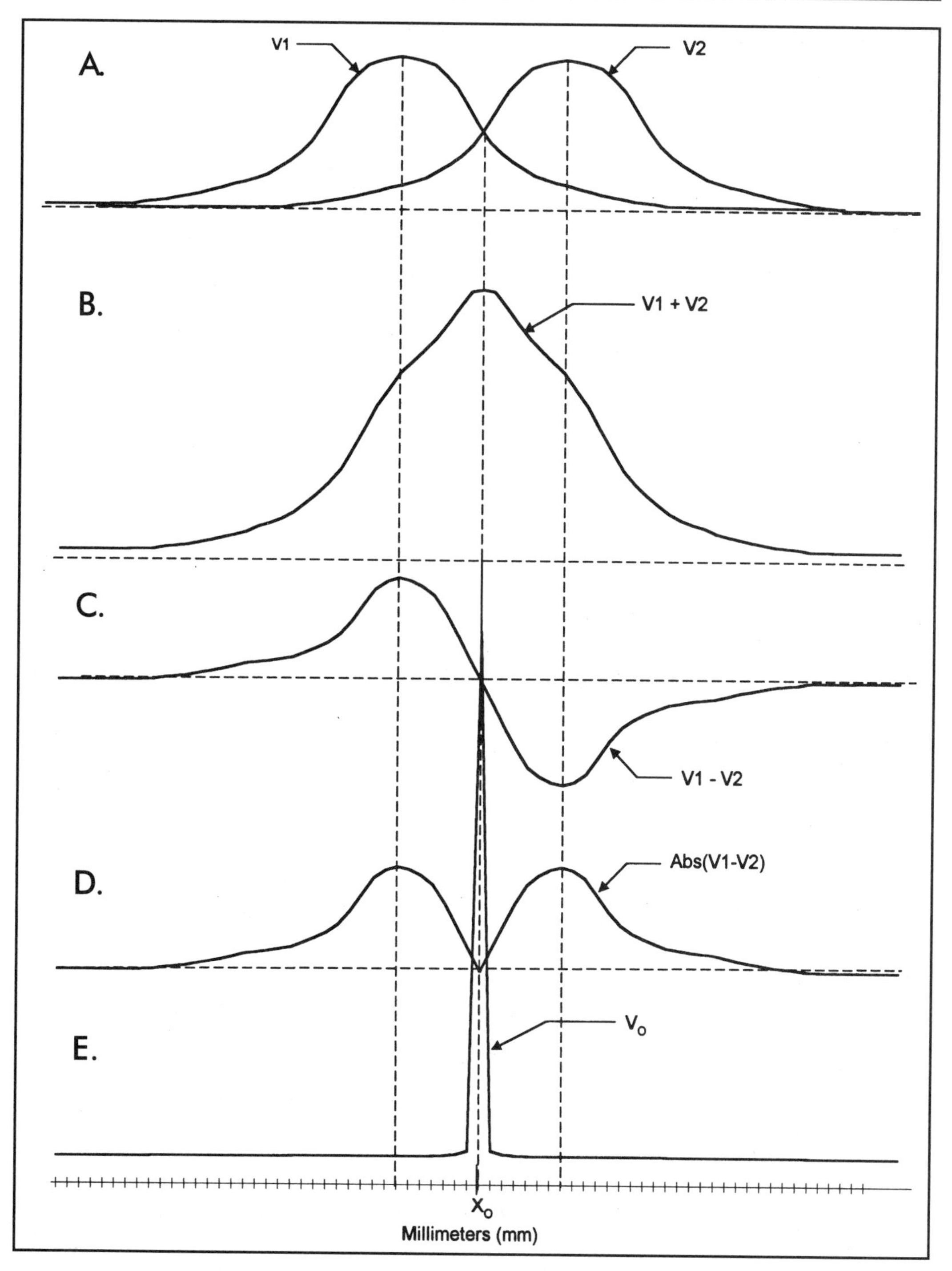

Figure 18-8. Sensor resolution improvement curves: A) V1 and V2; B) V1 + V2; C) V1 - V2; D) Abs (V1 - V2); E) V_o.

The final step is to take the ratio of the two signals, V3 and V4, solving **Equation 18-1** below to find V_O:

$$V_O = \frac{V1 + V2}{k + (V1 - V2)} \qquad \text{eq. (18-1)}$$

Where:

V_O is the output signal.

V1 is the signal from sensor S1.

V2 is the signal from sensor S2.

k is a small constant.

The original implementation did not use the k constant in the denominator, and that caused a problem that should have been foreseen. If there is no target present, which means S1 and S2 are equally illuminated by background light (optical sensors were used), or when the target is mid-way between S1 and S2 (which can occur with any form of sensor), then the value of V1-V2 goes to zero. This condition forces V_O to an extremely high value that is called "infinity" in the naive sense and "undefined" otherwise. In practical terms, this means that the amplifier or divider at the output will saturate, or the digital value tries to go to above full-scale. The constant k is set to whatever value will force V_O to exactly full-scale when V1 - V2 = 0.

The resolution-improved output VO is shown in **Figure 18-8e**. Notice the tremendous improvement in resolution compared with the beamwidths of V1 and V2, as shown in **Figure 18-8a**. The implication of this change to both position ambiguity and two-target discrimination is obvious. This waveform is an extremely high amplitude, narrow base, spike that accurately finds X_O.

In **Figure 18-7**, the block diagram assumes analog implementation, so the stages are labeled "amplifiers." The sum amplifier, the difference amplifier and the absolute value amplifier were operational amplifiers. The divider can be any of several analog multiplier/divider integrated circuits that are on the market, although I used the Burr-Brown DIV-100 module. The same block diagram can be used as the basis for a digital or computer implementation by programming the stages using math statements. Even ordinary BASIC can be used for this purpose.

Simulation Effort

The first step in demonstrating this principle involved a computer simulation on artificial data using an Excel 5.0 spreadsheet. A normal distribution ("bell-shaped") curve drawing template was used to draw curves representing V1 and V2 on fine-mesh linear graph paper with 1 mm (≈25/inch) spacing between lines.

Approximately 80 values for V1 and V2 were obtained by counting the amplitudes of the two curves at 1-mm intervals along the baseline. These values were entered into the "A" and "B" columns of the spreadsheet. Another column was programmed with a math function to solve **Equation 18-1**, and insert the result in the cell adjacent to the V1 and V2 cells, from which value was calculated. The chart function of Excel was then used to show V1, V2, V1+V2, V1-V2, ABS(V1-V2) and V_O. The result was consistent with the expectations of **Figure 18-8**. With that encouragement, a working model was produced.

Laboratory Model of the Two-Sensor Approach to SRI

In order to verify that this technique worked on sensor systems other than radar, an optical sensor pair was used in an experiment (**Figure 18-9**). The optical sensors were Burr-Brown OPT-101 devices. These sensors are operational amplifiers with an photodiode device built into the transparent 8-pin DIP IC package. The two OPT-101 devices were spaced 10-mm apart so that their cones of acceptance overlapped (which is also the minimum possible X-axis separation due to the size of the IC packages). The mid-point between the two OPT-101 devices corresponds to X_O, while the distance between corresponds to ΔX; any particular point along the path between S1 and S2 is designated X_i.

The target was a red LED mounted on a micrometer device that measured distances in millimeters. The initial position of the LED target was set so that it was outside the FOR of both S1 and S2. The LED was then advanced 1 mm at a time until it had traversed the entire distance from the left-most extent of the FOR of S1 to the right-most extent of the FOR of S2. The output voltages (V1 and V2) from S1 and S2 were measured with a 3½-digit digital voltmeter (DVM) at each 1 mm interval. Those data were then entered into an Excel spreadsheet and plotted on a chart. The curves in **Figure 18-8** were cut and pasted directly from the Excel chart to a Visio Technical 4.0 drawing, to represent actual results.

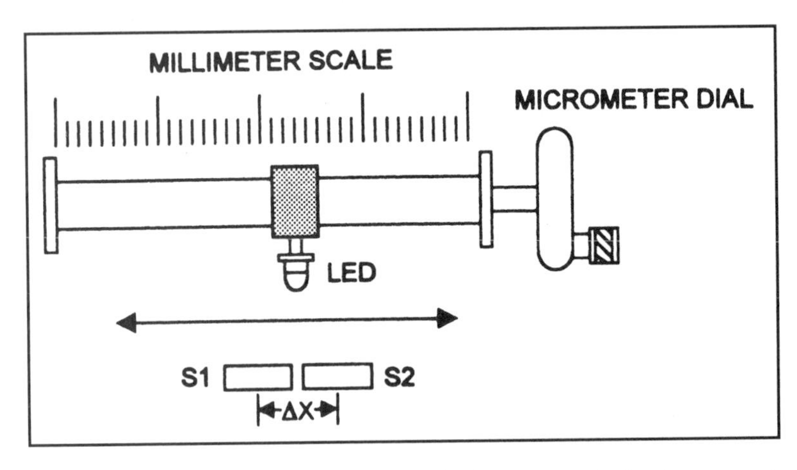

Figure 18-9. Test set-up using optical sensors to test concept.

Single-Sensor Method

The two-sensor method of SRI produces startling results, but at the cost of needing two sensors. It is easy to implement in analog circuitry, and can also be implemented in digital circuitry. Another approach uses a single sensor and a look-ahead calculation to synthesize V2. It is easily implemented digitally, but is somewhat difficult to implement in analog circuitry.

Assume that the values of V1 from the sensor are a series V_i in which each value represents the signal amplitude at sequential locations along the X-axis:

$$V1, V2, V3, V4, V5, V6, V7, V8, V9, V10, V11, \ldots, V_{ith}$$

Each value of V_i represents a value of V1 in **Equation 18-1**. The corresponding value of V2 is found by taking a subsequent value of V1 that is displaced a distance N (an integer) from V_i. Thus, in terms of **Equation 18-1**:

$$V1 = V_i$$

$$V2 = V_{i+N}$$

Equation 18-1 can be rewritten to the form:

$$V_O = \frac{V_i + V_{i+N}}{k + (V_i - V_{i+N})} \quad \text{eq. (18-2)}$$

When the V1 data from the experiment are plotted, the resultant curves are very similar to those of **Figure 18-8** and demonstrate very nearly the same degree of SRI. It was also found that a limited amount of "tuning" of resolution can be done by selecting values of N. However, with the 1 mm spacing used in the experiment, values $N > 5$ showed essentially the same curves.

In both the single-sensor and two-sensor methods there exists the possibility of creating a selectable beamwidth sensor system. Signal V_O could be the narrow beamwidth signal, V1 + V2 can be the wide beamwidth signal, and V1 (or V2) can be the medium beamwidth signal.

Summary

The sensor resolution improvement method is a variant on the monopulse resolution improvement method used for many years in radar technology. It appears to have application in any instrumentation problem where sensor resolution is an issue.

References

Carr Joseph J. and John M. Brown. Introduction to Biomedical Equipment Technology - 2nd Edition. New York: Prentice-Hall. (1993).

Carr, Joseph J. The Art of Science. San Diego: Hightext Publications, Inc. (1992).

Skolnik, Merrill. Introduction to Radar Systems. New York: McGraw-Hill. (1980).

Stimson, George W. Introduction to Airborne Radar Systems. El Segundo, CA: Hughes Aircraft Co. (1983).

Webster's New Collegiate Dictionary. Springfield, MA. G. & C. Merriam Co. (1979).

Index

Symbols

A

B

C

D

E

F

G

H

I

J

K

L

M

N

O

P

Q

R

S

T

U

V

W

X

Y

Z